Also in the Variorum Collected Studies Series:

WENDY DAVIES
Welsh History in the Early Middle Ages
Texts and Societies

RICHARD BRITNELL
Markets, Trade and Economic Development in England and Europe, 1050–1550

BRUCE M.S. CAMPBELL
Land and People in Late Medieval England

BRUCE M.S. CAMPBELL
Field Systems and Farming Systems in Late Medieval England

PAMELA NIGHTINGALE
Trade, Money, and Power in Medieval England

BRUCE M.S. CAMPBELL
The Medieval Antecedents of English Agricultural Progress

DAVID M. PALLISER
Towns and Local Communities in Medieval and Early Modern England

RODNEY M. THOMSON
England and the Twelfth-Century Renaissance

H.S. OFFLER
North of the Tees
Studies in Medieval British History

SUSAN REYNOLDS
Ideas and Solidarities of the Medieval Laity
England and Western Europe

ADRIAAN VERHULST
Rural and Urban Aspects of Early Medieval Northwest Europe

JOHN H. MUNRO
Bullion Flows and Monetary Policies in England and the Low Countries, 1350–1500

VARIORUM COLLECTED STUDIES SERIES

Manors and Maps in Rural England, from the Tenth Century to the Seventeenth

P.D.A. Harvey, Prato, 1982

P.D.A. Harvey

Manors and Maps in Rural England, from the Tenth Century to the Seventeenth

ASHGATE
VARIORUM

Published in the Variorum Collected Studies Series by

Ashgate Publishing Limited
Wey Court East
Union Road
Farnham, Surrey
GU9 7PT
England

Ashgate Publishing Company
Suite 420
101 Cherry Street
Burlington, VT 05401–4405
USA

Ashgate website: http://www.ashgate.com

ISBN 978–1–4094–0241–1

British Library Cataloguing in Publication Data
Harvey, P. D. A.
Manors and maps in rural England, from the tenth century to the seventeenth.
– (Variorum collected studies series ; CS950)
1. Manors – England – History – To 1500 – Sources.
2. Country life – England – History – To 1500 – Sources.
3. Land tenure – England – History – To 1500 – Sources.
4. Cartography – England – History – To 1500.
5. Cartography – England – History – 16th century.
I. Title II. Series
333.3'22'0942–dc22

ISBN 978–1–4094–0241–1

Library of Congress Control Number: 2010923095

VARIORUM COLLECTED STUDIES SERIES CS950

Mixed Sources
Product group from well-managed forests and other controlled sources
www.fsc.org Cert no. SGS-COC-2482
© 1996 Forest Stewardship Council
FSC

Printed and bound in Great Britain by
TJ International Ltd, Padstow, Cornwall

CONTENTS

This volume contains x + 340 pages

PUBLISHER'S NOTE

The articles in this volume, as in all others in the Variorum Collected Studies Series, have not been given a new, continuous pagination. In order to avoid confusion, and to facilitate their use where these same studies have been referred to elsewhere, the original pagination has been maintained wherever possible.

Each article has been given a Roman number in order of appearance, as listed in the Contents. This number is repeated on each page and is quoted in the index entries.

PREFACE

The eighteen articles reproduced or reprinted here were first published over a long period, between 1961 and 2004. They present no slowly evolving thought on my part, only the exploration of one topic after another, so they have been placed not in order of publication but in chronological order of their subject matter. The final article, covering a wide chronological range, forms an epilogue to the collection. As far as I can detect, no article contradicts another, but some answer questions raised in others and some work out more fully ideas that have been touched on in an earlier article.

Alongside research on rural England in the Middle Ages I have throughout my professional life worked on various aspects of the history of maps, not only medieval but also both earlier and later, not only from England but from all parts of the world. To give coherence to the present volume, the only articles here that stem from this work on maps are nos XV, XVI and XVII, which look at their use in running landed properties in early-modern England, and no. XIV, a necessary prelude to these, which shows how mapping to scale was introduced into England in the mid-sixteenth century.

Most of the articles are reproduced directly from their original publication, retaining their pagination but with silent correction of simple typographical errors. Short notes added to nos I, IV, IX, and XII draw attention to more significant mistakes. Direct reproduction of nos XI, XV, XVII and XVIII was not possible, so these have been reset with new pagination, and from nos XV and XVII some illustrations have been omitted, either because they add little to the text or because they are reproduced in another article.

I should like particularly to thank Miss Barbara Harvey, CBE, FBA (we share a surname but are not related), but for whose kindly urgings I should never have embarked on this project, and Mrs Yvonne Harvey (I am her husband), whose support, help and positive criticism are reflected on every page – with the exception only of no. XVI, published before we first met.

P.D.A. HARVEY

Durham
31 December 2009

ACKNOWLEDGEMENTS

For their kind and generous permission to reproduce or reprint my work in this way I am most grateful to the original publishers or their successors: Oxford University Press (nos I, VII, IX, X); Boydell and Brewer Ltd (nos II, IV, VIII); Shaun Tyas (no. III); Wiley-Blackwell (nos V, VI); Continuum International Publishing Group (no. XI); the British Agricultural History Society (no. XII); the British Archaeological Association (no. XIII); Portsmouth City Council (no. XIV); the Society for Landscape Studies (nos XV, XVIII); the Trustees of the British Museum (no. XVI); and the University of Chicago Press (no. XVII).

I am no less grateful to the copyright owners of the illustrations that have been reproduced: no. XIV fig. 5 by gracious permission of H.M. The Queen; nos XIII plate xxxii, XIV fig. 4, XV plate i, XVI plate xxix, XVIII plates i, ii, by permission of the British Library Board; no. XVII fig. 1 by permission of the Dorset History Centre; no. XV plate iii by permission of the Dean and Chapter of St Paul's Cathedral; no. XVII figs 7, 8, by permission of Northamptonshire Record Office; no. XV plates ii, iv, v, by permission of The National Archives; no. XVII figs 2, 3, 5, 6, 9, by permission of Warwickshire County Record Office; nos XVII fig. 4, XVIII plate iii, by permission of the Warden and Scholars of Winchester College.

I

Rectitudines Singularum Personarum and *Gerefa**

THE document known as the Rectitudines Singularum Personarum (henceforth RSP) is the only systematic record we have of the rights and obligations of workers and tenants on an English estate before the Norman Conquest. Its Old English text is known from one manuscript, Corpus Christi College, Cambridge, MS 383. This is a collection of Anglo-Saxon laws and similar texts, copied in the late eleventh or early twelfth century by a single compiler.[1] He worked probably at St Paul's Cathedral, London: a different but contemporary hand has added to the collection a list of places in south-east England with the number of seamen each had to provide for the king's service, and at most of these places St Paul's demonstrably held property. Later annotations suggest that the manuscript belonged to the Cathedral in the sixteenth century.[2] A Latin version of RSP, based on a slightly different (and slightly better) text, was included in the Quadripartitus. This was so called because, according to its preface, it was to be in four sections or books, but only the first two are known for certain. RSP is in its first book, again a collection of Anglo-Saxon laws and similar texts, but all translated into Latin. The preface can be dated between 1113 and 1118 and part of the second book between 1108 and 1114; we may reasonably suppose that the first book, with its translation of RSP, was compiled about the same time. Eleven manuscripts of the Quadripartitus are known, but each is selective, omitting some part or parts of the text, and RSP is found in only five of

* I am grateful to Professor C. C. Dyer for first pointing me towards Dorothy Bethurum's work; to Mrs C. P. Hall for alerting me to the impending removal of MS 383 from the Library of Corpus Christi College, Cambridge, on loan for exhibition; to Professor R. I. Page and Mrs G. Cannell who, at minimal notice, very kindly made the manuscript available to me before this occurred; and to Dr R. J. Faith, Dr Page, and Mr C. P. Wormald for reading this article in draft and for their helpful and encouraging comments.

1. F. Liebermann, *[Die] Gesetze [der Angelsachsen]* (3 vols., Halle, 1898–1916), provides the definitive edition of the text (i. 444–53), with introduction and notes (iii.244–52) and an account of the manuscript (i, p. xix). Further notices of the manuscript are in M. R. James, *A Descriptive Catalogue of the Manuscripts in the Library of Corpus Christi College, Cambridge* (2 vols., Cambridge, 1912), ii.230–1; N. R. Ker, *Catalogue of Manuscripts containing Anglo-Saxon* (Oxford, 1957), pp. 110–13; and M. P. Richards, 'The Manuscript Contexts of the Old English Laws', in *Studies in Earlier Old English Prose*, ed. P. E. Szarmach (Albany, 1986), pp. 171–92, at 181–3.

2. F. Liebermann, 'Matrosenstellung aus Landgütern der Kirche London', *Archiv für das Studium der neueren Sprachen und Litteraturen*, civ (1900), 17–24.

I

them; the earliest of these was copied in the mid-twelfth century, the latest about 1600.[1]

It was in its Latin version that RSP first came to the attention of scholars. In the second volume of his *General Introduction to Domesday Book*, published in 1833, Sir Henry Ellis printed some extracts from what he called 'a sort of glossary upon ancient services' with the comment that 'till the pages of the former volume had been printed off, this curious Tract was unknown to the present editor'. These extracts were taken from the early thirteenth-century copy of the Quadripartitus in Cotton MS Titus A.xxvii (cited once thus, once as Titus A.viii), but Ellis noted that the Old English text of the same document at Corpus Christi College pointed to an earlier origin.[2] What followed is of some historiographical interest, for it reflects the concern with Anglo-Saxon history then developing in Germany, and how closely work in this field ran parallel in the two countries. J. M. Lappenberg, archivist at Hamburg, read the extracts in Ellis's work, recognized that there was a copy of this document in a manuscript in the Hamburg city library (the copy of the Quadripartitus dating from about 1600), and at once, in 1834, published its full text in *Rheinisches Museum für Jurisprudenz*. He named it, from its heading in this manuscript, 'De Dignitate Hominum', and described it as a guide for itinerant Anglo-Norman justices; in passing he pointed out that Ellis (who as Principal Librarian of the British Museum had in fact the Cotton manuscripts in his keeping) had given the wrong reference.[3] The Old English text was first published in 1840 in the Record Commission's edition of *Ancient Laws and Institutes of England*; the editor, Benjamin Thorpe, in place of a modern English translation 'of which it seems hardly susceptible', printed beside it the Latin of the Quadripartitus, again citing Cotton MS Titus A.xxvii incorrectly (this time it became Tiberius A.xxvii). However, he described the document more convincingly than either Ellis or Lappenberg, as 'an enumeration of the several classes of persons employed on a domain, of the services to be rendered by each, and of the reciprocal duty of the lord to those engaged on his land'. He also gave it its lasting name, RSP, its heading in two of the five manuscripts of its Quadripartitus text.[4]

1. F. Liebermann discusses Quadripartitus and prints the full text both in *Quadripartitus, ein englisches Rechtsbuch von 1114* (Halle, 1892) and in *Gesetze*. T. F. T. Plucknett, *Early English Legal Literature* (Cambridge, 1958), pp. 24–30, and L. J. Downer, *Leges Henrici Primi* (Oxford, 1972), pp. 21–8, 34–7, confirm Liebermann's date for Quadripartitus and support his suggestion that the Leges Henrici Primi, written by the same author, are its missing books.

2. H. Ellis, *A General Introduction to Domesday Book* (2 vols., Record Comm., 1833), ii. 425–6n., 427n., 435–6n., 453n.

3. J. M. Lappenberg, 'De dignitate hominum Anglo-Saxonum', *Rheinisches Museum für Jurisprudenz*, vi (1834), 145–53. C. P. Cooper, *Appendices to a Report on Rymer's Foedera* (printed 1837, distributed 1869), provides a notice of the same manuscript text (App. A, p. 109) and a full transcript (App. C, pp. 48–9) which, had it been published when printed, would have been the first English edition of RSP.

4. [B. Thorpe], *Ancient Laws and Institutes of England* (Record Comm., 1840), pp. iv–v, xi, 185–9 (or, in the two-vol. edn., also 1840, i, pp. xi, xxv–xxvi, 431–41).

Thus introduced to historical scholarship, the text's importance was at once recognized in both England and Germany: it made an immediate contribution to knowledge of early English institutions and social structure, then under learned discussion.[1] In 1842 Heinrich Leo published at Halle a book entitled *Rectitudines Singularum Personarum*. It dealt with various aspects of Anglo-Saxon society – its first hundred pages are on place-names – but one section used RSP as the basis for an account of rural economy, and at the end of the book the document was printed in four parallel texts: Old English as printed by Thorpe, Old English with spelling and punctuation standardized, the Latin of the Quadripartitus, and a German translation.[2] Lappenberg cited RSP in the first volume of his *Geschichte von England*, published in 1834 and translated into English in 1845;[3] J. M. Kemble wrote of it, as a 'valuable document', in his two-volume work *The Saxons in England*, published in 1849 and translated into German in 1853–4;[4] and since then there have been few general accounts of Anglo-Saxon England that do not at least mention it. In 1858 Reinhold Schmid printed the text (Old English, Latin, and German translation) among the appendices of his comprehensive text of Anglo-Saxon laws, *Die Gesetze der Angelsachsen*; the document had not come to light in time to be included in the first edition in 1832.[5] Schmid's work was the precursor of the monumental *Gesetze der Angelsachsen* by Felix Liebermann; its first volume, containing the text of the laws, was published in 1898 and includes what is still the definitive edition of RSP (again Old English, Latin, and German translation).[6]

By then RSP had acquired a companion, Liebermann's own discovery. In Corpus Christi College MS 383, RSP is immediately followed by a further Old English text headed 'Bege sceadwisan gerefan': 'On the sensible reeve'. This heading need not be read as starting a new document; it looks analogous to the headings of the preceding sections of RSP ('Be wudewarde', 'Be bydele', 'Be berebrytte', and so on), but it should be said that the manuscript does not always clearly distinguish one document from the next, and that RSP itself is given no title beyond the heading of its first section, 'Ðegenes Lagu'. The text describes the qualifications and duties of the reeve in charge of a landed estate, including an account of the work to be done in the course of the year and lists of the tools and other equipment that had to be provided. It can be seen as comple-

1. S. Heywood, *A Dissertation upon the Distinctions in Society and Ranks of the People under the Anglo-Saxon Governments* (London, 1818), is an example of a work that would certainly have made substantial use of RSP had it come to light earlier.

2. Pp. 105–36, 222–47. H. Leo, *Treatise on the Local Nomenclature of the Anglo-Saxons* (London, 1852), translated from this work, omits the sections dealing with RSP.

3. J. M. Lappenberg, *Geschichte von England* (2 vols., Hamburg, 1834–7), i. 617; id., *A History of England under the Anglo-Saxon Kings*, trans. B. Thorpe (2 vols., London, 1845), ii. 357.

4. J. M. Kemble, *The Saxons in England* (2 vols., London, 1849), i. 38, 180, 214–15, 322–5; id., *Die Sachsen in England*, trans. H. B. C. Brandes (2 vols., Leipzig, 1853–4).

5. R. Schmid, *Die Gesetze der Angelsachsen* (2nd edn., Leipzig, 1858), pp. lxiii–lxiv, 370–83.

6. See *supra*, p. 1, n. 1.

menting RSP, which seems to have a section on everyone on the estate except the reeve himself, and this is how Liebermann saw it when he first published the text in *Anglia* in 1886, giving it the title Gerefa: it was simply a continuation of RSP. Its viewpoint was the same – that of the lord rather than the tenant, of the manager rather than the managed – its vocabulary similar, and some phrases of RSP recur in it: 'hede seðe scire healde', 'fela ðinga ðe ic ge tellan ne mæig', and others. However, it was easy to see why it had so far escaped notice, for it was through the Latin translation that RSP had first come to light and Gerefa is not included in the Quadripartitus. Its omission there, Liebermann explained, need not imply that they were taken to be two different documents in the early twelfth century. The author of the Quadripartitus may have felt that when the text passed from enumerating rights and obligations to describing the reeve's agricultural management it no longer fell within the scope of a collection of legal texts, or he may, reasonably enough, have been daunted by the thought of putting into Latin its formidable lists of implements. After all – and here Liebermann must have had his tongue slyly in cheek – Thorpe in 1840 had balked at the much easier task of translating RSP into English.[1]

In *Die Gesetze der Angelsachsen*, Liebermann placed Gerefa as a separate text immediately after RSP,[2] but his conclusion that they formed a single document had, if anything, strengthened. In the work's final volume, published in 1916, he provided a single introduction for them both and stated unequivocally that Gerefa is the second part of RSP. He gave an expanded list of the words and phrases common to the two texts, impressive evidence of their association, though the recurrence of some single words, as *began*, *fald*, or *ripan*, seems scarcely significant. Liebermann distinguished two layers in the grammar and spelling of RSP and Gerefa. The Old English text as we have it, in Corpus Christi College MS 383, must have been partly modernized in copying, not before about 1060 at the earliest and probably about 1070–1100. For instance, there is some confusion between the strong and weak declension of adjectives, and the paragraph headings, which were probably added on copying (they do not occur in the Quadripartitus version of RSP), sometimes include a word in later form than the same word in the text; thus the heading for the paragraph on the *wudu weard* reads 'Be wudewarde'. When the Quadripartitus leaves an English word untranslated it is sometimes in an older form than in the text of MS 383 – 'Cotsetle' instead of 'Kote setlan' is an example. Our Old English text still retains words that the copyist left unchanged in what must have seemed, by the late eleventh century, old-fashioned forms; they should not, however, be seen as older than about 970, and the work's alliterative style points to a date after 1000. Liebermann then adduced various historical indications

1. F. Liebermann, 'Gerefa', *Anglia*, ix (1886), 251–66.
2. Id., *Gesetze*, i. 453–5.

in the text which suggest an early eleventh-century date – some more convincing than others, but none of them claimed as conclusive – and finally proposed 1020–30 as the most likely time of its composition: its atmosphere of assured peace, free from Danish threat, suggested a time after 1018, whereas there is no mention of the heriots that appear in Cnut's secular laws, which he dated 1027–34. In deciding where the text was written, Liebermann proceeded by elimination. It shows no trace of Celtic influence, so did not come from the extreme west or south-west, nor of the dialects of Northumbria or Kent, nor of Kentish or Danelaw customs and terminology: the text therefore probably came from Wessex or the southern part of central Mercia.[1] It should be added that as the basis for his discussion Liebermann provided an impeccable transcription of the text from MS 383.[2] However, he imposed upon it a normalization of capital initials, breaks between words, and punctuation, procedures which at a few points, notably in Gerefa's lists of implements, have significant consequences.[3]

An important addition to Liebermann's conclusions was provided by Dorothy Bethurum in 1963. She concluded that RSP and its 'second part', Gerefa, were rewritten from an existing text by Wulfstan, Bishop of Worcester and Archbishop of York. This she argued on grounds of style: to some extent in RSP and much more in Gerefa, the vocabulary and particularly the alliteration and rhythm are characteristic of Wulfstan's prose. Moreover, the subject-matter accords with his interest in the ordered and orderly structure of society and its classes. Worcester neatly fitted Liebermann's view of the text's place of origin, and though Wulfstan resigned the see in 1016, this need not preclude Liebermann's date, for Wulfstan, who died in 1023, maintained contact with Worcester. The basis of his text was a document already at Worcester, 'probably dating from Oswald's episcopacy [i.e. 961–92] in the form in which he used it'.[4]

Bethurum's work should be seen in its own context of late Old English literary scholarship. Before 1932, as she herself pointed out, Wulfstan was little known as an author apart from the famous Sermo Lupi ad Anglos. There followed the work of, in particular, Karl Jost and Angus McIntosh which precisely defined Wulfstan's style. They were emphatic in their views: McIntosh that Wulfstan used his characteristic two-stress rhythm in all his formal writing; Jost that there were no imitators and that anything written in Wulfstan's style must be by him. Though Jost

1. Ibid. iii. 244–6.

2. I would query only one detail. Liebermann (ibid. i. 447n.) sees 'ræcan' in RSP para. 4.1b as altered in crayon (*Kreide*) to 'ræfan' or 'ræpan'. In fact the alteration is almost certainly in pale ink, analogous to the running correction of 'geoðeðe' in para. 14, and may well be a correction of 'ræpan' to 'ræcan'.

3. As pointed out by R. I. Page, '"The Proper Toil of Artless Industry": Toronto's Plan for an Old English Dictionary', *Notes and Queries*, ccxx (1975), 146–55, at 148–9. All words and passages cited here from RSP and Gerefa follow exactly the usage of the manuscript, not Liebermann's edition.

4. D. Bethurum, 'Episcopal Magnificence in the Eleventh Century', in *Studies in Old English Literature in Honor of Arthur G. Brodeur*, ed. S. B. Greenfield ([Eugene], 1963), pp. 162–70.

urged caution, their work opened the way to identifying various anonymous writings as Wulfstan's.[1] Among juridical texts, Dorothy Whitelock identified him as the compiler of the so-called laws of Edward and Guthrum and of Cnut's double law code, ecclesiastical and secular,[2] and Bethurum in 1950, looking at the group of six texts known as Grið, Geþyncðo, Hadbot, Norðleoda Laga, Mircna Laga, and Að, argued that he wrote or substantially rewrote the first three and put the whole group together. In 1949 McIntosh had considered that though the 'rhythmical texture' of RSP and Gerefa was very like Wulfstan's, 'a strict analysis of their two-stress phrases shows that they deviate significantly from those of Wulfstan'. Bethurum also commented that RSP and Gerefa contained much that was characteristic of Wulfstan; but 'they can scarcely be Wulfstan's work, for they display too detailed a knowledge of the lowliest duties of the peasant to come from the archbishop. It may be that they were written by Wulfstan's own reeve and were put into final form by Wulfstan.'[3] By 1963 she had changed her mind and was prepared to ascribe a larger role to Wulfstan. This was reasonable; after all, another bishop, Robert Grosseteste, turned his hand to a similar task some two hundred years later.[4] Historians are apt to be suspicious of arguments based largely on literary style, but Bethurum's credentials cannot be questioned: as the editor of Wulfstan's Homilies she was fully immersed in his writing and well qualified to recognize his hand at work in any context.

The conclusions of Liebermann and Bethurum have not been directly challenged, though one can detect some hesitation among scholars to accept them in detail. Thus most of those who have referred to RSP and Gerefa since 1963 have passed over in silence the question of Wulfstan's part in compiling them, but the only doubt raised over the general conclusions of Jost and McIntosh and the attributions stemming from them concerns Wulfstan's invariable use of the two-stress rhythm.[5] Before Wulfstan was brought into the picture there were some who favoured a later date for the text than Liebermann. In 1953 D. C. Douglas and Dr G. W. Greenway placed RSP in the 1042–1189 volume of *English Historical Documents*; in the second edition of 1981, without comment or explanation, it is dated '?1042–66', and in 1962 Professor H. R. Loyn referred

1. D. Bethurum, 'Wulfstan', in *Continuations and Beginnings: Studies in Old English Literature*, ed. E. G. Stanley (London, 1966), pp. 210–46, at 215; K. Jost, *Wulfstanstudien* (Berne, 1950), esp. pp. 5, 102–3, 116–17, 155–68, 270–1; A. McIntosh, 'Wulfstan's Prose', *Proceedings of the British Academy*, xxxv (1949), 109–42. Summarized accounts of Wulfstan's language and style and of the attribution of writings to him are in D. Bethurum, *The Homilies of Wulfstan* (Oxford, 1957), pp. 24–54, 87–98.

2. D. Whitelock, 'Wulfstan and the So-Called Laws of Edward and Guthrum', *ante*, lvi (1941), 1–21; ead., 'Wulfstan and the Laws of Cnut', *ante*, lxiii (1948), 433–52; ead., 'Wulfstan's Authorship of Cnut's Laws', *ante*, lxx (1955), 72–85.

3. D. Bethurum, 'Six Anonymous Old English Codes', *Journal of English and Germanic Philology*, xlix (1950), 449–63; McIntosh, 'Wulfstan's Prose', 126. Cf. Bethurum, *Homilies*, p. 46.

4. D. Oschinsky, *Walter of Henley and Other Treatises on Estate Management and Accounting* (Oxford, 1971), pp. 192–7.

5. I. M. Hollowell, 'On the Two-Stress Theory of Wulfstan's Rhythm', *Philological Quarterly*, lxi (1982), 1–11.

to it as 'a mid-eleventh-century document'.[1] Mostly, however, it has been assigned broadly to the early eleventh century; except by H. P. R. Finberg, Bethurum's suggestion of a tenth-century origin has been overlooked or ignored.[2] There is caution, too, in accepting Liebermann's view of Gerefa as an integral part of RSP, sundered from it by a mere accident of historiography; thus it is referred to as 'a second tract', 'a supplement as it were', or a 'companion piece'.[3] Professor R. I. Page has brought these implicit doubts into the open. Describing the relationship between RSP and Gerefa as 'unclear', he suggests that once it is accepted that the compiler of MS 383 drew his material from several sources, the 'compelling link' between the two documents is broken, and that (though he does not specify them) there are 'good philological, historical and archaeological reasons for treating *Gerefa* as a distinct text'.[4]

These doubts deserve to be taken further. What will be argued here is that RSP and Gerefa are documents of wholly different origin and purpose and probably of quite different dates; this argument is based simply on the nature and content of the two texts. They were brought together in the early eleventh century by someone who may well have been Wulfstan himself and who certainly had Wulfstan's interest in the ordered structure of society. Gerefa fills the apparent gap in RSP and, with its account of the reeve, makes the combined text a little treatise on the duties and mutual obligations of all the personnel of a country estate. There seems no doubt that the two texts were deliberately revised to weld them together; the repetition of phrasing pointed out by Liebermann cannot be coincidental. Nor does it seem possible that they were first united in MS 383. Nowhere else does its compiler appear as other than a simple copyist; the orthography of some of the repeated phrases is not of his time; it would be difficult to explain why he undertook this revision; and, if he had, the combined text would surely be headed or otherwise signalled in the manuscript as a distinct and single item. Nor need we suppose that, because the Quadripartitus omits Gerefa, the Latin translator was working from a text which comprised RSP alone; Liebermann's two possible explanations for the omission of Gerefa (and

1. D. C. Douglas and G. W. Greenway, *English Historical Documents, 1042–1189* (London, 1953), pp. 813–16; (2nd edn., London, 1981), p. 875 (it should be added, though, that in both editions the introductory note offers a wider date-span: probably the half-century before 1066); H. R. Loyn, *Anglo-Saxon England and the Norman Conquest* (London, 1962), p. 104.

2. H. P. R. Finberg, 'Anglo-Saxon England to 1042', in *The Agrarian History of England and Wales* (8 vols., Cambridge, 1967–), i, pt. 2, pp. 383–525, at 512. Finberg also accepts, without reservation, Bethurum's attribution to Wulfstan of the text's subsequent revision.

3. W. J. Corbett, 'England from A.D. 954 to the Death of Edward the Confessor', in *Cambridge Medieval History* (8 vols., Cambridge, 1911–36), iii. 371–408, at 401; Loyn, *Anglo-Saxon England*, p. 193; McIntosh, 'Wulfstan's Prose', 126; and Richards, 'Manuscript Contexts', p. 183.

4. R. I. Page, '*Gerefa*: Some Problems of Meaning', in *Problems of Old English Lexicography: Studies in Memory of Angus Cameron*, ed. A. Bammesberger (Regensburg, 1985), pp. 211–28, at 213–14.

they are not mutually exclusive) are entirely convincing.[1] Thus, where Liebermann saw two layers of writing in RSP and Gerefa, we should look for three: the original composition of each text, the revision that brought them together in the early eleventh century, and the partial modernization of grammar and spelling in the late eleventh or early twelfth century that gave us the Old English text of MS 383. We are unlikely to find any trace of the bottom layer, the original composition, in the spelling of the two texts as we have them, but their content and vocabulary offer some clues to their origins.

No one has suggested that Gerefa is a practical guide to running an estate, whether to instruct the overseer, the *gerefa* himself, or for the landlord to refer to as a check on his local manager's activities. There are too many omissions and inconsistencies. Thus, looking at its account of the work for each season of the year, we are shown an estate where grass was mown in the autumn but not in the early summer, where madder and vines were planted and beans, flax, and woad were sown, but not wheat, rye, barley, or oats. Over the last twenty years much philological work has been done on the two lists of implements, and reinterpretation has shown that at certain points they are more orderly than had been supposed.[2] Even so, they are neither systematic nor comprehensive. Professor C. E. Fell has detailed their shortcomings as inventories of domestic equipment,[3] and they can be criticized for their agricultural components too; thus ploughing equipment ('sulhgesidu') comes at the start of the second list, but ploughshare, coulter, and goad ('scear. culter. 7 eac gadiren') occur in the first list as individual items. In both lists there are groups of successive items with the same initial letter, suggesting a literary rather than a functional approach to the task. Nor is this confined to the lists of implements. Its use of alliteration and two-stress rhythm are primarily what led Bethurum to see Wulfstan's hand at work in Gerefa, and in some places alliteration has clearly dictated the content, not merely the form, of the tract. Thus the work to be done in May, June, and July concludes:

> sceap scyran. bytlian. boteatan. tynan. tymbrian. wudian. weodian. faldian. fiscwer. 7 mylne macian.[4]

We cannot suppose that alliterative effect did not influence the choice

1. While the copy of RSP translated for Quadripartitus must have differed slightly from the Old English text that survives, there can be no serious doubt that it included whatever changes were made to the text when it was linked with Gerefa.

2. Page, 'Proper Toil', 148–9; id., 'Two Problematic Old English Words', *Leeds Studies in English*, n.s. xvi (1985), 198–207, at 202–5; R. G. Poole, 'The Textile Industry in the Old English *Gerefa*', *Review of English Studies*, n.s. xl (1989), 469–78.

3. C. E. Fell, 'Some Domestic Problems', *Leeds Studies in English*, n.s. xvi (1985), 59–82, at 63.

4. Corpus Christi College, Cambridge, MS 383, fo. 67v.

of the occupations to be included. As Page has said: 'Whatever it is, *Gerefa* is not a straightforward account of running an estate.'[1]

We should thus see Gerefa as a literary exercise rather than as a didactic or administrative text, and the question then arises whether it had any model or precursor. Liebermann pointed out that in one passage in Gerefa there is an echo of Cato's De Agri Cultura (also called De Re Rustica); in 1886 he dismissed this as mere chance, saying that Gerefa was too slender to stand comparison with Cato's work, but in 1916 he suggested that the author had perhaps read Cato.[2] This suggestion has not found favour: Bethurum was 'not convinced' by it, and to Page Cato's possible influence 'looks slight', though he raises the possibility of a connection with Columella.[3] Both possibilities deserve serious consideration. There are six works of Classical Latin now known that might underlie a tract on agriculture. In chronological order they are: Cato, De Agri Cultura; Varro, Rerum Rusticarum Libri; Virgil, Georgicon; Columella, Rei Rusticae Libri; Pliny, Naturalis Historia, Book XVIII; and Palladius, Opus Agriculturae. Of these, only the works of Virgil, Palladius, and Pliny are believed from other evidence to have been read in late Anglo-Saxon England, but that need not be significant: our knowledge of what classical authors were available is bound to be patchy and is partly inferential.[4]

Gerefa falls into five sections:

1. The general qualities required in the reeve (paras. 1–5.1)
2. The management of workers (paras. 6, 7)
3. Work to be done in particular seasons and throughout the year (paras. 8–13)
4. Lists of equipment needed (paras. 14–17)
5. The need for the reeve to attend to every detail, and closing words (paras. 18–19).

In the works of Varro, Virgil, Pliny, and Palladius, there is nothing corresponding to more than two of these sections.[5] In Columella the first three sections are all represented, and at some length: the character and duties of the overseer (*villicus*), with his management of the workforce, are dealt with twice over, and the year's work is described in great detail month by month. There are, however, no lists of implements apart from a short account of the equipment needed for the olive har-

1. Page, '*Gerefa*: Some Problems', p. 217. Similarly Liebermann, 'Gerefa', 254: 'Der ökonom (*sic*) . . . konnte die schrift (*sic*) schwerlich praktisch brauchen.'

2. Loc. cit.; Liebermann, *Gesetze*, iii. 246.

3. Bethurum, 'Episcopal Magnificence', p. 168; Page, '*Gerefa*: Some Problems', p. 227.

4. J. D. A. Ogilvy, *Books Known to the English, 597–1066* (Cambridge, Mass., 1967), pp. 214, 222–3, 258–9, 261; H. Gneuss, 'A Preliminary List of Manuscripts Written or Owned in England up to 1100', *Anglo-Saxon England*, ix (1981), 1–60, at 53, 58.

5. Varro, book I, ch. xvii, xxvii, xxix–xxxvi, correspond to sections 2, 3; Virgil, book I, lines 160–75, 204–350, to sections 3, 4; Pliny, book XVIII, ch. viii, lix–lxxiv, to sections 2, 3; Palladius, book I, ch. xlii, and books II-XIII, to sections 3, 4.

vest.[1] In Cato, on the other hand, we find several long lists of implements: for an olive-yard, a vineyard, the pressing-room, and the oil cellar. The duties of the overseer (*vilicus*) and his management of the workers are described, and although there is no systematic account of each season's work, the start of such an account may perhaps be seen in a description of the work to be done in spring. There are also sections on work for winter evenings (as also in Columella) and wet weather, just as Gerefa describes indoor work for winter.[2]

Cato's text, though detailed and full of information on techniques of husbandry, is, in the form we know it, disjointed and unsystematic; Gerefa is orderly by comparison. However, if someone in the late tenth or early eleventh century, having read De Agri Cultura, felt moved to write a better arranged account for his own time, restricting it simply to the duties of an overseer, the result could well have been Gerefa. Columella's work is far longer than Cato's; it is systematic and comprehensive, though repetitious here and there. The first two chapters of Book XI deal first with the overseer's general character and duties, then with the work of each month. Gerefa could have originated as a contemporary version of this, vastly abbreviated, with the lists of implements added as the author's own idea. A late Anglo-Saxon writer would be no stranger to long lists of words and would need no classical model; indeed, one Anglo-Saxon glossary begins with a list 'De instrumentis agricolarum', while among the lists included in surviving scholastic colloquies, dialogues written as exercises in Latin, there is one of agricultural implements not unlike the first list in Gerefa.[3] It is hard to say which author, Cato or Columella, is more likely to lie behind Gerefa. The passage in Gerefa that Liebermann associated with Cato is not significant; the same precept – that the overseer should be familiar with all the operations of husbandry – can be found in Columella.[4] It is far from proved that either served as a model for Gerefa. All the same, it is difficult to believe that its author had not read one or other of these works.

Two points in Gerefa may well provide clues, not yet elucidated, to its literary background. One is the curious fact that it starts the cycle of the year's work in May. The starting-point we might reasonably expect on general grounds would be either the beginning of the agricultural year at the end of harvest, or else the turn of the year, the start of the calendar year in January. Classical precedent, if relevant, would also point to one or other of these times. Virgil and Pliny both start their

1. H. B. Ash *et al.*, *Lucius Junius Moderatus Columella on Agriculture* (3 vols., London/Cambridge, Mass., 1941–55), i. 78–95, iii. 50–127, 312–13.

2. W. D. Hooper and H. B. Ash, *Marcus Porcius Cato on Agriculture; Marcus Terentius Varro on Agriculture* (London/Cambridge, Mass., 1960), pp. 12–17, 22–9, 52–5, 56–9; Ash *et al.*, *Columella*, iii. 122–5.

3. T. Wright, *Anglo-Saxon and Old English Vocabularies*, 2nd edn., ed. R. P. Wülcker (2 vols., London, 1884), cols. 104–7; W. H. Stevenson, *Early Scholastic Colloquies* (Oxford, 1929), p. 4.

4. Hooper and Ash, *Cato*, pp. 14–15; Ash *et al.*, *Columella*, i. 88–9, iii. 52–7.

year after harvest; Varro, Columella, and Palladius in January or early February. If Cato's account of the work for spring represents the beginning of a year's cycle, he falls in the latter group, for his spring (*ver*) is the same as Lent (*onlængtene*), the last season of the cycle in Gerefa (indeed, setting vines and grafting occur in both). Discovery of a precedent for starting the year in May could well tell us something of the origin of Gerefa. It may, however, only reflect the Christian year: the first season after Easter, more likely to fall in late April in the Julian calendar than in the Gregorian. The other possible clue comes at the end of the work: the reeve should overlook nothing that might prove useful, not even a mouse-trap (*mus fella*) or, what is even less, a peg for a hasp (*hæpsan pinn*). This sounds like an allusion that the reader is expected to recognize, possibly a literary allusion. If so, it has eluded modern scholarship. It is conceivable, but unlikely, that the *mus fella* refers to the mouse-traps which Varro tells us were placed in the vineyards in Pandateria.[1] The *hæpsan pinn* would be harder to recognize in Latin guise; it might or might not be *gomphus*. But if these words are allusive, the allusion may not be literary at all, but topical or even personal, in which case it is lost forever.

Whether or not Gerefa had identifiable forebears, it clearly belongs to the literary, cultural, and scholastic world of the late tenth and early eleventh centuries, the world of Wulfstan, Ælfric, and Ælfric Bata, of glossaries and colloquies. It is unlikely to be much older. Its reference to building mills would fit this period better than an earlier one, and the word *gerefa*, meaning the manager of an estate, first appears in place of the specific *tungerefa* in the second half of the tenth century;[2] we may note in passing that the word does not appear in RSP. The most puzzling feature of Gerefa is that it is written in English. If what we had were a Latin text, it could be convincingly explained as an exercise for those learning the language. There would be obvious parallels for its use of an everyday subject to introduce the pupil to a wide vocabulary that would be genuinely useful in a monastic school where all conversation might have to be in Latin. One possibility is that the text was indeed first written in Latin, for just this purpose, and that what survives is an English gloss, just as we have a gloss as well as the original Latin for Ælfric's Colloquy.[3] Wulfstan (if it was he) would naturally have chosen the English, not the Latin, text to complement the English RSP. The difficulty with this explanation is the alliterative groups of words

1. Hooper and Ash, *Cato*, pp. 200–1.

2. P. Rahtz and D. Bullough, 'The Parts of an Anglo-Saxon Mill', *Anglo-Saxon England*, vi (1982), 15–37, at 18, 26–7; J. Bosworth and T. N. Toller, *An Anglo-Saxon Dictionary* (Oxford, 1898), and *Supplement* (Oxford, 1921), *sub* gerefa, tungerefa. The only occurrence of *gerefa* in this sense before the eleventh century seems to be in the (Northumbrian) gloss to the Lindisfarne Gospels (W. W. Skeat, *The Gospel According to St Luke in Anglo-Saxon and Northumbrian Versions* [Cambridge, 1874], p. 161; Ker, *Manuscripts containing Anglo-Saxon*, pp. 215–16).

3. Stevenson, *Colloquies*, pp. 75–102.

in the lists of implements. The alliteration elsewhere need be no more than a trick of style, possibly indeed imposed on the text at the Wulfstan revision, but the names of implements would have been hard enough to translate from what must have been a truly formidable Latin text, without introducing alliteration as well. These alliterative groups of words point strongly to original composition in English.

Another possibility is that the text was specially written as a supplement to RSP, that instead of taking an existing text to fill the gap in RSP's account of rural society, Wulfstan (or another) wrote one himself. This, too, is unlikely. The style of Wulfstan has not been detected uniformly throughout Gerefa, and it has not been suggested that he actually composed it, only that he revised a text which existed already. Moreover, RSP and Gerefa do not in fact fit together very well; each has a slightly different approach. RSP describes the obligations or perquisites of the various people on the estate, but says practically nothing about what they actually do; Gerefa describes the manager's job, but says nothing at all about what he specifically renders to his lord or receives in return. It is easy to see that someone might seize on Gerefa as an existing text, seeing it as a ready-made way of filling the gap in RSP's coverage. It is much harder to suppose that, if he sought to fill it with a text of his own composition, Gerefa would be the result. On the whole it seems most likely that Gerefa was originally written in English in the late tenth or early eleventh century as a literary exercise, perhaps as a simple *jeu d'esprit*, by someone who had read the agricultural work of either Cato or Columella. What was placed beside RSP may be simply an extract from a longer work which covered other topics dealt with by these authors. That is no more than a guess. All we can say for certain about Gerefa – and on this all are agreed – is that it was written as a literary work, not as a practical manual.

The contrast with RSP is striking. RSP is systematic, comprehensive, and thoroughly functional. Although there is occasional alliteration, at no point can we suppose that this has dictated the sense. There is less evidence here than in Gerefa of revision in Wulfstan's style, so what we have is presumably closer to the original text. RSP was a concise and practical statement of what an estate manager could expect to receive from other people on the property, or give out to them by way of perquisites. It begins (paras. 1–4.6) with those who held land of the estate owner as free men: the thegn, the *geneat*, the *kote setla*, the *gebur*. To the economic historian this order may seem strange: he would expect the *gebur*, who had a substantial amount of land, to precede the *kote setla*, who had only a small holding. But from the viewpoint of the estate manager the order is entirely logical. All the thegn's services are due to the king: the estate manager gets nothing from him. From the *geneat*, however, he gets occasional services and payments, from the *kote setla* some regular services as well, and from the *gebur* a great deal more. It is a straightforward progression. All the same, we may wonder whether the thegn really belonged to the original text. His obligations lay outside the estate and he might seem hardly relevant to it, though relevant to the general picture of rural society produced by the revision that added Gerefa to RSP. Whereas the obligations of the others are called *riht*, in singular or plural ('Geneat riht', 'Kote setlan riht', 'Geburgerihta'), those of the thegn are called *lagu* ('Đegen lagu'). *Lagu* is the only Scandinavian word in the whole text, and it is a word much used by Wulfstan.[1] There is a strong probability that the section on the thegn (paras. 1, 1.1) was added when the text was revised. Another passage which was almost certainly added then occurs at the end of the duties of the *gebur* (paras. 4.4–4.6). It consists of general precepts on the variety of local customs and the need for the manager to know those of his own area, and it stands out from its context of detailed obligations; it calls the manager 'seðe scire healde', a phrase that is used again in Gerefa.

After the free landholders, RSP turns to two parts of the estate's economy that may be farmed to local individuals apart from the demesne lands: the bees and the pigs. It is not difficult to see why the beekeeper and the swineherd were singled out like this. On some properties their work would have lain a long way from the estate centre – a very long way on some multiple estates – and effective supervision would have been difficult. A form of lease of the livestock in return for specific renders (and other obligations) would be a solution to this.[2] But alternatively the beekeeper and swineherd may be servile workers, like others on the demesne, and RSP, envisaging this possibility, adds two notes: the first is that on their death the same rules should apply to them both alike, and the second deals with the servile swineherd's perquisites. Text may well have been lost at this point, either at the Wulfstan revision or, more likely, in the copy that the reviser used; the first note seems inconsequential and leaves much to be explained, and with the second we would expect to be told also the perquisites of the servile beekeeper. The text then deals with other workers on the demesne. First it gives the perquisites that all servile workers are to have: the corn and other allowances in kind, first for the men, then for the women, and then the perquisites common to women and men alike – *feorm* at Midwinter and Easter, a strip of arable (*sulh æcer*), and at harvest an amount of corn (*hærfest handful*). There follows a section on the perquisites of the *folgere*. Then there is a series of sections giving the perquisites of each in turn of those doing specialized jobs on the demesne: the sower, the oxherd, the cowherd, the shepherd, the goatherd, and the cheesemaker. The only difficulty over the text's structure here is the section on the *folgere*.

1. Liebermann, *Gesetze*, iii. 244; Bethurum, 'Episcopal Magnificence', p. 165.
2. It may be relevant that RSP envisages the possibility that the lessee beekeeper and swineherd own horses.

The word *folgere* means, of course, simply 'follower' in general, but it must here be used in a specialized sense. The meaning generally attached to it derives from Cnut's secular law-code:[1] every free man, whether householder (*heorðfæst*) or *folgere*, should go twice a year to the hundred court. The contrast with *heorðfæst* is clear; and this, combined with the passage in RSP, led Thorpe, followed by Bosworth and Toller, and by others, to define *folgere* in this sense as 'a freeman who has no house or dwelling of his own, but is the follower or retainer of another (*heorðfæst*), for whom he performs certain predial services'.[2] But not only does that definition reflect a somewhat imprecise reading of RSP, which relates the perquisites of the *folgere*, not his services; it also reads more into the word's use in Cnut's law than the text justifies. Certainly the *folgere* there is specifically a free man, but there is no reason to suppose that the word itself is other than entirely neutral as between free and servile. The only other recorded instances of the word in this context seem to have no connotations of this sort; these are in the bishopric of Worcester estate survey of about 1170, where it occurs first in English, then in both masculine and feminine forms in Latin, and means simply the full-time workers of the manorial demesne.[3] If, like Liebermann and others, we take the *folgere* of RSP to be a free worker on the estate, the result is not nonsense: having described the perquisites of the servile demesne workers, the text now gives the perquisites of those who are free. But we should recognize that it really is no more than guesswork to assume that *folgere* implies a contrast with the workers who were servile: the supposed support from the law of Cnut is non-existent. On the other hand, we might justifiably take *folgere* to be any demesne worker, as in the Worcester survey, adding further perquisites to those just listed which they are all to have: the workers have already been referred to by three different names (*esne, ðeow, æhteman*), why not a fourth? But the text then does not make sense: we have just been told that the workers are each given a strip of arable; but the *folgere* gets two, one sown, one unsown. Clearly *folgere* is not just another name for the servile workers in general.

If there were a lacuna in the text at this point, so that the one word *folgere* was missing, we should unquestionably fill it with a word meaning 'ploughman', for three reasons. First, the ploughman, that is the holder of the plough, is an obvious and surprising omission from those named, with their particular perquisites, as doing specialized jobs on the demesne. Second, the position of the entry is exactly where, from later manorial records, one would expect the ploughman to be placed: the first of the demesne workers, followed by the sower, himself normally a ploughman

1. Whence the word passed to Quadripartitus, Leges Henrici Primi, and Bracton.

2. Thorpe, *Ancient Laws*, glossary, *sub* folgere.

3. M. Hollings, *The Red Book of Worcester* (4 pts., Worcestershire Historical Soc., 1934–50), pp. 37, 207 (date of survey from C. Dyer, *Lords and Peasants in a Changing Society* [Cambridge, 1980], p.4n.).

to whom this most responsible job was given, then by the oxherd who would have been the ploughman's partner, the man who guided the oxen.[1] Third, the perquisites of each of the specialized workers are related to the work they do, and those of the *folgere* would suit a ploughman well. It may well be that when the text was revised or copied, some word for ploughman was replaced by *folgere* through being misread or misunderstood, and that *folgere* itself never had this meaning. As other relevant instances of the word occur only in Cnut's laws, compiled by Wulfstan, and in a later record of Wulfstan's episcopal estates, we might reasonably suppose that the word was introduced into RSP at the Wulfstan revision. At the same time, it is not difficult to see how *folgere* could have come to mean ploughman. The most prominent, archetypal, demesne worker may have been given the name that really applied to them all, just as in later centuries *tascator* or tasker, meaning 'pieceworker' in general, came to mean 'thresher' in particular. Alternatively it may simply be the man who followed the plough (the *tentor* of later records), so called to distinguish him from his colleague who drove the oxen (the later *fugator*). This argument cannot be supported by any other instance where *folgere* does mean 'ploughman': it is simply a shot in the dark.[2] But so too is the explanation of *folgere* as a non-servile worker, while as 'ploughman' the word makes even better sense, not just in this passage but in the overall structure of the text. After the perquisites due to all the servile demesne workers we have the special gifts and payments given to workers doing special jobs, who are listed in just the order we would expect in a manorial document of the thirteenth or fourteenth century.

The text continues with the rewards of four men who in later parlance would be called manorial officers: the *bere brytta* (who presumably had charge of barn and granary and supervised the threshing), the beadle, the woodward, and the hayward. It is probably right to distinguish these from the demesne workers who precede them, though there is no clear distinction in the text. To do so would accord with later usage, and the beadle at least is said to be freer of work than other men ('weorces frigra ðonne oðer man'), implying that he is one of those who owe services, thus a landholding tenant, probably a *gebur*. At the same time, some or all of the specialized demesne workers whose perquisites are

1. Thus, for instance, 'Petrus bovarius tenet quinque acras et custodit boves domini. et vadit ad aratrum': J. E. Jackson, *An Inquisition of the Manors of Glastonbury Abbey of the Year MCLXXXIX* (Roxburghe Club, 1882), p. 37.

2. It is just worth noting that if *folgere* had and retained the meaning of 'ploughman' it would add a neat *double entendre* to the change in Langland's Piers Plowman from A version VI.33:

I haue ben his felawe this fiftene wynter

to B version V.549 (the C version is similar):

I haue ben his folwar al this fifty wyntre.

(W. W. Skeat, *The Vision of William concerning Piers the Plowman ... by William Langland* [2 vols., Oxford, 1886], i. 182–3). The phrase 'to follow the plough' is first recorded much later, but *sequi carectam*, 'to follow the cart', occurs in Wiltshire in 1278 and 1280 (M. W. Farr, *Accounts and Surveys of the Wiltshire Lands of Adam de Stratton* [Wiltshire Archaeological and Natural History Soc., Records Branch, vol. xiv, 1959], pp. 94, 111; I am much indebted to Dr R. Sharpe for this reference).

listed, from ploughman to cheesemaker, may also have been tenants, for if, as we are told happened in some places, the *kote setlan* did Monday works, the implication is that they were full-time workers on the demesne.[1] We may suppose that these *kote setlan* were distinct from the specifically servile demesne workers whose general perquisites are described before those of the specialized workers, and that the demesne labour force (including the specialists) comprised both cottage tenants and slave workers. This would accord with Domesday Book's consistent distinction between *bordarii* and *cottarii* on the one hand, *servi* and *bovarii* on the other. It is not unreasonable to use later records to help to interpret RSP: it was, after all, from the Anglo-Saxon structure that the later medieval forms developed. None the less, just because so much in RSP fits in well with what we find later, we should be particularly wary of reading beyond the strict letter of the text and of carrying back into earlier ages the general structure of rural society, of family and community, that we find in, say, the thirteenth century. The formal distinction between demesne worker and manorial officer, between the cheesemaker in one sentence and the *bere brytta* in the next, may have been quite different (and much less clear) in the age when RSP was written from what it came to be subsequently.[2]

After the woodward, the last of the four officers, RSP concludes with some general comments on the variety of local customs and the need to know and observe them. This passage (paras. 21–21.5) was almost certainly added when the text was revised and combined with Gerefa, like the passages, already noted, on the duties of the thegn and concluding the section on the *gebur*. With these should be placed two shorter comments which may be seen as interpolations in the text at this time: the comment that the beekeeper should do many things that cannot be set out in full (para. 5.4) and the note, 'swa ic ær be beo cere cwæð', in describing the obligations of the swineherd (para. 6.3).[3] These include all occurrences of the words *lagu* and *scir* in the text and all occurrences of the first person singular. They are mostly concerned with generalities – the variety of customs, the impossibility of enumerating them all – and when they refer to specific obligations or rights they simply list their names with little or no further definition (paras 1, 1.1, 4.5, 21.4).[4] They also include several of the phrases that recur in Gerefa – several but not all. The work of revision was certainly not confined to these

1. M. M. Postan, *The Famulus* (Economic History Review Suppl. 2, [1954]), pp. 31–2.

2. Both Kemble, *Saxons in England*, i. 214–15, and Corbett, 'England from A.D. 954', pp. 403–4, see no distinction between these manorial officers and the other demesne workers.

3. I strongly suspect, from their content and phrasing, that paras. 4.3b, c and 5.5 also date from the Wulfstan revision, but reserve the unavoidably lengthy discussion of this for a future occasion.

4. It is fair to add, though, that this last might equally be said of para. 2, the duties of the *geneat*, which I do not doubt belonged to the original text.

sections, though they seem in their contents to stand out most clearly from the rest.[1] It is possible, too, that in the process of bringing together and revising the two texts, phrases already in RSP as originally written were introduced into Gerefa. The occurrence of the same phrase in the two texts need not mean that it was an interpolation in both. Detailed work on the language may establish more precisely the extent and limits of the changes made on revision.

Removal from RSP only of what seem the most obvious additions to the text leaves us with a document that is even more businesslike than before. It is exact, detailed, and wastes no words. When was it written? We have seen that the revision which linked it with Gerefa, whether or not by Wulfstan himself, was probably made in the early eleventh century; that Gerefa was probably written not long before; and that RSP, whether or not regarded as a single text with Gerefa, has commonly been dated to the same period. For well over a hundred years after its discovery in 1833, historians were reluctant to consider that RSP might be any older than this – indeed, it came to be seen as interestingly early evidence of the conditions it describes. It was generally accepted that Viking attacks and other external pressures changed a society of free Germanic settlers into a subject peasantry bound to local lords by labour services and other demands, and that the last stages of this 'manorialization' came with the Norman Conquest. Into this scheme an early eleventh-century RSP could only just be fitted: it shows a rural society, far advanced along the manorializing road, where the local tenants – *gebur* and *kote setla* – owe heavy predial services. No wonder that Maitland called it 'that mysterious document', or that to Douglas and Greenway it was 'a description of agrarian conditions in the time of Edward the Confessor'.[2] H. P. R. Finberg blamed 'Maitland's polemic' for relegating the Hurstbourne Priors statement of services to as late a date as about 1050.[3] Maitland, however, was no polemicist but a great historian; and it was a much wider consensus which maintained the broad model, with significant variants, of an Anglo-Saxon society that developed from general freedom to widespread serfdom, a model which required the latest possible date for Hurstbourne services and RSP alike. Traces of this model can be found in English and German historians long before Maitland

1. In trying to identify passages added at the Wulfstan revision I have been guided purely by content and language, and at no point consulted the details of the conclusions that Bethurum reached from a consideration of style. The passages where she particularly sees Wulfstan's hand at work are, in fact, paras. 3.4, 4.2, 4.4, 5.1, 5.4, 6.3, and 21–21.5 (Bethurum, 'Episcopal Magnificence', p. 165). It is interesting that our two lists overlap as much as they do and that we both (independently) see the use of the first person singular as a sign of the reviser. That the two lists do not coincide does not reflect on either approach: the reviser may have adjusted the style of a passage with little change to its language and none to its content, and Wulfstan himself did not invariably write in idiosyncratic style (Hollowell, 'On the Two-Stress Theory').

2. F. W. Maitland, *Domesday Book and Beyond* (Cambridge, 1897), p. 37; Douglas and Greenway, *English Historical Documents*, p. 813.

3. Finberg, 'Anglo-Saxon England', p. 452n.

– the historians who first discovered and discussed RSP – and it continued long after his time, most notably in the work of F. M. Stenton.[1] Work that ran counter was apt to be ignored or dismissed as eccentric and irrelevant to scholarly understanding of this period. Finberg cites Seebohm as a victim of this,[2] but in work on RSP a sadder case of neglect was C. M. Andrews. His book, *The Old English Manor: A Study in English Economic History*, published in 1892, takes RSP as the basis of its account of Anglo-Saxon society. It is erudite, contains much of interest, and anticipates some of the conclusions of later scholarship. It argues that manorial organization originated in tribal settlement, under hereditary leaders, in the first stages of English occupation, though with some reservations: 'We look for no singleness of origin nor uniformity of custom in the study of the Saxon manor.'[3] The implications of the last two words excluded the work from contemporary mainstream scholarship in Anglo-Saxon history. It deserves reassessment, now that Aston, Finberg, and others have long since successfully challenged the old certainties. Now too, we can re-examine the date of RSP without calling in question entrenched views on broad development.

The closest indication Liebermann offers of a *terminus a quo* for RSP lies in its references to money for ships and to the *trimoda necessitas* of service on bridges, fortifications, and in the *fyrd*, which are first mentioned in Æthelræd II's law-codes;[4] but this need not preclude an earlier date, and in any case they are mentioned only in the services of the thegn, likely on other grounds to have been added to the text at the Wulfstan revision. Some features of RSP point to an earlier origin. It makes no mention of church, smith, or (unlike Gerefa) mill, though this need not be inconsistent with the early eleventh century: mills and churches were only then reaching the local community and we cannot formally demonstrate that the smith was there until the twelfth century, though we may reasonably guess that he was the first of the three to arrive.[5] Two words in RSP seem to point to a date before the eleventh century. One is *ealdor man* in the sense of overseer or manager, here the overseer of the estate, the *tungerefa*; it is otherwise known with this

1. Thus, Lappenberg, *History of England*, ii. 318–20, 332; Kemble, *Saxons in England*, i. 53–6; Maitland, *Domesday Book and Beyond*, pp. 221–3, 318–26; F. M. Stenton, *Anglo-Saxon England* (2nd edn., Oxford, 1947), pp. 463–5. Three essays by Stenton, collected in his *Preparatory to Anglo-Saxon England*, ed. D. M. Stenton (Oxford, 1970), are of especial interest on the historiography of this question and on Stenton's own view: 'Frederic Seebohm' (1912; pp. 16–22), 'Early English History, 1895–1920' (1946; pp. 346–56), and 'The Thriving of the Anglo-Saxon Ceorl' (1958; pp. 383–93).

2. H. P. R. Finberg, *Lucerna: Studies of some Problems in the Early History of England* (London, 1964), pp. 131, 136.

3. C. M. Andrews, *The Old English Manor: A Study in English Economic History* (Baltimore, 1892), p. 60.

4. Liebermann, *Gesetze*, iii. 245.

5. Rahtz and Bullough, 'Anglo-Saxon Mill', 18, 26–7; J. Blair, 'Introduction: From Minster to Parish Church', in *Minsters and Parish Churches: The Local Church in Transition, 950–1200*, ed. Blair (Oxford U. Comm. for Archaeology, Monograph xvii, 1988), pp. 1–19, at 7–8; P. D. A. Harvey, 'Non-Agrarian Activities in Twelfth-Century English Estate Surveys', in *England in the Twelfth Century*, ed. D. Williams (Woodbridge, 1990), pp. 101–11, at 108, 110–11.

meaning only from King Alfred's English translation of Bede's Ecclesiastical History, where Cædmon went 'to þam tungerefan, se þe his ealdorman wæs'.[1] The other is *pund* as a measure of capacity, the meaning it almost certainly bears in RSP, where the servile workers' corn allowance is 12 *pund* for a man, 8 *pund* for a woman. This usage occurs in a manuscript of medical recipes in the mid-tenth century, where we are told, for instance, that a *pund* of ale weighs six pennies more than a *pund* of water.[2] A mid-tenth-century date for RSP seems reasonable and is not so far from Bethurum's reference to Bishop Oswald of Worcester (961–92). All the same, we should not overlook the possibility that it may be significantly older.

What was its purpose? There is a clear and interesting distinction in the text. On the one hand are the landed tenants, and those who held the bees and the pigs at farm: in every case their obligations are said to vary from place to place according to local custom, so that only what is the norm or typical can be described. On the other hand are the servile beekeeper and swineherd, the specialized demesne workers, and the four officials entered at the end: their perquisites are all defined precisely and with no proviso of that sort. This strongly suggests that the document was written for a single estate, where local custom might produce variations in tenure and farm, but where a single authority could produce uniformity in perquisites. A text written about estates in general would find variety in perquisites as well. It follows that RSP is a functional, administrative text, not a descriptive one; it tells the local manager what he might expect to get from the landed tenants and the farmers of stock, and exactly what he may or should allow as perquisites to the demesne workers and local officers. Against this might be argued the references to *his hlaford* in the obligations of *kote setla* (twice) and *gebur* (once) and the perquisites of the hayward (once): they imply that the lord might differ from one case to another. But elsewhere the text reads simply *hlaford* or (once) *se hlaford*;[3] we may reasonably assume that 'his' was added at the Wulfstan revision, widening the text's meaning so as to give it general application. Almost certainly RSP was composed to guide the local managers of a single estate-owner, a single *hlaford*, perhaps to be read to them, perhaps to be referred to as a standard rule. It says nothing of their own obligations, the amount of farm they are to render to the lord, for this may have varied from place to place and was in

1. Bosworth and Toller, *Dictionary: Supplement*, *sub* ealdorman.

2. Eid., *Dictionary*, *sub* pund III; O. Cockayne, *Leechdoms, Wortcunning, and Starcraft of Early England* (3 vols., Rolls Ser., 1864–6), ii. 298 (the text, from B[ritish] L[ibrary] Royal MS 12 D. xvii, is dated mid-tenth century by Ker, *Manuscripts Containing Anglo-Saxon*, p. 332); C. E. Fell, 'Old English *Beor*', *Leeds Studies in English*, n.s. viii (1975), 76–95, at 84. *Pund* is also used of liquids in A. S. Napier, *The Old English Version of the Enlarged Rule of Chrodegang* (Early English Text Soc. o.s. vol. cl, 1916), p. 15, but as it is simply translating *libra* this need not be significant.

3. The one reference to *se hlaford* occurs in para. 5.5, which I suspect (on other grounds) was added at the Wulfstan revision (supra, p. 16, n.3).

any case quite a different matter. What RSP does is to set standards for the internal regulation of the manors of the estate.

Some passages in the Carolingian capitularies deal with the same sort of topic as RSP or Gerefa,[1] and we may well wonder whether RSP was written for royal estates. Certainly RSP has no textual link with the capitularies, nor in fact is there any conceptual link: none served the same function. None of the relevant capitularies looks at what is to be given or received from the viewpoint of the local official; all are concerned with higher levels of oversight.[2] There is a clear distinction in RSP between *cynig* and *hlaford*, and, unless it underwent more change on its revision than we need suppose, it was written not for royal administration but for the estate of some magnate. Where did this estate lie? Liebermann, we have seen, determined reasonably enough that RSP came from Wessex or southern central Mercia, but we may be a little more precise. The name *bere brytta* for the officer in charge of the granary is an unusual one; it is used, occasionally, down to the fifteenth century, but is known only in Hampshire, Wiltshire, and Somerset – particularly east Somerset and west Wiltshire.[3] Liebermann and others have drawn attention to the connection between RSP and the Old English estate survey from Tidenham in Gloucestershire; in both the *geneat*, among other things, is to 'ridan 7 auerian 7 lade lædan'. This survey is preserved in a cartulary of Bath Abbey, and Tidenham belonged to Bath from 956 until a date between 1061 and 1065; the survey cannot be more precisely

1. As has been pointed out, with reference to Gerefa, by Bethurum, 'Episcopal Magnificence', p. 169, and by Rahtz and Bullough, 'Anglo-Saxon Mill', 20n.

2. Relevant passages are in A. Boretius and V. Krause, *Capitularia regum francorum* (2 vols., Monumenta Germaniae Historica, Legum sectio II, Hanover, 1883–97): i.61 (a summary, but comprehensive, list of rural works forbidden on Sundays), 81–2 (labour services due from the tenant of a quarter-*factus* of land or less), 83–91 (Capitulare de Villis: rules for managing royal estates), 172 (character and duties of *vilicus bonus*), 254–6 (lists of goods on royal demesnes), 286 (agricultural services due), ii.437–8 (character and conduct of those in charge of royal manors).

3. H. Kurath and S. M. Kuhn, *Middle English Dictionary* (Ann Arbor, 1956–), *sub* berebret; R. E. Latham *et al.*, *Dictionary of Medieval Latin from British Sources* (London, 1975–), *sub* berebretus. Of the references there that do not identify the place, 'PR Winch. 1209, 23/56, 80/16' relates to Cheriton and Overton (Hants); '*Pat. R. Edw I* 1277' to a suit where the Abbot of Keynsham (Som.) was a defendant; '*Ac. Cornw.* i.62' to Mere (Wilts.); 'PRO SC 12 [*recte* 11] /700 m.1' to Ashton (Wilts.); 'Pipe Winc. 159405, r 24d' to Cheriton (Hants; I am grateful to Miss H. M. Forbes for this information). The word also occurs at South Wraxall (Wilts.) about 1180 (BL, Harley MS 61, fo. 85r; I am grateful to Dr Ann Williams for advice on the date); at Pilton (Som.) in 1189 (Jackson, *Inquisition*, p. 32); at Downton (Wilts.) in 1209 and 1211 (H. Hall, *The Pipe Roll of the Bishopric of Winchester, 1208–1209* [London, 1903], p. 20; N. R. Holt, *The Pipe Roll of the Bishopric of Winchester, 1210–11* [Manchester, 1964], p. 37); at Alresford (Hants) in 1211 (ibid., p. 144); at Overton (Hants) in 1232 and 1252 (Hants Record Office, Eccl. Com. 2/159282, 159447); at East Knoyle (Wilts.) in 1232, 1252, 1401, and 1426 (ibid. 159282, 159447, 159405, 159427); in an accounting treatise copied at Luffield Priory (Northants) in the late thirteenth century but of unidentified origin (Cambridge University Library, MS Ee. I 1, fo. 231v; cf. Oschinsky, *Walter of Henley*, pp. 29, 32); in the fourteenth-century estate rules of the Mohun family of Dunster (Som.; ibid., pp. 476–7); and at Rimpton (Som.) in 1341 and 1358 (Hants Record Office, Eccl. Com. 2/159351, 159368; I am grateful to Dr C. Thornton for this information). The Hampshire references are all from the estates of the bishopric of Winchester, and the word may well have been transferred from its manors in the other two counties.

dated.[1] The simplest explanation of the connection between the two texts is that RSP was composed for the estates of St Peter's, Bath, which resumed its earlier role as a monastery about 963, but which probably then already had a long history as a property-owning minster church.[2] Bath will thus have given us not one but two of our few surviving written relics of Anglo-Saxon estate administration. Nothing in either document is inconsistent with this conclusion, and we may even conjecture that the general principles embodied in RSP were supplemented for each manor by the more specific statement of services represented by the Tidenham survey. There are, however, other possible explanations for the connection between the survey and RSP. A text could easily move, especially from one monastic house to another, and all that we can be reasonably certain of is that RSP originated in south-west England, very likely in west Wiltshire or east Somerset.[3] Glastonbury comes at once to mind as a possible alternative place of origin.

That the text at some point did move is shown not only by the Wulfstan revision (if this really was by Wulfstan), but by what is in some ways the most puzzling passage of the whole document. It tells us, of the farmer of the bees (para. 5.1), that 'mid us is ge ræd þæt he sylle V sustras huniges to gafole'. This is the only point where the text uses the first person plural. If our interpretation is correct it cannot be part of the original text, for self-evidently everything described was 'mid us', not just one custom. On the other hand, it cannot have been added on the Wulfstan revision, for if we have correctly judged both the purpose and the effect of this revision, it broadened and extended the text's application, substituting the general for the particular, and not introducing new, purely local, detail. We can only assume that at some time before it reached Wulfstan – or whoever else revised it – the text, composed for one estate, had already been copied and annotated for use on another, perhaps even the bishop of Worcester's. The Wulfstan revision may well have changed other similar annotations to give them general application, simply leaving this one unaltered by an oversight; they may well lie

1. W. de G. Birch, *Cartularium Saxonicum* (3 vols., London, 1885–93), iii. 102–3; Liebermann, *Gesetze*, iii. 245; T. H. Aston, 'The Origins of the Manor in England', in *Social Relations and Ideas: Essays in Honour of R. H. Hilton*, ed. Aston *et al.* (Cambridge, 1983), pp. 1–43, at 9n.

2. *The Victoria History of Somerset* (London, 1906–), ii. 69–70. I am grateful to Mr C. P. Wormald for pointing out the interesting fact that in MS 383, RSP immediately follows the text known as Dunsaete, which relates to arrangements in an area on the River Wye that may have adjoined Tidenham. In his forthcoming *The Making of English Law: King Alfred to the Norman Conquest*, chs. iv and v, he discusses the possibility that the treaty known as the second code of Æthelræd II, Dunsaete, and RSP were brought together at Bath as a small collection of texts.

3. It may seem inconsistent to attach importance for the date of RSP to the words *ealdorman* and *pund* and for its place of origin to *bere brytta*, while ignoring its use of the word *metecu*, otherwise known only in the will of Thurketel of Palgrave (Suff.), which dates from 'before 1038' (D. Whitelock, *Anglo-Saxon Wills* [Cambridge, 1930], pp. 68–9, 179–80). But in *ealdorman* and *pund* (as with *gerefa*) it is not the word itself, but the more changeable usage, that is in question, and the regional use of *bere brytta* is attested by not one but many examples. *Metecu*, however rare, is a self-explanatory compound, and it is not unreasonable to suppose that the two documents in which it appears might be separated by two or more generations in time and by the breadth of England in space.

behind some of the customs said to obtain 'on suman landum' (or similar phrase).[1] It is interesting that the author of Quadripartitus made just such a change in this passage, rendering 'mid us' as 'In quibusdam locis'.[2]

RSP is thus a complex text, comprising at least three, probably four, levels of composition and adaptation. Even if the general outline suggested here is accepted, there is still much to be done by way of refinement and correction through detailed work on the text's language and content. It will be especially interesting if Old English scholars can either confirm or challenge Bethurum's attribution to Wulfstan of the text as we have it. In a sense we have two quite different documents in one: on the one hand, it forms the core of an eleventh-century treatise on the ordering of society in a rural community and, on the other, it is a tenth-century record of estate administration. In both roles it is a record of great importance. Over the past fifty years we have become increasingly aware of the part played by written records in the practical work of estate management in Anglo-Saxon England, but we have no more than a few fragments of relevant texts. RSP may reasonably be seen as the most informative survivor.

NOTE: Two mistakes have been corrected. On p. 5, n. 2, 'ræƿan' (with a wen) replaces 'ræþan' (with a thorn), a printer's error introduced after page proofs had been returned; and on p. 8, line 5, 'late eleventh or early twelfth' replaces 'late tenth or early eleventh', an error entirely the fault of the author.

1. Paras. 2, 3, 4, 4.3, 4.5, 5.1, 6.1.

2. Just as, by way of generalizing the precepts of RSP, Quadripartitus often translates not only *his hlaford*, but *hlaford* by itself, as *dominus suus*.

II

The Manorial Reeve in Twelfth-Century England

We know a great deal about the manorial reeve from the first half of the thirteenth century onward. He stands before us in surveys, in court rolls and, above all, in manorial accounts: the villein tenant who ran the lord's demesne for him under the successive scrutiny of bailiff, steward and auditor and who had to account to his lord for every penny received, every peck of corn harvested, every egg laid. We may see him as an oppressed victim of the contemporary economic system or as peculiarly well placed to profit from it: this depends on our own point of view, on the particular records, perhaps on the particular reeve. But we are not short of the information we need in order to form our opinion.

This is not the case when we look at his twelfth-century predecessor. The document we now call *Rectitudines singularum personarum*, which originated perhaps in the mid-tenth century and which gives an account of the rights and duties of the various workers on a rural estate, may well have been written for the guidance of local reeves; the reeve is the one person left out. In the early eleventh century it was read by someone, possibly Bishop Wulfstan of Worcester, who saw that by supplying this omission it could be turned into a little treatise on rural society in general and he added a postscript on the duties of the reeve that is now known as *Gerefa*, the Anglo-Saxon word from which 'reeve' derives. This text, however, seems to owe more to classical literature than to contemporary English practice and we arguably learn more of the Anglo-Saxon reeve from the gap in the *Rectitudines* than from Wulfstan's – if it was his – attempt to fill it.[1] But this is the fullest, indeed the only, detailed picture of the rural reeve before the records of the thirteenth century: for the intervening period we have only casual references.

If we look at these as far back as the late eleventh century and domesday book we are largely breaking new ground. Vinogradoff makes some interesting comments, but he never looks at the reeve in detail.[2] Stenton devotes a paragraph of the introduction to his Danelaw charters to the reeves that they mention, and more recently DE Greenway has usefully brought together the references to reeves in the Mowbray charters she edited.[3] However, the only general discussion is in five pages of Lennard's *Rural England 1086–1135*, though for the reeves in domesday book this

1 PDA Harvey, '*Rectitudines singularum personarum* and *Gerefa*', *English Historical Review* cviii (1993).

2 P Vinogradoff, *Villainage in England* (1892), 356–8; P Vinogradoff, *The growth of the manor* (1904, 2nd edn 1911), 191–2, 271–2.

3 *Documents illustrative of the social and economic history of the Danelaw*, ed FM Stenton (1920), cvi–vii; *Charters of the honour of Mowbray 1107–1191*, ed DE Greenway (1972), xlv–vi.

has been valuably supplemented by James Campbell in his contribution to the essays commemorating the ninth centenary of 1086.[4] It is time to look at these reeves again in the light of other recent work on the period, in the light especially of Rosamond Faith's book which brings together so much recent exploratory work on the Anglo-Saxon and Anglo-Norman periods; with its new interpretations and insights it is a real starting-point for looking at the period afresh, through new eyes.[5]

The Latin records in which we meet the thirteenth-century manorial reeve refer to him as *prepositus*, or occasionally by the all-purpose word *serviens* (employee). *Prepositus* corresponded to one aspect of his work – he was set over his fellow villagers to command their services for their lord – but if he had first been given a Latin name in the thirteenth century he would more appropriately have been called *villicus*, a manager of rural property. But the translation of the English *gerefa* as *prepositus* goes back well before the thirteenth century and it is nearly always as *prepositus* that we meet the rural reeve in earlier Latin texts. Nonetheless in the Shaftesbury Abbey surveys of about 1130 and 1180 he is more often *prefectus* and is probably twice called *pretor*.[6] *Prefectus* occasionally occurs in other sources and in the so-called laws of Henry I at the beginning of the century we find *tungrevius*, a direct Latinization of the English *tungerefa*, reeve of the vill.[7]

The word occurs in the laws of Henry I in a list of those who are to attend the shire court, from the bishop downwards. The rural reeves have to be called *tungrevii* because both *prepositi* and *prefecti* have already been included in the list.[8] This introduces a problem in looking for these rural reeves in twelfth-century sources: they may have been nearly always *prepositi*, but so were other people as well.[9] The *prepositi* of the laws of Henry I were probably royal reeves, agents of the sheriff; they are often mentioned, usually *prepositi* sometimes *prefecti*, among them the reeves of hundreds who are sometimes distinguished from the others.[10] By the end of the twelfth century *prepositi*, royal reeves, were often included in the list of officers to which royal letters patent were addressed.[11] To ask whether the reeve of a royal manor is to be seen as one of these royal reeves or simply as a manorial reeve like any other is to make a distinction that might not have been understood in the twelfth century; the royal manor was after all in the sheriff's keeping. There were other *prepositi* as well. There were the reeves of a borough, royal or seigneurial.[12] The second-in-command of an abbey, the

4 R Lennard, *Rural England 1086–1135* (1959), 271–6; J Campbell, 'Some agents and agencies of the late Anglo-Saxon state' in JC Holt (ed), *Domesday studies* (1987), 205–7, 215–16.

5 RJ Faith, *The English peasantry and the growth of lordship* (1997).

6 London, British Library, Harley MS 61, fos 37r–89r passim (*prefectus*), 80r (*prepositus*), 59v, 69v (*pretor*; though Henry *pretor de operacione augusti* on fo 59v might suggest the *budel* of other entries rather than a reeve, this is presumably the same Henry who appears in the list of jurors on fo 59r as *prefectus*).

7 Thus *Domesday book, seu liber censualis Wilhelmi primi regis Angliae* [ed A Farley] (2 vols 1783) i, fo 218v and *Sir Christopher Hatton's book of seals*, ed LC Loyd and DM Stenton (1950), 243; *Leges Henrici primi*, ed LJ Downer (1972), 98.

8 'centenarii, aldermanni, prefecti, prepositi, barones, vavassores, tungrevii et ceteri terrarum domini'.

9 As Lennard, *Rural England*, 274 points out.

10 WA Morris, *The medieval English sheriff to 1300* (1927), 53–6; *Domesday book* i, fo 218v.

11 Thus *A formula book of English official historical documents*, ed H Hall (2 vols 1908–9) i.21 (letters patent of 1194).

12 Thus *The great roll of the pipe for the eleventh year of the reign of King Henry the second* (Pipe Roll Society old series viii 1887), 12 (Gloucester), 39 (Lincoln); *Earldom of Gloucester charters*, ed RB Patterson (1973), 37 (Bristol), 157 (Cardiff).

prior, might be called *prepositus* following the usage of the rule of St Benedict and so might the provost of a cathedral chapter.[13] A local official with some particular function might be called reeve, like the reeve of the market (*prepositus fori*) at Glastonbury.[14] Usually it is fairly clear what sort of reeve is meant, but some instances are ambiguous and it may be easier than it seems to fall into error.

Prepositus means the man set before or above his fellows and the laws of Henry I make the point by contrasting him with the *subditi* or *subjecti*, those who are under his authority.[15] Was the title *prepositus* more appropriate to the eleventh-century manorial reeve than it was to his thirteenth-century successor, or did he acquire it only by analogy because it was applied more realistically to other kinds of *gerefa*? This is an interesting question to bear in mind: the title may or may not reflect actual change in status. It is a question that underlies the three questions about the twelfth-century manorial reeve to be specifically addressed here successively in ascending order of difficulty: who was he? whose was he? and what did he do?

Who was he?

The reeve was probably always a manorial tenant. We meet him in this guise on every one of the thirteen private estates from which we have twelfth-century surveys.[16] He was a peasant, *villanus*, with usually a standard holding: a yardland (virgate), a half-yardland, two oxgangs (bovates). Occasionally he held more – two yardlands, half a hide – or he might supplement his holding by taking over the lord's mill or further lands and by the late twelfth century we see reeves participating in the local land market, acquiring and disposing of pieces of free land by charter.[17] None of this set the reeve apart from his fellow tenants. At Pittington, county Durham, about 1150 a reeve was put in place by a written charter: Baldwin was granted land and a mill in return for serving as reeve or, if he relinquished office, for 10*s* a year and two sestars of honey.[18] The surveys often acquit the reeve of part or all of the obligations due for his holding in return for his service; it would accord with thirteenth-century practice for this to happen even if not actually mentioned in the survey.[19] On three manors in

13 *RB 1980: the rule of St Benedict, in Latin and English*, ed T Fry (1981), 284–6; *Liber eliensis*, ed EO Blake (Camden 3rd series xcii 1962), 219, 290, 293, etc; EU Crosby, *Bishop and chapter in twelfth-century England* (1994), 346.

14 *An inquisition of the manors of Glastonbury Abbey of the year M.C.LXXXIX*, ed JE Jackson (Roxburghe Club 1882), 13.

15 *Leges Henrici primi*, ed Downer, 174–7, 194–5, 363.

16 Two additions should be made to the eleven non-royal estates listed in PDA Harvey, 'Non-agrarian activities in twelfth-century English estate surveys' in D Williams (ed), *England in the twelfth century* (1990), 101 n2. The first is Bury St Edmunds Abbey, whose early or mid-twelfth century surveys are in *Feudal documents from the Abbey of Bury St Edmunds*, ed DC Douglas (1932), 25–44; the second is Abingdon Abbey, whose surveys of 1185–90 are in *Chronicon monasterii de Abingdon*, ed J Stevenson (2 vols Rolls Series 1858) ii.299–309, 324–34. Douglas (*Feudal documents*, lxv) dated these Bury surveys to the late eleventh century, connecting them with documents that underlie entries in domesday book, but I do not doubt that they are significantly later.

17 Lennard, *Rural England*, 273; *Cartularium monasterii de Rameseia*, ed WH Hart and PA Lyons (3 vols Rolls Series 1884–93) iii.243 (two yardlands), 270 (demesne meadow); E Miller, *The abbey and bishopric of Ely* (1951), 284–5; *Documents of the Danelaw*, ed Stenton, cvii n1.

18 *Feodarium prioratus dunelmensis*, ed W Greenwell (Surtees Society lviii 1872), 129 n1.

19 Thus *Boldon buke*, ed W Greenwell (Surtees Society xxv 1852), 7, 10, 12 (fully remitted); *Inquisition*

the surveys of about 1195 from Ramsey Abbey the reeve's holding and services are the paradigm for the rest; on three manors in the Templars' survey of 1185 the reeve is the first tenant named.[20] In many other cases too the reeve is probably placed at the head of the lists of tenants, but without being identified; on many of the manors described in the surveys no reeve is named among the tenants but this does not mean that he was not there. On several manors of the Glastonbury estates in 1189 there was a reeve among the jurors who provided the information for the survey but not among the tenants – where, however, he probably appears without being thus identified.[21] Thus, to take an unusual name, the Sericus *prepositus* who was a juror at Christian Malford was presumably the same as the Sericus who had a cottage holding and one acre of demesne.[22]

In charters too we must meet many reeves without recognizing them. In a charter of 1174–81 quitclaiming three tenants, presumably at Maplestead in Essex, to the canons of St Paul's the witnesses include Hugh the reeve of Maplestead but in the actual grant of the same three tenants, a charter with a very similar witness list, we have instead simply Hugh of Maplestead.[23] We need not doubt that it was the same Hugh. That the reeve was sometimes so called, sometimes not, itself tells us something of the office and how it was regarded: it was important, significant in some contexts, but not so important that it must always be mentioned as its holder's outstanding characteristic. The reeve may occasionally emerge from anonymity, but he returns to it again all too easily.

Along with the estate surveys, charters are our best source of information on twelfth-century reeves. As in charters of the thirteenth century, the reeve is sometimes mentioned as the tenant of property adjoining the free holdings that are being conveyed but also, quite unlike the thirteenth century, he appears in the lists of witnesses to the transaction. There is no more cogent contrast between the manorial reeve of the twelfth century and his thirteenth-century successor than the fact that whereas he often witnessed charters his successor did not, even if this is no more than a simple but telling illustration of the development of the law of villeinage: not just the reeve but any villein tenant was now unable to attest a conveyance of free property. It is only rarely that we find someone who is apparently a manorial reeve witnessing a charter that can be firmly dated even a few years after 1200.[24]

Demonstrably neither surveys nor charters can be relied on to tell us when they are speaking of a reeve, so we can reasonably suppose that there may have been far more of them than these sources suggest. The note about the making of domesday book at the start of the related Ely survey, the *inquisitio eliensis*, tells how inquiries were made of the sheriff, the barons and their French retainers, the whole of each hundred and the

of Glastonbury, ed Jackson, 46–7 (half remitted); *Cartularium de Rameseia*, ed Hart and Lyons, 245, 246, 250 (no remission recorded).

20 *Cartularium de Rameseia*, ed Hart and Lyons, 245, 246, 250; *Records of the Templars in England in the twelfth century*, ed BA Lees (1935), 39, 46, 47.

21 *Inquisition of Glastonbury*, ed Jackson, 103, 112, 120, 127.

22 Ibid 112, 114, 115.

23 *Early charters of the cathedral church of St Paul, London*, ed M Gibbs (Camden 3rd series lviii 1939), 202–3.

24 Examples are in *Registrum antiquissimum of the cathedral church of Lincoln*, ed CW Foster and K Major (Lincoln Record Society 10 vols 1931–73) iii.325–6 (1203–4), vii.44 (1205–6); where, ibid vii.57, a *prepositus* witnesses a grant of c1215–19 we may wonder whether this can be a manorial reeve or even whether the date is correctly assigned.

priest, the reeve and six inhabitants of each vill.[25] Likewise the laws of Henry I rule that if the lord or his steward cannot attend the shire court or the hundred court the reeve, the priest and four of the best persons of the vill should go to represent everyone not summoned by name.[26] The assumption is that every vill will have a reeve. That this ruling was not mere theory is implied by a charter, probably of the late 1150s, in which Roger de Mowbray gave to the church of South Cave in the east riding of Yorkshire a house in the marketplace there. Fifteen important witnesses – the archdeacon of York, the abbot of Selby, canons and knights – are followed by six more 'of the vill of Cave' (*et de villa de Cave*): first was Hubert, the vicar, then four other names and finally Reginald, reeve of the same vill (*prepositus eiusdem ville*), that is the priest, the reeve and four others, just as the laws of Henry I required.[27]

David Postles has recently shown how seriously we should take the lists of witnesses to twelfth-century charters: they were not just names snatched from the air.[28] In this grant from Cave we can reasonably assume that we are looking at a transaction concluded in the hundred court where the local representatives were, sensibly enough, included among the named witnesses and in this case actually identified as such. How often they occur in other witness lists without being identified is almost anyone's guess: the conveyances recorded in charters must often have been made in the court and the reeve named among the other witnesses, whether or not he is actually called *prepositus*. Where a reeve does appear on witness lists he is often either right at the end or the last named but one, the last being sometimes specifically the writer of the charter, sometimes a clerk (*clericus*) who was probably the scribe.[29] In two late twelfth-century grants of land in Abney, Derbyshire, the first witness is a clerk who probably wrote the charters, the second a reeve.[30] This is unusual, but less unusual is the fact that the reeve, and the clerk too, came not from Abney but from Eyam, two miles away: a reeve who witnessed a conveyance may be the reeve not of the vill where the property lay but of a neighbouring vill. This example comes from the charters of Rufford Abbey and two others from the same collection are grants in Ollerton and Eakring, both Nottinghamshire, witnessed respectively by the reeves of Laxton and Winkburn, three or four miles away.[31]

However, we can reasonably follow Campbell in assuming that the *inquisitio eliensis* and the laws of Henry I were right to assume that in principle every vill had a reeve whom we often fail to recognize because our sources simply do not tell us that this was what he was.[32] Was this so throughout England? Witness lists to charters from eastern England name far more reeves than those from other parts of the country. Twenty-six of the Danelaw charters edited by Stenton name a rural reeve as witness, at

25 *Libri censualis vocati domesday-book additamenta* [ed H Ellis] (Record Commission 1816), 497.

26 *Leges Henrici primi*, ed Downer, 100–1; GC Homans, *English villagers of the thirteenth century* (1941), 334–6 discusses the occasions when the reeve and four or six men might represent the vill in the thirteenth century.

27 *Early Yorkshire charters*, ed W Farrer and CT Clay (Yorkshire Archaeological Society 12 vols 1914–65) iii.438.

28 D Postles, 'Choosing witnesses in twelfth century England', *Irish Jurist* new series xxiii (1988).

29 Examples are in *Documents of the Danelaw*, ed Stenton, 219 (last name), 346 (last name apart from the writer of the charter), 294 (last name apart from a *clericus*), 214 (last name but one).

30 *Rufford charters*, ed CJ Holdsworth (Thoroton Society 4 vols 1972–81) i.64–5. A similar case, where the first witness was specifically the writer, the second the reeve, is in *Documents of the Danelaw*, ed Stenton, 236.

31 *Rufford charters*, ed Holdsworth ii.220, iii.430.

32 Campbell, 'Agents and agencies', 205.

least 5 per cent of the relevant documents.[33] In collections from other parts of the country, such as the charters of the Redvers family or the earls of Gloucester, they are named much less often, though even here they appear and in these two collections we meet the reeves of Fleet in Dorset and of Winkleigh in Devon as addressees.[34] The preponderance of identified reeves among witnesses in eastern England may have no significance beyond regional scribal custom, which was clearly not very marked even there. Any suspicion that it points to an actual preponderance of reeves can be set at rest by domesday book. There reeves are seldom mentioned, but they do occur and nearly seventy are entered as manorial tenants in 1086 – of whom half are in Herefordshire.[35] Within Herefordshire their distribution follows no clear pattern: they are mentioned in manors of the king, the bishop and six other lords and occur in thirteen of the county's twenty domesday hundreds.[36] All this seems to point to little more than random naming of reeves in surviving records. It allows us to accept that there probably was a reeve in every vill and that his position, though significant within the local community, was not of such weight that one would expect it to be mentioned every time his name was recorded.

Might there be more than one reeve in every vill? Sometimes more than one reeve attests a charter. The last two witnesses to a grant to St Peter's hospital, York, of land at Swaythorpe near Bridlington in the 1160s are Swain the reeve and Godwin the reeve; in the 1180s or 1190s the witnesses to a grant in Hensall near Snaith include Reginald and William, named together as being then the reeves (*tunc prepositi*).[37] These pairs of reeves may, of course, have come from different manors or even from different vills: we have seen that charters might be witnessed by reeves of neighbouring places. However, the earl of Gloucester's letter to his officials at Winkleigh in the mid-twelfth century was not to his reeve but to his reeves.[38] A large manorial complex might well include a number of vills and hamlets, each with its own officials: Leominster, Herefordshire, with over two hundred ploughs, had eight reeves in 1066 and seven in 1086, while the possibly two reeves mentioned at Hanbury, Worcestershire, in the survey of about 1170 may reflect contemporary expansion of settlement there.[39] But we find more than one reeve as tenants of a single lord in what seem unexceptional manors of modest size. In the surveys they occur at Hemingford Grey (Huntingdonshire), Caddington (Bedfordshire), East Brent (Somerset), Idmiston and Tilshead (both Wiltshire).[40] To look further at this we have to turn to our second question.

33 *Documents of the Danelaw*, ed Stenton, 60, 90, 98, 122, 157 (x2), 184, 187, 201, 205, 207, 212, 214, 218, 219, 228, 236, 274, 285, 294, 327, 331, 346, 365, 369, 372.

34 *Charters of the Redvers family and the earldom of Devon 1090–1217*, ed R Bearman (Devon and Cornwall Record Society new series xxxvii 1994), 149–50; *Gloucester charters*, ed Patterson, 74.

35 See the five regional volumes of the *Domesday geography* of England, ed HC Darby, IB Terrett, IS Maxwell, EMJ Campbell and RAW Finn (1952–67), passim.

36 *Domesday book* i, fos 179r–87v.

37 *Early Yorkshire charters*, ed Farrer and Clay ii.189, i.383.

38 *Gloucester charters*, ed Patterson, 74.

39 *Domesday book* i, fo 180r; *The red book of Worcester*, ed M Hollings (Worcestershire Historical Society 1 vol in 4 1934–50), 186, 188; C Dyer, *Hanbury: settlement and society in a woodland landscape* (1991), 28–9.

40 *Cartularium de Rameseia*, ed Hart and Lyons, 243–4; *The domesday of St Paul's*, ed WH Hale (Camden Society old series lxix 1858), 113; *Inquisition of Glastonbury*, ed Jackson, 73–4, 76, 125–6; *Charters and custumals of the abbey of holy trinity, Caen*, ed M Chibnall (1982), 46–7. An intriguing

Whose was he?

Need the rural reeve have belonged to anyone? In an age when any office might become hereditary, from parish priest to sheriff, when the knightly tenant or the manorial lessee could effectively turn into the owner of the property, was the twelfth-century reeve at least some way along the road to making himself independent? Probably not. Two possible cases of a reeve being succeeded in office by his son are both ambiguous and there seems to be no recorded case of a woman serving as reeve, as one might expect would occasionally happen if a hereditary principle had become established.[41] Even in the fourteenth century a reeve might be succeeded by his son without the remotest hint of independent office.[42] No case has emerged of a rural reeve who paid to be admitted to office as the reeve of Abingdon did, and land that went with the office of reeve is rarely mentioned: the seven acres of *refland* and a small piece of meadow that belonged to the offices of reeve and forester at Hampton Lucy, Warwickshire, about 1170 are quite exceptional.[43] In the eleventh and twelfth centuries reveland more often meant an estate's inland, the area under the reeve's control.[44] We may be fairly certain that the reeve was not his own master.

If he was not, to whom did he answer? Was he the reeve of the vill or the reeve of the manorial lord? Stenton, Lennard and Campbell have all posed this question and all are agreed that there is no clear answer.[45] Oddly it is not, as one might expect, that there is insufficient evidence. It is rather that the evidence is clear and wholly contradictory. Even at the simplest level we find that those named as reeves among the witnesses to charters are sometimes called reeve of such-and-such a place, sometimes reeve of a particular person, usually the donor.[46] In a vill of single lordship the same reeve could and would indeed be both – it would be meaningless to distinguish him as one or the other – and he might well have been called either even in a divided vill, where he will have been the tenant of one or another manorial lord. But other, more cogent evidence points in both directions.

case appears in two late twelfth-century charters of Nigel de Mowbray, each granting away the lands and persons of some of his tenants at South Cave. The first included Gamel the reeve with his 2 oxgangs, the second, to a different recipient probably not long afterwards, included Gregory the reeve also with two oxgangs, but we cannot be sure they were in office simultaneously, and more than one explanation is possible. *Charters of Mowbray*, ed Greenway, 220–1, 231.

41 Swain the reeve of Nettleham, Lincolnshire, appears in *Pipe roll 12 Henry II* (Pipe Roll Society old series ix 1888), 10 and in *Pipe roll 13 Henry II* (ibid xi 1889), 48, while *Documents of the Danelaw*, ed Stenton, xxiii n4 cites a charter referring to land of Richard son of Swain the reeve of Nettleham: was 'the reeve' here Richard or Swain? William the reeve and William son of the reeve at Tilshead, Wiltshire, about 1170 are probably a single individual, but this may signify no more than loose description, the first step towards creating a surname separate from the office: *Charters and custumals of Caen*, ed Chibnall, 46–7.

42 PDA Harvey, *A medieval Oxfordshire village: Cuxham 1240 to 1400* (1965), 64–5.

43 *Chronicon de Abingdon*, ed Stevenson ii.306 (reeve may give gersum *pro tenenda villa*); *Red book of Worcester*, ed Hollings, 277–8; P Vinogradoff, *English society in the eleventh century* (1908), 372 n7 gives mid-twelfth-century examples of reveland with this meaning.

44 Vinogradoff, *Growth of the manor*, 225, 283–4; Vinogradoff, *English society*, 372–3; a late twelfth-century example of reveland with this meaning is in 'Charters of the earldom of Hereford', ed D Walker in *Camden miscellany* xxii (Camden 4th series i 1964), 71.

45 *Documents of the Danelaw*, ed Stenton, cvi–cvii; Lennard, *Rural England*, 274–6; Campbell, 'Agents and agencies', 206.

46 Thus on one Lincolnshire charter of Henry II's reign the witnesses include Alan the reeve of Walter and Ranulf the donor's reeve and on another Elias the reeve of Blyth: *Documents of the Danelaw*, ed Stenton, 212, 369.

There is no difficulty in finding cases where the rural reeve acts clearly as the manorial lord's local agent. We find instances of rural reeves exercising a good deal of authority in this role. In the mid-twelfth century an annual gift by Roger de Mowbray to the canons of Hirst of 30*s* and malt from the Isle of Axholme, Lincolnshire, was to be given to them by the reeves – who failed in this, so the canons were told they could collect it themselves from the Mowbray tenants. Other gifts from Mowbray manors, of sheaves and of wood, were likewise under the surveillance of the local reeves. In another Lincolnshire charter it was the reeve of Bourne, named as a witness, who gave seisin of a grant of land to its new owners.[47] At Kington St Michael, Wiltshire, in the 1180s Garwi, reeve of Robert de Courtenay, took 'by force and without judgement' a half-acre belonging to a yardlander tenant of Glastonbury Abbey and kept it for two years.[48] There is yet more telling evidence. A charter recording the grant to Rufford Abbey of land in Averham, near Newark, in the mid-twelfth century names as the last witness Gilbert, reeve of Ella, fifty miles away in Yorkshire, near Hull. The explanation is that the grantor at Averham, William Tison, was also the owner of Ella; although named there as reeve of Ella, Gilbert's appearance among the witnesses shows that he was Tison's reeve.[49] When Geoffrey de Mandeville gave Westminster Abbey the vill and church of Hurley in Berkshire, in 1085 or 1086, the land of Edric the reeve was excluded from the grant; a few years later William Warenne's grant of Heacham, Norfolk, to Lewes Priory went out of its way to say that the land of Payn, the reeve, was to be included.[50] Excluded – presumably to deal with residuary rights and neighbouring properties – or included in these grants of entire vills, the reeve is implicitly understood as the reeve of the grantor, his local agent: there would be no reason to make special mention of the reeve of the vill itself.

Some references are more ambiguous than may appear at first sight. Peterborough Abbey's chronicler tells us that William I 'granted his land on the hardest terms and at the highest possible price . . . He did not care at all how very wrongfully the reeves (*ða gerefan*) got possession of it from wretched men, nor how many illegal acts they did.'[51] These reeves may well have included those of individual royal manors, like Ælfsi, reeve of Sutton, Berkshire, whose claims to rights in local woods in William's reign were violently disputed by the abbot of Abingdon.[52] A little later, between 1107 and 1111, Swain, reeve of the royal manor of Fawsley, Northamptonshire, was ordered to return what he had taken from the adjacent lands of Thorney Abbey in Charwelton, and, as Campbell points out, the reeves whom Archbishop Anselm feared would oppress his tenants, if he himself did not visit the manors to hear their complaints, were clearly rural reeves.[53] These might seem obvious instances of local

47 *Charters of Mowbray*, ed Greenway, 152–5 (Hirst), 200, 132–3; *Documents of the Danelaw*, ed Stenton, 345–6.

48 Robert de Courtenay held two and a half hides of the abbey in Kington St Michael, five hides in neighbouring Grittleton: *Inquisition of Glastonbury*, ed Jackson, 107, 109, 111.

49 *Rufford charters*, ed Holdsworth ii.167; *Early Yorkshire charters*, ed Farrer and Clay xii.134.

50 *Westminster Abbey charters*, ed E Mason (London Record Society xxv 1988), 296–8; *Early Yorkshire charters*, ed Farrer and Clay viii.57–8.

51 *The Peterborough chronicle 1070–1154*, ed C Clark (1958, 2nd edn 1970), 10; translation from *The Anglo-Saxon chronicle*, ed GN Garmonsway (1953), 218.

52 *Chronicon de Abingdon*, ed Stevenson ii.10.

53 *Facsimiles of early charters from Northamptonshire collections*, ed FM Stenton (Northamptonshire Record Society iv 1930), 14–15; *The life of St Anselm, archbishop of Canterbury, by Eadmer*, ed RW Southern (1962), 71; Campbell, 'Agents and agencies', 216.

reeves acting firmly and all too zealously in their lords' interests. However, they can equally, perhaps more convincingly, be seen as acting in the interests of the vill, extending its common rights at the expense of its neighbours, even – as Anselm hints – of their lord's other tenants.

Both the *inquisitio eliensis* and the laws of Henry I unequivocally speak of the reeve of the vill, not the reeve of the manorial lord. In the late thirteenth century the reeve was often chosen by the manorial tenants, not simply appointed by the lord, and this may well have been the ancient custom: as early as 1218–19 the tenants of the bishop of Winchester's manor of Burghclere were amerced for having made a bad choice.[54] We may see this as a vestige of a time when he was an officer of the community rather than of the manorial lord, and this same picture is given by some twelfth-century charters. We have seen the reeve being included or excluded in grants of entire manors. In the 1150s Peter Boterel gave to the church of St Melaine at Rennes just the reeve of Nettlestead, Suffolk, his heirs and all his lands; that the reeve was his tenant there can be no doubt, but it seems scarcely credible that he would want to give his own agent to a Breton church or that the clergy of St Melaine would find the gift a useful one without the rest of the manor.[55] This reeve was the reeve of the vill. But that he and his holding should have been singled out as the gift is interesting when we look at another grant of a single local tenant in Yorkshire: William Paynel gave the Hospitallers Gamel son of Arthur, his wife and all his boys, with warranty, failing which he would give them instead William the reeve with his family.[56] There is a hint here that the risk of losing the reeve as a tenant was a particular guarantee of the actual grant – that in ways that we can only guess, to have the vill's reeve among one's own tenants was a particularly valuable asset to a manorial lord.

Since the evidence is so clear both that the rural reeve was the reeve of the lord of the manor and that he was the reeve of the local community, should we envisage a dual system in which there were both? When a mid-twelfth-century grant of land at Hellifield, west riding of Yorkshire, is witnessed both by Richard the reeve of Peter des Arches (the donor) and Swain the reeve of Kettlewell (twelve miles away) we may well be looking at two different kinds of reeve.[57] This might explain how we sometimes find two reeves on a single small manor, as at Hemingford Grey or Caddington or East Brent: one was the reeve of the lord, the other the reeve of the vill. However, nothing we know of twelfth-century England could lead us to expect uniformity, so this does not preclude a norm that combined the two roles in a single person. When we look at the reeves of the thirteenth century we find not even residual traces of a dual system; the reeve who still acted for the vill in certain circumstances was apparently the reeve of the lord of the manor.[58] Vinogradoff saw the reeve as an officer of the community on whom manorial duties came to be imposed.[59] To pursue this further we have to consider what the reeve's function on the manor was, what he actually had to do.

54 Hampshire Record Office, 11M59/B1/7. I am grateful to Dr Katharine Stocks for telling me of this.

55 DC Douglas, *The social structure of medieval East Anglia* (1927), 232. Another instance of the grant of a reeve is mentioned in *Registrum antiquissimum*, ed Foster and Major vi.175.

56 *Early Yorkshire charters*, ed Farrer and Clay vi.243.

57 Ibid xi.159.

58 HM Cam, 'The community of the vill' in V Ruffer and AJ Taylor (eds), *Medieval studies presented to Rose Graham* (1950), 10–13; WO Ault, 'The vill in medieval England', *Proceedings of the American Philosophical Society* cxxvi (1982), 193–4, 205–6.

59 Vinogradoff, *Villainage*, 356–7; Vinogradoff, *Growth of the manor*, 191.

What did he do?

We have already had several glimpses of the reeve at work: representing the vill at the hundred court, collecting rents of cash and of corn, supervising the lord's harvest and his woodlands, acting for the lord in giving seisin of land, extending his vill's or his lord's rights, whether legitimately or not, in neighbouring areas. These glimpses could be multiplied: here and there we see quite a lot of things that the reeve did. He was responsible for the waste land of the vill, let out lands and paid the vill's rents or amercements. He ordered the ploughing and winnowing, looked after the barn and granary and travelled on the lord's business. He held scotales for the tenants.[60] What we do not see so easily is the context in which he did all this: the seigneurial, the tenurial, the communal setting for his activities, and thus the overall picture.

It is reasonable to start from the premiss that this overall picture is unlikely to have been radically different in the mid-eleventh century from what it was to be in the mid-thirteenth: that there was no revolution in the reeve's position and function beyond what external circumstances dictated. The reeve-shaped gap in the *Rectitudines singularum personarum* points this way, so do what indications we have from domesday book onward. Starting, then, with the reeves of the mid-thirteenth century we see them acting as their lords' agents in operating manorial demesnes as home farms, extracting from the husbandry the maximum profit for the lord, profit that was handed over in its entirety to the lord himself, his receiver or steward. The reeve was the key local figure in the system of direct exploitation called demesne farming; in early manorial accounts he often accounts jointly with his superior, the bailiff, but it is always the reeve alone who carries the financial responsibility.[61]

This way of managing large estates became the fashion only in the late twelfth and early thirteenth centuries; it replaced what seems to have been widespread leasing of individual manors, with leases that got shorter in the course of the second half of the twelfth century, so that running the manor in demesne through the reeve, with annual accounting for profits, need not have seemed so big a change from a lease renewed annually.[62] What is not clear at all is how lords drew profits from their manors before the mid-twelfth century. For long it was an all but undiscussed question: historians wrote about manorial demesnes, their grant or sale, their expansion or contraction, as though everyone knew how they were run, how manorial demesnes were translated into seigneurial profits, when in fact no one really knew at all. Vinogradoff by implication and Postan more explicitly both assumed some form of demesne farming, seeing the frequent leasing of the later twelfth century as a temporary aberration.[63] On the other hand Lennard and, more recently, Edward Miller and John Hatcher have seen this leasing as the continuation of earlier systems, which one way or another

60 *Chronicon petroburgense*, ed T Stapleton (Camden Society old series xlvii 1849), 162, 164 (waste, letting land); *Pipe roll 14 Henry II* (Pipe Roll Society old series xii 1890), 202 (waste); *Feudal documents*, ed Douglas, 120, 128 (rent); *Pipe roll 17 Henry II* (Pipe Roll Society old series xvi 1893), 16 (amercement); *Red book of Worcester*, ed Hollings, 83, 148 (ploughing); *Chronicon petroburgense*, ed Stapleton, 163 (winnowing); *Inquisition of Glastonbury*, ed Jackson, 38 (barn and granary); London, British Library, Harley MS 61, fo 85r (travel); *Red book of Worcester*, ed Hollings, 37, 60, 85 and *Domesday of St Paul's*, ed Hale, cvii–ix, 144 (scotales).

61 PDA Harvey, *Manorial records* (1984, 2nd edn 1999), 30.

62 PDA Harvey, 'The pipe rolls and the adoption of demesne farming in England', *Economic History Review* 2.xxvii (1974), 353–4.

63 Vinogradoff, *English society*, 229–32; M Postan, *The medieval economy and society* (1972), 88–99.

drew a predetermined income from the demesne, leaving the hope of greater profit, the risk of loss, to someone closer to the locality.[64] Sally Harvey brought this question into the open in 1983 when she concluded, looking at domesday book, that we simply do not know whether or not lords mostly used some form of leasing to draw profits from their demesnes.[65]

Where there were relatively small estates consisting of a single manor or a single vill there need be no problem; it was presumably run by the owner who himself bore all risk and took all profit. The problem arises with larger estates. It is easy to see why some past historians assumed a system of demesne farming. So much of what we see before the mid-twelfth century is entirely at one with what we see in the thirteenth and early fourteenth: the listing of tenants' obligations in twelfth-century surveys, no different from the custumals of the thirteenth; domesday book's emphasis on the size of manorial demesnes; and the account of everyone's activities in the *Rectitudines singularum personarum* which with minimal alteration might apply to any late thirteenth-century manorial demesne. However, although we should keep our minds open to the possibility, it seems most unlikely that distant demesnes on large estates were run by direct management before the late twelfth century. We know so much about demesne farming in the thirteenth and fourteenth centuries because it was recorded in formal and elaborate manorial accounts; these were not generally adopted until the second half of the thirteenth century, though they appear on some of the largest estates much earlier. But this is not to say that demesne farming was otherwise conducted without written records; it is difficult to see how this would have been possible and some mid-thirteenth-century accounts from Bury St Edmunds provide a unique insight into the management of an estate by demesne farming before it adopted a system of fully formalized accounting, revealing a situation of chaotic complexity.[66] They show that the introduction of regular accounting, far from imposing bureaucratic forms, actually brought in a streamlined efficiency which was essential if the whole system was to work properly. Though it has become increasingly clear that written records played a significant part in Anglo-Saxon private administration, it seems unlikely that this can have been on a scale to make demesne farming remotely possible.

We have to look then to some form of leasing. Certainly domesday book often tells us that a particular manor was at farm, often indeed to a named lessee. This of course had the effect of converting the manor of the large estate into a single unit of production, as though the lessee was the owner of a single manor. But that domesday book particularly draws attention to this leads us to suppose that most manors were not at farm in this way. If we assume that some system of rendering predetermined amounts of cash or produce to the lord was everywhere the norm, what exactly happened on these other manors? How did they differ from those that were at farm?

There is plenty of evidence to show that on some large estates a well organized system of food-farms operated, a rota by which each manor of the estate contributed a

64 Lennard, *Rural England*, 106; E Miller and J Hatcher, *Medieval England: rural economy and social change 1086–1348* (1978), 204.

65 SJP Harvey, 'The extent and profitability of demesne agriculture in England in the later eleventh century' in TH Aston, PR Coss, CC Dyer and JI Thirsk (eds), *Social relations and ideas: essays in honour of RH Hilton* (1983), 45–9.

66 PDA Harvey, 'Mid-13th-century accounts from Bury St Edmunds Abbey' in A Gransden (ed), *Bury St Edmunds: medieval art, architecture, archaeology and economy* (1998).

fixed amount of produce in specified weeks, so that the central household was continually supplied week by week throughout the year. We find full details of the system on several monastic and other corporate estates – Bury St Edmunds Abbey, Rochester cathedral priory, St Paul's cathedral and others – and the 'farm of one night' that we find in domesday book is seen as a vestige of a similar system on the royal estates.[67] Some estates preserved a memory of the origin of the system: at Ely it was attributed to Abbot Leofsige (1029–42), at Canterbury cathedral priory to Archbishop Lanfranc (1093–1109).[68] But even if these memories were reliable, we need not suppose that these reformers did more than create the rota. What they organized into a rational system was what already existed: a fixed render of produce from each manor.

Whether organized into a rota or unsystematically delivered on demand or at local convenience, food-farms raise another question that has scarcely been addressed, though Maitland touched on it long since.[69] What happened if the demesne failed to produce the required farm? Who made up the balance? Or, if in a good year more was produced than was needed for the farm, who kept the surplus? The answer might be the reeve: if the lord had not brought in another farmer from outside should we see the reeve as the effective lessee of the demesne, doing well in a good year when the crops far exceeded the requirements of the farm, impoverished and starving in a bad year when the farm took all that the demesne produced? This is possible. Certainly we are sometimes shown the reeve as the alternative to a farmer: the early twelfth-century Burton Abbey surveys mention either a reeve or a farmer on each manor.[70] This need not imply that if there was a farmer there would be no reeve – there is far too much evidence elsewhere to the contrary – but rather that if a particular manor had not been let to farm the abbey would be dealing directly with the reeve. All the same it is difficult to see the reeve as an entrepreneur elected by the vill to make what he could from the lord's demesne. It is much easier to see him certainly as responsible for supplying the lord with the set produce due from the manor but to see this food-farm as an obligation not on him personally but on the vill as a whole. Much early evidence of communal obligation of one kind or another points in this direction. In a good year the whole vill would benefit, in a bad year the whole vill would suffer. Few will have been as lucky as the tenants of Arlington and Chedworth, Gloucestershire, whose reeves' recorded obligation in 1066 was, almost incredibly, to render 'what they wished'.[71]

One objection to this can be dealt with briefly. There are references in domesday book to manors being farmed by the local inhabitants, the men of the vill, and this would be meaningless if the farm was already their communal responsibility. But Hoyt showed that these references are few and more ambiguous than might appear, and where the Burton Abbey survey of 1114–18 tells us that Abbots Bromley is at

67 Ibid 134–5; RAL Smith, 'The financial system of Rochester cathedral priory' in *Collected papers by RAL Smith* (1947), 43–4; *Domesday of St Paul's*, ed Hale, xxxix; JH Round, *Feudal England* (1895, new edn 1964), 96–100.

68 *Liber eliensis*, ed Blake, 152–3, 411–12; Miller, *Ely*, 37–9; RAL Smith, *Canterbury cathedral priory* (1943), 129.

69 FW Maitland, *Domesday book and beyond* (1897, new edn 1960), 183–4.

70 CGO Bridgeman (ed), 'The Burton Abbey twelfth century surveys', *Collections for a History of Staffordshire* 3.1916 (William Salt Archaeological Society 1918), 212–47. Both are mentioned only at Winshill in the 1114–18 survey (where the reference to Bristwi holding the manor for £4 may well be an addition) and at Willington in the 1116–26 survey: ibid 240–3, 236–8.

71 *Domesday book* i, fo 164r.

farm to 'the men' (*homines*) the later survey of 1116–26 defines and names them as a group of five tenants – a small consortium, not a communal enterprise.[72]

A more serious and more obvious objection calls for longer consideration. If the annual renders were a communal obligation – the responsibility of the vill as a whole headed by the reeve, so that the food-farm was simply a charge on the local tenants – why should the lord cultivate lands himself as a manorial demesne? Here the new ideas that Faith brings together are particularly helpful. First is her demonstration, so convincing once it has been so clearly pointed out, that we should not equate the inland of the eleventh century with the manorial demesne of the thirteenth: it included tenant holdings as well, the lands of the peasant tenants whose services provided the labour, specialized or unskilled, for working the demesne. The inland was this entire complex; it was the reveland, the whole area that lay in the reeve's control.[73] Second, at separate points in discussing the Anglo-Norman countryside Faith quotes from two other historians whose comments have particular significance in this context. One is Nellie Neilson who wrote that heavy food-farms were of 'great influence in creating the rudiments of the demesne . . . the distinction between land used to produce the lord's food and lord's land being easily lost'.[74] The other is Barbara Harvey, writing of the Westminster Abbey estates in the twelfth century: 'The earlier the period, the more blurred does this distinction between demesne and tenant-land become . . . the notion that these two parts of the manorial estate were readily separable or that the one rather than the other should supply them with this or that kind of income lay in the future.'[75] Faith also mentions the suggestion that the mid-twelfth century saw peasant holdings 'territorialized', for the first time defined as specific quantities of arable and so on rather than as a simple share in the vill's resources.[76] If we bring these thoughts together we may wonder whether not only the holdings of the tenants but the holdings of the manorial lords as well were first defined on the ground between perhaps the late eleventh century and the mid-twelfth. It is all too easy to see earlier records through eyes that know the thirteenth century. When domesday book speaks of the lord's ploughteams *in dominio* we translate this as 'on the demesne': but are they not really the ploughteams 'in his control'? Is it not possible that domesday book expresses the lord's share in the manorial arable resources in terms of ploughteams simply because in many, even in most, cases this is the only way it could possibly be expressed?

Let us return to the reeve. If we assume the development of the manor on something like these lines many of the problems disappear. He did in the eleventh century just what he did in the thirteenth, look after the lord's resources in the vill, as part of the general oversight and organization of the inland. He was neither the vill's reeve nor the lord's but both: the distinction would be meaningless, and could be made only in the changing conditions of the twelfth century. What happened then in the vill of divided lordship? Presumably it had more than one reeve, but we may wonder just when and how the vill of divided lordship emerged. It is now generally accepted that

72 RS Hoyt, 'Farm of the manor and community of the vill in domesday book', *Speculum* xxx (1955), 147–69; 'Burton Abbey surveys', ed Bridgeman, 223.

73 Faith, *English peasantry*, 49–50 and passim.

74 Faith, 186, quoting N Neilson, *Customary rents* (in Oxford Studies in Social and Legal History ii 1910), 18–19.

75 Faith, 202, quoting BF Harvey, *Westminster Abbey and its estates in the middle ages* (1977), 129–30.

76 Faith, 219, referring to PDA Harvey (ed), *The peasant land market in medieval England* (1984), introduction, 12–13.

parishes, nucleated settlements and common fields all developed late in the Anglo-Saxon period from an earlier pattern of dispersed homesteads or hamlets.[77] In a landscape of dispersed settlement, where parish, village and common fields were all unknown, what in fact determined whether a particular house and holding belonged to one vill or another if there was no dominating physical boundary? What defined the vill? The only possible answer is single lordship.

To sum up, the model proposed takes as its starting point a vill of single lordship where the inland, with its arable held in common just as its pasture was held in common, was worked by its tenants under the leadership of their reeve. He, on their behalf, delivered to the lord an amount of produce that was fixed in advance, unchanging from year to year. Convenience may often have dictated that cash replaced the food-farm, but in either form the render was fixed in tradition and in local custom. If the lord thought he might reasonably get a better return from any of his manors, we know from later experience that the strength of local custom would create difficulties in the introduction of any change. The answer was to hand the vill over to a farmer who, on the spot and with a direct personal interest in maximizing profit, would find it easier to achieve the increased returns that eluded the manorial lord.[78] It is likely that it was at this point that the demesne lands came to be separated from the tenants' lands, though the division of these among individual tenants may well have come later; simple fashion and pressure from tenants could have extended the separation of demesne lands to vills that were never set to farm. Thinking of the tenth-century leases of the bishop of Worcester's lands we may decide that the process began early, but if so it moved slowly and had still a long way to go by the time of domesday book.[79] Defining the demesne lands on the ground facilitated the division of vills into separate manors and opened the way to a peasant land market, though this could really take off only when the peasant holdings themselves had been defined on the ground. Later of course, in the second half of the twelfth century, long-term leasing gave way to short-term, then to demesne farming. Through all of this the work of the reeve will have undergone curiously little change, whether he was organizing the vill's husbandry to render the traditional farm to the manorial lord or running the now defined demesne lands to produce maximum profit, at first for the lessee on the spot, later for the distant lord.

This model needs to be tested against much more of the detailed evidence of what happened in individual manors before the mid-twelfth century. It cannot be said too strongly that we cannot expect to find great uniformity and it may well be that there was no such thing as a norm in the manorial development of this period. But however we interpret the evidence at our disposal we should never lose sight of the reeve. He is a key to our understanding of much that happened on landed estates in eleventh- and twelfth-century England.

77 Faith, 165–7, 149–52, 143–7.

78 *Domesday book* i, fo 16v records that the reeve of Tangmere, Sussex, was given 20*s* a year as a payment from the farmer; it is tempting to see this as a douceur to the vill for a break with custom that was to the community's disadvantage and it would be interesting to find any analogous case.

79 C Dyer, *Lords and peasants in a changing society: the estates of the bishopric of Worcester*, 680–1540 (1980), 17.

III

English Cathedral Estates in the Twelfth Century

Introduction

Since Professor Everett Crosby published his magisterial work, *Bishop and Chapter in Twelfth-Century England*, in 1994 it might seem both unnecessary and foolhardy to write about cathedral estates in the twelfth century.[1] His book's sub-title is *A study of the mensa episcopalis* and it is a detailed account of the way the properties of bishopric or cathedral were divided to form separate estates – separate *mensae* – for the bishop and for the chapter, whether of monks or of secular canons, and an account too of the way this process affected relations between bishop and chapter. Professor Crosby's thesis, compellingly argued, is that whereas traces of the division of estates can be found in some sees in the eleventh century and even earlier, it was only in the course of the twelfth that this became a general development, with all its implications for the bishop's authority in his cathedral church. It was not until the thirteenth century that chapters gained effective independence, so that chapter and bishop would be seen as separate entities, in church administration and in the law of property. Professor Crosby recounts this with a wealth of detail, looking at what exactly happened in each see in turn, and as a result his book is a wonderful source of information and insight, a reference work for all aspects of the way cathedrals and their properties were run in the twelfth century.

However, despite Professor Crosby's labours, some points can still usefully be made in focusing on the estates rather than on their owners. Some are implicit in what he writes but, because they do not relate directly to his theme, he does not discuss them; others lie outside his scope altogether, but are still relevant to cathedral estates at this time. Their estates were of course crucial to the history of the cathedrals in any period, not least in this period of great achievement. They provided the wealth that paid for every activity. Most obviously, they supported the routine of the cathedral, its clergy and its services, but just as certainly they underlay all building works, all artistic and scholarly work, every charitable initiative. The present paper looks at three questions basic to these estates in the twelfth century: how they moved with the sees, how they came to be split up and how they were managed.

1 E. U. Crosby, *Bishop and Chapter in Twelfth-Century England* (Cambridge, 1994).

First, though, it is worth reiterating what Professor Crosby shows very clearly: there was no such thing as a typical English cathedral in the twelfth century – they were all different. There were, after all, no more than fifteen of them at the start of the century, seventeen at the end, ten of them served by monastic chapters and seven by secular canons. Their recent histories were very varied.[2] Between 1050 and 1102 two dioceses had merged,[3] and six had changed the site of their cathedral, two of them twice,[4] while in the process three had moved from a secular chapter to a monastic one, one from monastic to secular.[5] Possibly there were no fully secular chapters before the Norman Conquest: the non-monastic chapters lived a common life under one or other canonical rule, and how and when this broke down must have differed from one secular cathedral to another.[6] In 1109 and 1133 two new sees were created, one in a long-established Benedictine abbey, the other in a newly founded house of Augustinians.[7] This variety of experience clearly made great differences between one cathedral and another, between one chapter and another, and it contrasts with what was to follow: there were hardly any further changes to the dioceses or to the sites of their cathedrals until the sixteenth century.

Nor, in the twelfth century, was there such a thing as a typical cathedral estate. Domesday Book shows that Chichester and Wells had lands only in a single county; at the other end of the scale Canterbury's estates lay in ten counties, Winchester's and Lincoln's in nine. The annual income that it records from Canterbury's properties was £1750, from Chester's £85.[8] As we shall see, on a few sees Domesday Book enters some lands as specifically earmarked for the monks of the cathedral priories or for the canons, but there is no hint of an allocation of this sort on the other bishoprics. By 1200, within each great division of monastic and secular, cathedrals and their estates were probably a good deal more alike, more homogeneous, than they had been a hundred years before.

2 The changes are outlined in D. Knowles, *The Monastic Order in England* (2nd edn, Cambridge, 1963), pp. 129–34; K. Edwards, *The English Secular Cathedrals in the Middle Ages* (2nd edn, Manchester, 1967), pp. 10–11; Crosby, *Bishop and Chapter*, pp. 31–5.

3 Ramsbury and Sherborne.

4 Crediton to Exeter; Dorchester to Lincoln; Elmham to Thetford, then Norwich; Lichfield to Chester, then Coventry; Selsey to Chichester; Sherborne to Salisbury; Wells to Bath.

5 Norwich, Coventry, Bath; Salisbury.

6 Edwards, *English Secular Cathedrals*, pp. 8–14, 19–20.

7 Ely, Carlisle.

8 Crosby, *Bishop and Chapter*, p. 367.

ENGLISH CATHEDRAL ESTATES IN THE TWELFTH CENTURY

Movement of Estates

The fluidity of sees in the late eleventh and early twelfth centuries clearly had profound effects on the properties that ultimately supported the cathedral church but, typically, just how far seems to have varied from one case to the next. We must remember that at this point any distinction between the estates of the bishop and the estates of his cathedral mostly simply did not exist: property of the bishop, property of the cathedral priory or canons, was all one. Thus Domesday Book enters the see of Hereford's lands as the bishop's in its index for Herefordshire, but as 'land of the church of Hereford' in its text; Henry I granted rights of tithes and wood in the New Forest to the canons of Salisbury, but his subsequent confirmation was to the bishop and chapter together.[9] What then happened when the bishop moved his see from one church to another? Did he take the estate with him, leaving the erstwhile cathedral bereft, or did he plant himself unendowed on the charity and property of the new cathedral?

Tradition pointed to the former. The patrimony of St Cuthbert seems to have followed the saint and his bishop throughout their wanderings from Lindisfarne to Durham.[10] In this case, certainly, the community of Lindisfarne moved with the bishop, but we mostly find the same pattern when the bishop simply changed his see from one church to another. The lands may have been seen as property of the cathedral as well as of the bishop, but the cathedral clergy were the bishop's community and he was their head. It was he who controlled the property and who determined what he left for the support of his former cathedral, what he allowed for the support of his new one. Thus, when he moved the see to Lincoln in 1072, Remigius granted the canons of the former cathedral at Dorchester some small properties in and around Dorchester itself: two mills, ten bordars with their lands, 100 acres of arable, six chapels in south Oxfordshire.[11] The bishop made the grant; there was no suggestion that the canons were merely being given what was already their own or that they had any rights over the vast estate that Remigius took with him to Lincoln. Again, when John de Villula transferred his see from Wells to Bath Abbey in 1090 he secured gifts of important properties in the area but also, thinking 'that as he had increased the revenues of the church in one direction, so he might be at liberty to diminish them in another', he gave away rents that had been used to support the canons of Wells.[12] One result might be a significant shift in the estate's centre of gravity. Looking again at Lincoln, an extreme case, the bishop of Dorchester's 90 hides at Dorchester

9 Domesday Book, vol. i, fols 179r, 181v; Crosby, *Bishop and* Chapter, pp. 341–2.

10 E. Craster, 'The Patrimony of St Cuthbert', *English Historical Review*, 69 (1954), pp. 177–99.

11 *Registrum Antiquissimum of the Cathedral Church of Lincoln*, ed. C. W. Foster and K. Major (Lincoln Record Society, 10 vols, 1931–73), i. p. 247.

12 *Ecclesiastical Documents*, ed. J. Hunter (Camden Society, old series VIII, 1840), pp. 21–2.

will effectively have provided a home farm for his cathedral clergy, who will also have drawn heavily on the demesne at Thame, 10 miles away, but when the see moved over a hundred miles to Lincoln and these properties with it, their function and their management are bound to have changed.[13]

In some cases the bishop had no option: his new cathedral had little or no endowment, so he had to bring with him the lands to support himself and the church. This is what happened when Leofric moved from Crediton to Exeter, taking the manor of Crediton with him.[14] On the other hand, a bishop might increase his revenue by moving his cathedral to a church already well provided; Robert Limesey nearly doubled his income through Coventry's elevation to a see, and this reasoning underlay abortive attempts to move sees to Bury St Edmunds and Malmesbury Abbeys.[15] Equally, a bishopric and cathedral might gain by a bishop's gift of properties he held personally; Wells benefited from estates that Duduc had been given in hereditary right, and Robert Limesey brought to Coventry some properties that were his own, not the bishopric's.[16]

Division of Estates

It was the movement of sees that produced most changes in cathedral estates at the start of our period, but it was internal re-organisation that produced continuing change until at least the mid-twelfth century. Remembering that in the late eleventh century the bishop, not the cathedral, was in practice in charge of the estates, we can see these changes simply as successive reductions in the property under his control. They occurred through three distinct processes. All three can be traced before the end of the eleventh century and all three advanced simultaneously; the first to be effectively achieved was enfeoffment for knight service, the second was the creation of prebends and the third was the division of the estates between bishop and cathedral. It was only this last that led to permanent separation of properties, recognised by the law of church and state, but the other two processes equally withdrew properties from the bishop's control and redirected the income they produced.

The obligation to provide knights for the king's service was laid on tenants-in-chief, bishops and monasteries among them, soon after the Norman Conquest, and this obligation came to be met by enfeoffment, granting out lands in return for the service required. This service varied greatly, from the sixty knights demanded from Canterbury (with Rochester), Lincoln and

13 Domesday Book, vol. 1, f. 155r; S. P. J. Harvey, 'The Extent and Profitability of Demesne Agriculture in England in the Later Eleventh Century', in *Social Relations and Ideas: Essays in Honour of R. H. Hilton*, ed. T. H. Aston *et al.* (Cambridge, 1983), pp. 54–5.

14 Crosby, *Bishop and Chapter*, p. 272.

15 Ibid., p. 118; Knowles, *Monastic Order*, pp. 131–2.

16 *Ecclesiastical Documents*, p. 15; Crosby, *Bishop and Chapter*, p. 117.

Winchester to the two from Chichester, and political and other bargaining must have played some part in the assessments: from York, for instance, only seven knights were required.[17] The process was well advanced at Canterbury before 1100; here, as at Ely, Worcester and probably on other estates as well, some at least of the requirements were met by changing the terms of tenure of tenants who were already in occupation: military service replaced payments in cash or produce.[18] This makes it difficult to trace the detailed chronology of enfeoffment. A further complication is that most bishops granted out far more lands for knight service than they needed to: 105 knights' fees at Lincoln by 1166, for instance, some forty-seven at York. As the archbishop of York explained, 'Our predecessors ... wished to provide for their relatives and employees'.[19] The obligation was the bishop's and it was to the bishops' estates, not the cathedrals', that these enfeoffed lands passed. Thus, unlike other large, long-established monasteries, the cathedral priories had no need to let out lands for knight service. The lands of course were not entirely lost to the bishop: he continued as feudal lord and drew from his tenants the appropriate feudal incidents. But unless they reverted to the bishop through a tenant's failure to produce an heir or other escheat, they were lost for ever to the estate's regular economic potential.

The prebends that were created in the secular cathedrals came to be included, naturally enough, in the chapter's share of the estate when the division between bishop and cathedral finally took effect, but they seem to have originated before this division began to take formal shape. Thus Canterbury prebends at Newington-by-Sittingbourne were described in about 1100 as belonging to the archbishop and to the cathedral.[20] The prebend was a separate, independently held estate for each canon, whose rights might even continue after his death: at Lincoln the following year's income went to the dead prebendary's estate, but at Salisbury it passed to the other canons and the poor.[21] The beginnings of prebends have been traced in Domesday Book at St Paul's and at Hereford;[22] passages in Salisbury documents dated 1091 that

17 *Red Book of the Exchequer*, ed. H. Hall (3 vols, Rolls Series, 1896), i, pp. 13–16, 22–8.

18 Crosby, *Bishop and Chapter*, p. 73; E. Miller, *The Abbey and Bishopric of Ely* (Cambridge, 1951), pp. 67–9, 165–6; C. Dyer, *Lords and Peasants in a Changing Society: the Estates of the Bishopric of Worcester, 680–1540* (Cambridge, 1980), p. 47.

19 Crosby, *Bishop and Chapter*, p. 369; *Red Book of the Exchequer*, pp. 374–6, 412–15.

20 F. R. H. Du Boulay, *The Lordship of Canterbury* (London, 1966), p. 21.

21 *Statutes of Lincoln Cathedral*, ed. H. Bradshaw and C. Wordsworth (2 vols in 3, Cambridge, 1892–7), ii, p. 140; D. Greenway, 'The False *Institutio* of St. Osmund', in *Tradition and Change: Essays in Honour of Marjorie Chibnall*, ed. D. Greenway *et al.* (Cambridge, 1985), pp. 80, 97–100.

22 *Early Charters of the Cathedral Church of St. Paul, London*, ed. M. Gibbs (Camden Society 3rd Series, LVIII, 1939), pp. xxii–xxiii; M. Jones, 'The Estates of the Cathedral Church of Hereford, 1066–1317' (Oxford B.Litt. thesis, 1958), pp. 40–1. I am grateful to Mrs Marjorie Maslen (Miss M. Jones) for permission to cite her

imply a well-established system were probably written in the mid-twelfth century,[23] but prebends may well have developed early in the century in some at least of the secular cathedrals. At York the first prebends were set up by Archbishop Thomas (1070–1100) and at St Paul's they have been associated with Bishop Maurice (1085–1107), though he is more likely simply to have regulated their development than to have completed the pattern.[24] Later at Hereford some prebends lay among the chapter's properties, some in areas that otherwise lay in the bishop's estate, pointing to their creation before the estate had been effectively divided.[25] Indeed, we may even see the creation of prebends as an important step in bringing about the separation of the cathedrals' estates from those of the bishops.

When this came about, the proportion of the chapter's properties that had been allocated to prebends differed from one cathedral to another. At St Paul's the bulk of the estate remained in the hands of the chapter corporately, and the thirty prebends were relatively modest: nine distinct prebends were formed from the single manor of Willesden, which had been valued in 1086 at £6 6*s*. 6*d*. a year.[26] At Lincoln, on the other hand, the substantial properties of the cathedral were all divided between the prebendaries, while the common property listed in 1146, though in total a significant amount, might reasonably be described as a lot of odds and ends: half the offerings at the cathedral altar, an orchard where the king's vineyard used to be, St Nicholas's church at Newport, and so on.[27] The value of the individual prebends in a cathedral might differ considerably; Exeter, where they were all worth the same, was exceptional.[28] Later in the century the cathedral dignitaries might have properties allocated to them from the chapter's estates; at York the treasurer and the precentor had properties attached to their offices by the mid-twelfth century and at Lincoln the vicars-choral had properties of their own by 1190.[29] These allocations of property may normally have developed from simply earmarking certain sources of income for particular offices,

---

thesis.

23 Greenway, 'The False *Institutio'*, pp. 88–91.

24 Hugh the Chantor, *The History of the Church of York 1066–1127*, ed. C. Johnson (London, 1961), p. 11; *Early Charters of St. Paul's*, pp. xxiv–xxv; Greenway, 'The False *Institutio'*, p. 89.

25 Jones, 'Estates of Hereford', p. 43.

26 C. N. L. Brooke, 'The Composition of the Chapter of St Paul's, 1086–1163', *Cambridge Historical Journal*, 10 (1950–2), pp. 127–8; *A History of St Paul's Cathedral and the Men Associated with it*, ed. W. R. Matthews and W. M. Atkins (London, 1957), p. 61.

27 K. Major, 'The Finances of the Dean and Chapter of Lincoln from the Twelfth to the Fourteenth Century', *Journal of Ecclesiastical History*, 5 (1954), pp. 149–50.

28 Edwards, *English Secular Cathedrals*, pp. 39–41.

29 C. T. Clay, 'The Early Treasurers of York', *Yorkshire Archaeological Journal*, 35 (1943), p. 18; C. T. Clay, 'The Early Precentors and Chancellors of York', *ibid.*, p. 120; Edwards, *English Secular Cathedrals*, p. 274.

particular departments, within the cathedral, the process that probably gave rise to the earliest prebends.

In the monastic cathedrals we see the same process of allocating properties to particular obedientiaries. In some non-cathedral abbeys the process was well advanced by the end of the twelfth century; here, of course, the separation of the abbot's estates from the convent's corresponded to the division of the bishop's from the cathedral priory's, but with the important difference that abbot and convent never came to be seen, outside the monastery, as completely independent property-owning entities. In the cathedral priories further subdivision failed to take hold. At Canterbury most of the priory's sources of income were allocated to one or other of three obedientiaries, but only until the 1160s, when a treasurer was appointed to receive and distribute all income;[30] at Rochester the sacrist and chamberlain had small properties of their own but under close central control;[31] at Winchester Bishop Henry of Blois introduced what may have been the first system of central auditing in any English monastery.[32] There is a difference between receiving revenue from specified sources and actually administering these properties oneself, and it seems that in most monastic cathedrals administration of the vast bulk of the estates continued in the priory's centralised control. Even at Norwich, where the obedientiaries by the late thirteenth century were in control of the properties allocated to them, a single steward representing the prior apparently had wide powers over the administration of all the manors.[33]

In investigating just when the chapters' estates were divided from the bishops', Professor Crosby is concerned to establish at what date they became completely separate entities in the eyes of the law and of the church. It is easy to overlook that this is by no means the same as the date when they first began to be administered separately by their respective owners, when the chapter took charge of its own estates for day-to-day management and for year-on-year auditing. Yet in looking at the estates from the cathedrals' viewpoint this is an interesting question, of great practical, indeed crucial, importance at the time, and it is important too to the historian concerned with landed property in medieval England, its organisation and development.

It is not an easy question to answer. Relevant evidence is sparse and involves some reading between the lines. Here again we have to make a clear distinction between drawing revenue from property earmarked for a

30 R. A. L. Smith, *Canterbury Cathedral Priory* (Cambridge, 1943), p. 14.

31 R. A. L. Smith, 'The Financial System of Rochester Cathedral Priory', in his *Collected Papers* (London, 1947), p. 44; C. Flight, *The Bishops and Monks of Rochester 1076–1214* (Kent Archaeological Society, monograph 6, 1997), p. 199.

32 R. A. L. Smith, 'The *regimen scaccarii* in English Monasteries', in his *Collected Papers*, pp. 55–6.

33 H. W. Saunders, *An Introduction to the Obedientiary and Manor Rolls of Norwich Cathedral Priory* (Norwich, 1930), pp. 42–3.

particular purpose and having that property in one's own administrative control. We find particular sources of income allocated to the clergy of some cathedrals at a very early date. At Worcester they had a specific endowment from the ninth century, and at Canterbury, indeed, from the end of the eighth, though whether this had a continuous history to the Norman Conquest is uncertain.[34] On several bishoprics Domesday Book distinguishes properties set aside for the cathedrals. At Winchester it separates the bishop's manors from those of 'his monks' in Hampshire and the description of parts of these properties as for the monks' food or clothing points to more detailed allocations within the cathedral priory; it has special headings for properties allocated to Canterbury Cathedral Priory in Kent and Essex and identifies those of the canons of St Paul's in Middlesex and three other counties; at Worcester it distinguishes the bishop's manors in Oswaldslow Hundred from those of the cathedral, and two other Worcestershire manors are specifically 'for the monks' food'.[35] But we should not see this as more than a first step towards separating the properties: Domesday Book's headings, in its county indexes and text, mostly make it clear that it views them as single estates.

We are not helped by some apparent early evidence that we must reject. The twelfth-century history of the see of Wells, which moved to Bath, tells how Bishop Gyso (1061–88) made a clear division between the lands that supported the bishop and those that supported the chapter, so that posterity would be in no doubt which was which, but the writer's interest in claiming for Wells the properties that Bishop John de Villula took with him to Bath is all too obvious.[36] At Rochester and at Ely the division came to be attributed to a distinguished figure – Bishop Gundulf (1077–1108) and Bishop Hervey (1109–31) respectively – simply to lend it impugnable authority and antiquity,[37] and there is no evidence to support the later tradition that it was Lanfranc (1070–89) who divided the estates at Canterbury.[38] We are looking at a gradual and irregular process, which may have receded as well as advanced, and these notable personalities may all have advanced it in one way or another; but there is too much evidence of bishops' later involvement in these cathedrals' properties to suppose that they gave the chapter or priory effective control of its own estates.

From St Paul's we have more, and more varied, evidence than from most cathedrals. A lease of the manor of Twyford in Middlesex in 1102 was made by the canons alone, but we must discount this, for a lease of the same manor

34 Dyer, *Lords and Peasants*, p. 18; Du Boulay, *Lordship of Canterbury*, pp. 18–20.

35 Domesday Book, vol. i, fols 37v, 40r–41v, 43r; vol. i, fol. 4v, vol. ii, fol. 8r; vol. i, fols 127v–128r, 136r, 211r, vol. ii, fols 12v–14r; vol. i, fols 172v–174r.

36 *Ecclesiastical Documents*, pp. 19–20.

37 Flight, *Bishops and Monks of Rochester*, pp. 42–3; *Liber Eliensis*, ed. E. O. Blake (Camden 3rd Series, XCII, 1962), pp. l–li, 261–3.

38 B. W. Kissan, 'Lanfranc's Alleged Division of Lands Between Archbishop and Community', *English Historical Review*, 54 (1939), pp. 285–93.

in 1114 was made by the bishop of London, the dean and the canons.[39] However, a lease of Ardleigh in Essex in 1141, like other leases of 1152 and 1155, was by the chapter alone.[40] In the leases made by the canons themselves we can be more confident of seeing the actual situation than in documents addressed to them by the king and others, so we may reasonably deduce that between 1114 and 1141 the dean and chapter gained lasting administrative control of the properties assigned to them. We may see confirmation of this in the bishop's charter of about 1127 restoring to the canons a wood he had enclosed in his park at Clacton in Essex and, more convincingly, in the dean and chapter's foundation of Markyate nunnery in 1145, confirmed by the bishop of Lincoln in whose see it lay but without any reference to the bishop of London.[41] Certainly in the statutes drawn up when Ralph de Diceto was dean, in the last twenty years of the century, there is no suggestion that the bishop or any of his officers was involved in running the estates. They include a vivid vignette of the chamberlain reciting his accounts for audit before the assembled chapter: so much received from Belchamp, so much for such and such a place, this amount lacking in the rent from Drayton, this amount from somewhere else and so on – all of them places quite unknown to anyone in the chapter except the chamberlain himself, and if it should happen – which God forbid! – that the chamberlain should die, no one would have the least idea where to go to collect the rents, with the result that the chapter would be deprived and distressed (*privatum et desolatum*); so it should all be written down to safeguard the future.[42] Again, not a hint that anyone was involved but the chapter itself.

The date when this administrative independence was achieved perhaps differed a good deal from one cathedral to another. St Paul's may well have been especially precocious, remembering how its canons' lands are already distinguished from the bishop's in Domesday Book. But all cathedrals seem to have achieved this degree of independence in the course of the twelfth century, probably earlier in the century rather than later. Thus at Hereford the bishop in 1135–6 granted churches to Llanthony Priory, reserving certain tithes to the dean and cantor, while his successor in about 1180 formally granted a church to the cathedral canons in view of their poverty;[43] this suggests – it does not prove – that the chapter acquired administrative control of its estates between these dates. There may seldom have been a simple progression towards administrative independence; it is worth remembering that after they had achieved control of their estates the monks of Canterbury

39 *Early Charters of St. Paul's*, p. 137; *The Domesday of St. Paul's*, ed. W. H. Hale (Camden Society, old series LXIX, 1858), pp. 127–8.

40 *Domesday of St. Paul's*, pp. 135–6, 128–9, 132–3, 134–5.

41 *Early Charters of St. Paul's*, pp. 43–4, 119–22.

42 *Registrum statutorum et consuetudinum ecclesiae cathedralis sancti Pauli Londinensis*, ed. W. Sparrow Simpson (London, 1873), p. 133.

43 Jones, 'Estates of Hereford Cathedral', pp. 234, 239.

Cathedral Priory twice relinquished it to the archbishop, once voluntarily in 1150, once involuntarily in 1186.[44]

From Worcester in about 1170 and from Durham in 1183 or 1184 we have detailed surveys of the bishop's estate alone; no corresponding survey of the cathedral's estates survives.[45] At Durham we can be fairly sure that none was made, for two of the four known copies of the survey are in the cathedral's archives, and while this points to close interest in the bishop's affairs – what H. S. Offler calls 'capitular responsibility and neighbourly inquisitiveness' – it seems unlikely that the monks would have preserved the record of his estates more carefully than one of their own.[46] The contents of cathedral archives and cartularies might well repay detailed analysis to discover just when chapters kept their records apart from the bishop's – though neighbourly inquisitiveness, a continuing interest in the bishop's properties, might obscure the point when independence was achieved, and so too might less than efficient record management: the records of Canterbury Cathedral Priory and the archbishop were kept together until the thirteenth century.[47] Estate accounts from the archbishopric and the cathedral priory at Canterbury in the 1220s and 1230s are set out on quite different principles, pointing to an administrative separation of long standing,[48] and though those of the first half of the thirteenth century from Winchester Cathedral Priory and bishopric are strangely alike, so are those from other contemporary Hampshire estates, suggesting a regional style rather than a common administrative origin.[49] The likelihood is that most cathedrals were running their own estates, without reference to the bishop, by the mid-twelfth century or not long after, even though it was not until later that this separation was formally recognised by the church and by the law of the land.

Overall the bishops probably lost out by separating the cathedral estates. Charitable gifts, of lands, houses, rents, tithes and so on, continued to be

44 *The Historical Works of Gervase of Canterbury*, ed. W. Stubbs (2 vols, Rolls Series, 1879–80), I, pp. 143–5, 35.

45 *The Red Book of Worcester*, ed. M. Hollings (Worcestershire Historical Society, 4 pts, 1934–50); *Boldon Buke*, ed. W. Greenwell (Surtees Society, XXV, 1852).

46 H. S. Offler, 'Re-reading Boldon Book', in his *North of the Tees* (Aldershot, 1996), no.XII, pp. 6, 22.

47 Crosby, *Bishop and Chapter*, p. 16. The dates when cathedrals acquired their own corporate seals, while of great interest, is probably not relevant in this connection: seals from what were probably eleventh-century matrices are known from Canterbury, Durham, Exeter and Sherborne (T.A. Heslop, 'English Seals from the Mid Ninth Century to 1100', *Journal of the British Archaeological Association*, 133 (1980), p. 7).

48 Lambeth Palace Library, estate documents 1193; Canterbury Cathedral Archives, Assisa Scaccarii rolls 1–5.

49 *Manorial Records of Cuxham, Oxfordshire, circa 1200–1359*, ed. P. D. A. Harvey (Oxfordshire Record Society, 50; Historical Manuscripts Commission, Joint Publication 23, 1976), pp. 18–19, 79.

made throughout the century to the cathedral clergy, monastic and secular;[50] no one was going to give anything to an already wealthy prelate. These gifts were often very small, but cumulatively they contributed significantly to the changes that cathedral estates underwent in the twelfth century. All in all the composition and structure of the estates that supported the cathedrals in 1200 were very different from what was recorded in Domesday Book in 1086. The period was one of great fluidity and change, in marked contrast to the three hundred years that followed, when the stability of the cathedrals' estates mirrored the stability of the sees themselves.

Management of Estates

How they ran their estates, how the results of tilling the soil reached the cathedral, and in what form they arrived there, are questions that, quite properly, Professor Crosby sees as lying outside his brief. Here we might expect greater difference between one cathedral and another, reflecting the history of the estate and the chapter's internal organisation, but evidence is limited. It is fullest by far from St Paul's, but this very fact, as well as the nature of some of the evidence, makes it untypical, and for the others we must proceed mostly on the basis of probability and analogy with what happened on similar estates of other large landowners.

In one point, however, we can be fairly certain that what happened at St Paul's happened also at most or all of the other cathedrals with properties that they had long held. It is likely that in the eleventh century the usual way of taking profit from a manor on a substantial estate was by the food farm, an annual fixed render of corn and other produce. At St Paul's fifteen manors were organised in a regular rota, each providing its food farm in particular weeks so as to ensure a constant flow of produce for the chapter's use.[51] We know the same system operated at Rochester,[52] as well as at Ely, where it was thought to date from the reign of Cnut (1016–35), and at Canterbury, where it was ascribed to Lanfranc (1070–89),[53] and we would expect to find it at the other cathedrals where there was a stable nucleus of long-held properties; certainly it operated on other large monastic estates. A food-farm rota may in some cases underlie the division of the estate between bishop and chapter. It is a commonplace that the static cathedral kept the properties close at hand, while the peripatetic bishop took those further away,[54] though at Canterbury

50 At Hereford gifts were particularly to found prebends (Jones, 'Estates of Hereford Cathedral', pp. 44–5), whereas at Lincoln there were many small gifts to the common fund (Major, 'Finances of the Dean and Chapter of Lincoln', p. 151).

51 *Domesday of St. Paul's*, p. xxxix.

52 *Custumale Roffense*, ed. J. Thorpe (London, 1788), pp. 12, 33; Smith, 'Financial System of Rochester Cathedral Priory', pp. 43–4.

53 *Liber Eliensis*, pp. 152–3; Miller, *Abbey and Bishopric of Ely*, pp. 37–9; Smith, *Canterbury Cathedral Priory*, p. 129.

54 Thus, Miller, *Abbey and Bishopric of Ely*, pp. 76–7; Dyer, *Lords and Peasants*, p. 19.

the cathedral priory held properties not only in Kent but as far afield as Devonshire, Oxfordshire and Norfolk.[55] It could well be that estates were divided in such a way as to leave the cathedral's food farm undisturbed – it would tend to be the nearer manors that supplied it. Certainly the structure of food farms, which affected the organisation of both estate and household, was remarkably resistant to change and on some large estates, St Paul's among them, it continued in place long after it had become a palpable archaism.[56]

It was already old-fashioned by the early twelfth century. At St Paul's our earliest – thirteenth-century – detailed accounts of the food farm show that it comprised only corn, supplied for ale and bread; the rest of the farm was paid in cash, a fixed payment from each manor, which had probably replaced earlier renders of meat, cheese and other produce.[57] Cash was by the twelfth century the usual way of bringing the profits of an estate to a static landowner, and we can assume that it was cash that was rendered by the manors that lay outside a cathedral's food-farm rota: when Domesday Book recorded that the St Paul's manor of Willesden produced £6 6*s.* 6*d.* a year, it probably meant exactly what it said.[58] This explains why the St Paul's prebends were relatively modest, formed from those parts of the estate that could be hived off without disturbing the food farms. At Lincoln, whatever arrangements for food farms had been in place when the see was at Dorchester must have been entirely disrupted by the long-distance migration of the bishopric; significantly, they were not reinstated on the re-assembled estate which, unencumbered in this way, could be entirely split up among the canons.

Historians have never fully addressed the basic problem that arises from manors held for traditional farms in food or in cash: who locally carried the risk or kept the profits? The answer, arguably, is the local community as a whole, led by its reeve.[59] But where the landlord felt that, properly managed, his lands could yield a higher return, what did he do? The answer here is in no doubt: he handed the property to a lessee or farmer, *firmarius*. This lessee, who might pay a significant sum to get the lease,[60] would answer for the

55 Smith, *Canterbury Cathedral Priory*, p. 100.

56 *Domesday of St. Paul's*, p. lii.

57 *Ibid.*, pp. xlvi–xlvii, 154–65*.

58 My understanding of what was happening at Willesden differs from that of R. S. Hoyt, 'Farm of the Manor and Community of the Vill in Domesday Book', *Speculum*, 30 (1955), pp. 164–5.

59 I suggest this in a forthcoming article, 'The Manorial Reeve in Twelfth-Century England', in *Lordship and Learning: Studies in Memory of T. H. Aston*, ed. T. Charles-Edwards *et al.* (Woodbridge, 2004).

60 We learn, almost incidentally, that Richard, archdeacon, offered 20 marks for the lease of Belchamp St Paul (*Domesday of St. Paul's*, p. 129); Professor F. R. H. Du Boulay sees payments 'to the archbishop' in a late-eleventh-century list of Canterbury manors as entry-fines for leases (*The Domesday Monachorum of Christ Church Canterbury*, ed. D. C. Douglas (London, 1944), pp. 98–9; Du Boulay,

traditional farm or more and would have every incentive to develop the land's potential since he would keep any profit for himself. This was not a new idea in the early twelfth century, though leasing may have become more widespread then. There is an important series of written leases from Worcester in the late tenth century,[61] and there are many references to lessees in Domesday Book, on episcopal as on other estates. For example, at Rochester, a priest named Anchitillus held Langfield in Kent for £5 a year, not necessarily paid in cash; at Winchester, Ralph son of Sefrid rendered £7 10*s*. to the bishop for Polhampton in Hampshire, which in 1066 had been producing £8 for the monks' food; at Chichester, the manor of Preston in Sussex, worth £18 a year, had once been let for £25 to a lessee who had been unable to meet this commitment.[62]

In three ways the St Paul's records throw a valuable light on leasing manors in the twelfth century. Uniquely, they include a series of written leases, from the early to mid-twelfth century.[63] Although a few manorial leases of this period are known from other estates, no group like this has survived elsewhere and, indeed, it is likely that local lessees would normally be set in place without a written agreement. Second, these leases show that in the mid-twelfth century individual canons were themselves acting as lessees on particular manors, a very neat arrangement which ensured the lessee's probity and commitment – though not necessarily his efficiency – while enabling the canon to supplement the income from his prebend. Third, a list of manors and lessees in 1181 shows that whereas some manors were leased to two lessees jointly, some lessees had leased several manors; interestingly, it was noted in the same year that it was inadvisable to allow the lessee of a manor to lease the church – that is, the tithes and glebe lands – as well.[64]

How far this corresponded to what happened in other cathedrals we do not know. Before the end of the twelfth century the system of management changed radically. The tenth-century Worcester leases were for three lives and the St Paul's leases seem all to have been for the life of the lessee, but from the mid-twelfth century very short leases, even for a single year, became more usual.[65] This, however, proved a short-lived development, generally replaced in the late twelfth and early thirteenth centuries by direct management: running each manor by the lord's own officers, so that the landlord himself bore all the risk and took all the profits. We can trace the introduction of the system from the accounts entered on the Exchequer's pipe rolls whenever an estate came into the king's hands through vacancy or escheat, and these show

---

Lordship of Canterbury, p. 200).

61 Dyer, *Lords and Peasants*, pp. 17–18.

62 Domesday Book, vol. i, fols 5v, 41v, 17r.

63 Most are printed in *Domesday of St. Paul's*, pp. 122–39.

64 *Domesday of St. Paul's*, pp. 110–12, 146.

65 P. D. A. Harvey, 'The Pipe Rolls and the Adoption of Demesne Farming in England', *Economic History Review*, 2nd series 27 (1974), p. 353.

that it was starting to be adopted in the mid-1180s and was all but universal on large estates by the end of John's reign.[66] Here, though, the historian of cathedral estates is at an unusual disadvantage, for their separation from the estates of the bishop meant that they no longer came into the king's hands. However, we know that demesne farming was in operation on manors of Canterbury Cathedral Priory in 1207–8 and there is no reason to doubt that other cathedrals followed this general fashion on their estates.[67]

Conclusion

We have taken no more than a quick glance at three aspects of the estates that paid for all activities of the cathedrals in the twelfth century. In their composition and management there was nothing to set them apart from other large estates of the period. As a group, however, they are particularly well documented, and they shared some peculiarities in their development: their movement with the sees, the creation of prebends in the secular cathedrals and, above all, their division between chapter and bishop. Of the chronology of that division in its effects on the day-to-day workings of the estates, we probably have much still to learn, though we now have the firm overall chronology that Professor Crosby has given us. What he has shown us clearly, and what underlies much of what is discussed here, is that we underestimate at our peril the continuing power of the bishops in the twelfth-century cathedrals. They held the purse-strings for much longer than we might have supposed and for much longer than some cathedral chapters would have wished us to think.

[66] *Ibid.*, pp. 345–59.

[67] *Interdict Documents*, ed. P. M. Barnes and W. R. Powell (Pipe Roll Society, new series XXXIV, 1960), pp. 69–80.

IV

Non-Agrarian Activities in Twelfth-Century English Estate Surveys

In 1959 Reginald Lennard's important work, *Rural England 1086–1135*, brought together many years' work on the English countryside, its society and economy, as revealed by Domesday Book and by the records of the succeeding generation. In one chapter, 'Local and regional variety and the "normal" village', he explores the non-agrarian occupations of the countryside: what economic activities went on besides growing crops and raising livestock.[1] This paper continues this story where Lennard left off, looking at the non-agrarian activities revealed by estate surveys down to the end of the twelfth century, when much fuller records begin to give us a picture we can view with real confidence of the detail and variety of rural life. It cannot be viewed exactly as a sequel to Lennard's work: it will look rather less at the variations, regional or local, rather more at the 'normal' village or other rural community. This is partly dictated by our sources: while much more detailed than Domesday Book, twelfth-century manorial surveys are less uniform in their treatment, and between them cover only a minute proportion of the country.

These surveys survive from only twelve estates;[2] from one, St Paul's

1 R. Lennard, *Rural England 1086–1135* (Oxford, 1959), pp. 237–87.

2 The surveys are as follows (in brackets are the abbreviations used for succeeding references): (1) Trinity Abbey, Caen (Caen), 1106–31, c. 1170, c. 1200 – *Charters and custumals of the abbey of Holy Trinity, Caen*, ed. Marjorie Chibnall (British Academy, Records of Social and Economic History, new ser. v; 1982), pp. 33–86; (2) Burton Abbey (Burt.), 1114–18, 1116–26 – C. G. O. Bridgeman, 'The Burton Abbey twelfth century surveys', *Collections for a history of Staffordshire 1916* (William Salt Archaeological Soc.; 1918), pp. 212–47; (3) Peterborough Abbey (Pet.), 1125–8 (date from E. King, *Peterborough Abbey 1086–1310* (Cambridge, 1973), pp. 140–5) – *Chronicon Petroburgense*, ed. T. Stapleton (Camden Soc., old ser. xlvii; 1849), pp. 157–66; (4) Shaftesbury Abbey (Shaft.), c.1130, c.1180 (I am grateful to Dr Ann Williams for advice on these dates) – BL Harley MS 61, ff. 37–89; (5) royal demesne in Bedfordshire and Buckinghamshire, 1154–7 – R. Richmond, 'Three records of the alien priory of Grove and the manor of Leighton Buzzard', *Bedfordshire Historical Record Society*, viii (1924), pp. 22–4, and G. H. Fowler, 'Extents of the royal manors of Aylesbury and Brill, circa 1155', *Records of Buckinghamshire*, xi, no. 7 (1926), pp. 402–5; (6) Ramsey Abbey (Ram.), c.1160, c.1195, 1201 (for different manors at each date: dates from J. A. Raftis, *The estates of Ramsey Abbey* (Toronto, 1957), pp. 305–8) – *Cartularium monsterii de Rameseia*, ed. W. H. Hunt and P. A. Lyons

Cathedral's, we have no more than fragments, while the only survey from the royal demesne covers just three manors in Buckinghamshire and Bedfordshire. The other surveys, however, cover most or all of what are mostly very large estates, all ecclesiastically owned, comprising some four hundred manors in all. The Trinity Abbey, Caen, and, even more, the Templars' estates were widely scattered. The rest were concentrated in limited areas, but these areas are fairly well distributed throughout the country: their lords included the abbots of Glastonbury in Somerset, Ramsey in Huntingdonshire, Burton in Staffordshire and the bishop of Durham. Chronologically too the surveys are well distributed, from the Burton and early Caen surveys in the early years of the century to the late Glastonbury and Evesham surveys right at the end. From five of the estates we have two or more successive surveys of some or all of the manors, while from a sixth estate, Ramsey's, the mid-century survey tells us so much of earlier conditions, 'in the time of King Henry', that we have in effect two successive surveys here too. Likewise at Glastonbury the earliest of the several surveys refers back extensively to a still earlier one, now lost, also of Henry I's reign.

Some economic historians have found these successive surveys of the same manors invaluable; they show what changes were occurring and point to general trends. In looking at non-agrarian activities it is harder to use them in this way, even though long-term changes were taking place. This is partly because the surveys, even those from a single estate, were clearly drawn up for different purposes and thus give different information. The earliest Glastonbury survey does little more than list the livestock on the demesne of each manor. Most do more than this and describe the services of the local tenants, but some merely give the number of virgaters, cottagers and so on, naming only those whose tenure differs in

(Rolls Ser.; 3v., 1884-93), iii, pp. 241-314; (7) bishopric of Worcester (Worc.), c.1170 (date from C. Dyer, *Lords and peasants in a changing society* (Cambridge, 1980), p. 4n) - *The Red Book of Worcester*, ed. Marjory Hollings (Worcestershire Historical Soc.; 4 parts, 1934-50), *passim*; (8) St Paul's Cathedral, 1170-81 - *The Domesday of St Paul's of the year M.CC.XXII*, ed. W. H. Hale (Camden Soc., old ser. lxix; 1858), pp. 109-17, 140-52; (9) Glastonbury Abbey (Glast.), 1171, 1198-9, 1201 (dates from M. M. Postan, 'Glastonbury estates in the twelfth century', *Economic History Review*, 2nd ser. v (1952-3), pp. 358-9) - Trinity College, Cambridge, MS R.5.33, ff. 110v-116 *bis*v; and 1189 - *An inquisition of the manors of Glastonbury Abbey of the year M.C.LXXXIX*, ed. J. E. Jackson (Roxburghe Club; 1882); (10) bishopric of Durham (Durh.), 1183-4 - *Boldon Buke*, ed. W. Greenwell (Surtees Soc., xxv; 1852); (11) Knights Templars (Temp.), 1185 - *Records of the Templars in England in the twelfth century*, ed. Beatrice A. Lees (British Academy, Records of Social and Economic History, old ser. ix; 1935), pp. 1-135; (12) Evesham Abbey (Eve.: all references are to Harley MS 3763), late 12th century - BL Cotton MS Vespasian B.xxiv, ff.22v-73r *passim*, and BL Harley MS 3763, ff. 72r-81v.

some way from the norm.[3] Others name all the tenants, even those with entirely regular holdings.[4] Sometimes the description of the holding itself tells us of non-agrarian activity, as in the case of a mill. Sometimes particular services are a guide, like the obligation to do smith's work – shoe horses and maintain the ironwork on ploughs – or to pay rent in the form of fish or even wild fowl. But in general it is the surveys with most tenants' names that are the most informative, for these names often include non-agrarian occupational names. In later records such names may mislead, for they became fixed as family surnames that might or might not describe an individual's actual occupation. Mrs Le Patourel has found as early as 1242 a man named Potter who was really a butcher.[5] It is unlikely that this happened in the twelfth century, and it is reassuring that a tenant is sometimes called son of the turner, son of the smith, son of the weaver and so on[6] – the first step, but no more, in turning the descriptive craft name into a hereditary family surname. But if such names can safely be taken as a guide to occupations, they are certainly not used with a consistency that permits argument from silence. It is only in the later twelfth-century surveys that more than a very few tenants are identified by more than a simple Christian name, while the Templars' survey in 1185 shows a regional difference: in the sections for Lincolnshire and Yorkshire there are significantly fewer occupational names and most of the bynames used are patronymics.[7]

However, one thing that occupational names show very clearly in these surveys is the distinction between village and town. Professor Hilton has shown in the later Middle Ages a fundamental difference between the settlements we call villages and the settlements we call towns: in the towns a sizable proportion of the inhabitants followed a great variety of non-agrarian occupations.[8] The towns may or may not have been important as local market centres; but they certainly served as centres that supplied the wholly agricultural vills around with the specialised crafts and trades that no one vill could support on its own. The surveys show the same pattern throughout the twelfth century. In the Templars' survey of 1185 the entry for Baldock (Hertfordshire) is about twice as long as the entry for Finchingfield (Essex), twenty-five miles away; but whereas at Finchingfield the only tenant named by his craft is a single smith, at Baldock we have

3 E.g. Durh.
4 E.g. Glast. 1189.
5 H. E. Jean Le Patourel, 'Documentary evidence and the medieval pottery industry', *Medieval Archaeology*, xii (1968), p. 102n.
6 Thus Godric *filius tornatoris* (Burt., pp. 233–4), Edmund *filius fabri* (Ram., p. 248), William *filius textoris* (Glast. 1189, p. 131).
7 Temp., pp. 78–135.
8 R. H. Hilton, *The English peasantry in the later middle ages* (Oxford, 1975), pp. 77–88.

9 tailors, 6 tanners, 5 smiths, 4 masons, 4 clerks, 2 cooks, 2 farriers, 2 mercers, 2 shoemakers, 1 fowler, 1 goldsmith, 1 merchant, 1 saddler, 1 vintner, 1 weaver.[9] The same survey gives a similar picture of many varied occupations among the Templars' relatively few tenants in towns of national or regional importance - London, Bristol, Oxford - and also in the other local centres: Witham (Essex) and Strood (Kent).[10] The late-twelfth-century Shaftesbury survey gives very few craft names and the only manor to mention more than one (apart from millers and smiths) is Bradford on Avon where a fuller, a goldsmith, a merchant and a tailor (*scharere*) were former tenants and a (textually corrupt) note records a local by-law for tanners and brewers.[11] Among the Caen surveys of about 1170 the entries for Felsted (Essex) and Minchinhampton (Gloucestershire) show the same concentration of craft names, while on the contemporary bishopric of Worcester survey they appear not only at Worcester itself but also on the adjacent rural manor of Northwick.[12] We find just the same phenomenon in the two Burton surveys of the early twelfth century: few tenants are given more than a Christian name, but at Burton-upon-Trent we find (taking both surveys together) 2 bakers, 2 cooks, 1 dyer, 1 gardener, 1 huntsman, 1 mason, 1 salter(?), 1 shoemaker, and, in neighbouring manors, a goldsmith and a shoemaker at Stapenhill, a baker, a brewer, a cook and a fuller at Wetmore, a saddler at Winshill.[13] This spillover of craftsmen may be evidence of suburban growth, or it may be that those living in the town were adding to their resources by taking additional holdings outside it: Aluric the baker who held two bovates at Wetmore may very well have been the same Aluric the baker who held two bovates at Burton.[14] From twelfth-century towns we have (very notable) overall urban surveys only from Winchester, where there was a great variety of crafts.[15] Of towns entered in Domesday Book only Bury St Edmunds gives us a list, probably more impressionistic than complete, of the many crafts practised there.[16] The estate surveys show that this pattern held good for towns in general, including those much less important than Bury or Winchester.

The towns, then, were settlements where a variety of crafts was concentrated to serve the purely agricultural communities around. In a few

9 Temp., pp. 10–12, 65–9.
10 Temp., pp. 1–5, 13–16, 17–20, 42–3, 58.
11 Shaft., ff. 76r, 77r.
12 Caen, pp. 40–4, 56–62, 67–72; Worc., pp. 32, 37–8.
13 Burt., pp. 212–14, 221, 238–9, 241 (*celarius*, saddler; the editor, p. 273, takes it to be cellarer, but this is unlikely to be correct).
14 Burt., pp. 213, 221.
15 *Winchester in the early middle ages*, ed. M. Biddle (Winchester Studies 1; Oxford, 1976), pp. 428–39.
16 *Domesday Book, seu liber censualis Willelmi primi regis Angliae*, [ed. A. Farley] (London, 2v., 1783), ii, p. 372.

other settlements a single non-agrarian activity seems to have predominated – they were akin to the industrial villages of the later Middle Ages. There are two clear examples in the estate surveys. One was Wells-next-the-Sea, beside the mud-flats of the north Norfolk coast. The Ramsey survey lists some fifteen holdings of fisheries, some held on a timeshare basis – the fourth, the sixth or the eighth night in a particular fishing area – and all paying rent in kind: from 500 to 7000 eels a year and from 2 to 36 *steilinges*, an unidentified but clearly much larger fish. But many other holdings at Wells, both simple messuages and what were called lands (*terre*), also paid rents of so many *stiche* of (perhaps 25) eels. Fishing must have been at the very least a widespread activity at Wells.[17] The other vill of this sort was Arne in Dorset. This too lay beside coastal mud-flats, on the shore of Poole Harbour, but here the dominant activity was not fishing but salt-making. Shaftesbury Abbey's 21 tenants here each held from one to four saltpans (*plumba*) for rents that included a *wikeworc salis*, probably a day's work each week making salt for the abbey itself.[18] It seems likely that fishing or salt-making was not the only activity at Wells and Arne and that here too corn was grown, stock was raised – though in each case the holdings of what may or may not have been agricultural land are described in unusual terms. We would expect a landlord to exploit as far as he could any peculiar but useful local product and thus for it to loom unduly large in his records – corn and livestock, after all, he could get anywhere. But there is a clear contrast between these two places and, say, Henbury (Gloucestershire) where the Worcester survey gives no more than a passing reference to fisheries and salt-pans.[19]

On many manors, in fact, we have just such passing reference to local products that supplemented the basic agricultural economy. Some are crops other than the usual cereals. It is interesting that on half the manors in Evesham Abbey's survey tithe obligations refer to flax as well as (mostly) to corn and lambs. More definite, and rather more surprising, are the references to viticulture. Work in the vineyard was a normal service for tenants at Glastonbury itself and on some other manors of the abbey's estate, and on about a third of the manors in the Evesham survey; vine-growing is also mentioned on manors of Shaftesbury Abbey and the bishopric of Worcester.[20] This agrees with other evidence that more vines were grown in southern England now than in later centuries, when the Gascon wine trade had fully developed to meet English demand.[21] There are quite a number of references to fisheries and fishermen; the Templars

17 Ram., pp. 296–9.
18 Shaft., ff. 60v–61r.
19 Worc., p. 411.
20 E.g. Glast. 1189, pp. 22, 31; Eve., ff. 72v, 80r; Shaft., f. 82v; Worc., pp. 170-1, 406.
21 E. M. Carus-Wilson, 'The effects of the acquisition and of the loss of Gascony on

even drew 7s annual rent from a fishing-boat (*navicula piscarii*) that they leased out.[22] At Hartlebury (Worcestershire) an enclosure in the Severn produced annual rent of 1000 eels or 5s and at Zoy (Somerset) on the edge of Sedgemoor a fishery rendered 3000 eels, 3000 *exherell'*, another unidentified fish.[23] These were obvious large-scale fishing-grounds, but there are many cases of mills rendering smaller numbers of eels as part of their rent,[24] and on three of the bishop of Durham's manors the mill-pond was let out apart from the mill, presumably for fishing.[25] Wild fowl are mentioned only at Glinton (Northamptonshire), a Peterborough manor on the edge of the Fens, where two tenants, one specifically called a fowler, each paid rents that included ten wild geese;[26] persons named as fowlers occur in a few other places as well.[27] We see other local resources being exploited. One was building-stone: in 1183–4 Lambert the marble-worker (*marmorarius*) held 30 acres at Stanhope (co. Durham) for his service to the bishop of Durham; the so-called Frosterley marble comes from just this part of Weardale.[28] There are more references to exploitation of woodlands: charcoal-burners (*wodecokeres, carbonarii*) are mentioned,[29] and at Temple Balsall (Warwickshire) the rent paid by the smith gave him the right to take dead wood for charcoal, while again in Weardale three turners held 17 acres in return for making 3100 wooden dishes (*scutelle*).[30]

Like Lambert the marble-worker these turners held what was presumably agricultural land, and in all these cases the surveys show us these other forms of production as subsidiary or supplementary to the agriculture that was the community's main business. Here again a note of caution is needed. At Arne and at Wells-next-the-Sea the surveys may exaggerate the dominance of salt-making and fishing. Elsewhere they almost certainly understate the extent of non-agricultural production. Land and the tenure of land are what the surveys are about. If the tenant did some other kind of work this need come to light only in the unusual case of his paying rent in the form of the products of his craft, like the Weardale turners. Otherwise it was the merest chance if the survey mentioned his

the English wine trade', *Bulletin of the Institute of Historical Research*, xxi (1946–8), pp. 145–8.

22 Temp., p. 32.

23 Worc., p. 207; Glast. 1201, f. 111v. *Exherell'* were clearly a small fish, and the word is a diminutive of Anglo-Norman *esche*, bait; I am grateful to Dr Laura Wright for this information.

24 E.g. Pet., p. 158; Worc., p. 37; Eve., f. 74r.

25 Durh., pp. 11, 18, 35.

26 Pet., p. 162.

27 E.g. Caen, p. 57; Temp., p. 67.

28 Durh., p. 30.

29 E.g. Burt., p. 216; Glast. 1189, p. 41.

30 Temp., p. 34; Durh., p. 28.

occupation in naming him. On the Burton and Caen estates, where we have successive surveys only a few years apart, the same tenants may be given an occupational name in one survey, not in the other.[31] The contrast between Arne and Wells-next-the-Sea on the one hand and many predominantly agrarian communities on the other may well have been less extreme than a simple reading of the surveys might suggest. At the same time the natural resources for non-agricultural production – the woods, the quarries, the water – may have been widespread but were not ubiquitous. Most communities of twelfth-century England must have been overwhelmingly agricultural, their economy based on crops and livestock. In such a 'normal' vill what non-agrarian services were performed? For what craftsmen was demand so great that a single community could support them by itself, without having to go to the local town to seek their services? The answer is simple and straightforward: millers and smiths.

Mills are one of the subjects on which the officials of 1086 had to collect information, and some six thousand are recorded in Domesday Book. Their distribution raises problems: the very few mills in Devon and, especially, Cornwall, the relative shortage of mills in Middlesex, the occasional mill or group of mills of great value, like the seven mills at Battersea worth over £42 a year. One possibility is that Domesday Book records a late stage in the general replacement of household querns by village mills, a process that had not quite reached the farthest West Country; another is that village mills were replacing mills that had served wider areas.[32] The twelfth-century surveys offer less direct help in these problems than one might expect. For one thing they do not necessarily enter all mills. At Whickham (co. Durham) we are told that the mill had once rendered 3 marks a year but was now, in 1183–4, included in the farm of the manor.[33] There were probably many manors where the mill was simply counted as part of the demesne, was included in the property let out to the manorial farmer and then left out of a survey that was concerned only with local tenants and their rents. At Temple Balsall (Warwickshire) in 1185 three tenants paid 3s 4d each 'for land of the mill', but there is no mention of the mill itself.[34] Where we find that no mill is mentioned on about half the manors in the Glastonbury survey of 1189 it need not follow that these manors had no mills. When the second Burton survey records mills at Abbots Bromley and Potlock that do not appear on the first survey, some ten years earlier, we cannot safely argue that new mills had been built.[35] There are problems

31 E.g. Caen, pp. 58, 70 (no. 44: William Smedbred, William *fullo*); Burt., p. 216 (B: Alured *parmentarius*; A: Alured).
32 Lennard, *Rural England*, pp. 278–87.
33 Durh., p. 34.
34 Temp., p. 34.
35 Burt., pp. 223, 236.

even in comparing the value of mills on successive surveys from a single estate: a mill was seldom farmed out without an amount of land that might vary from one lease to the next, and the surveys make clear, what Domesday Book does not, that various services might or might not be required of the lessee as well as the rent in cash. However, for what it is worth, successive surveys give the impression that mills were generally increasing in number and in value, though with local exceptions: at Burwell (Cambridgeshire) Ramsey Abbey's Hulme Mill paid 4d rent and provided the manorial farmer with his food in the time of Henry I, but by the 1160s lack of water meant that it rendered only 6d a year in all.[36] Most of the mills recorded must have been water mills, though usually it is only where they paid rents of fish that we can be sure of this - and we are specifically told of a horse mill at Darlington (co. Durham) in 1183–4 and of a windmill at Dunwich (Suffolk) in 1185.[37] Almost certainly these surveys have more to tell us about the mills of twelfth-century England; but the difficulties that arise call for careful consideration of what we know of the local circumstances in each case.

Domesday Book records mills but does not record smiths, and it is the twelfth-century surveys that first show how widely they were distributed; there was probably a smith in most vills at this date. We cannot suppose that they can all be identified in the surveys. Many had service holdings, land held in return for specified smith's work, usually seeing to the ironwork of the demesne ploughs. But others held for cash or agricultural works like the other tenants and we recognise them as smiths only by the chance of the survey's calling them *faber*. Like the miller the smith normally had agricultural land for his support as well as his craft; at Burwell (Cambridgeshire) he held as a service holding a croft and 22 acres that had been his father's and also held a croft and 24 acres for a cash rent with services of ploughing and reaping.[38] At Kingston (Dorset) the smith for his half-virgate not only had to maintain the ironwork of the demesne's buildings and ploughs and shoe one draught animal on all four feet; he also had to castrate animals on the demesne whenever this was needed or else find someone else to do it, and this may have been part of a smith's normal work.[39]

Carpenters are the only other craftsmen to appear at all regularly in agricultural vills and they are far less frequent than millers or smiths, though in the Ramsey surveys curiously few smiths are mentioned and there are almost as many carpenters: six as against seven. More typically, in the Worcester survey there are only two carpenters but seventeen smiths.

36 Ram., p. 310.
37 Durh., p. 17; Temp., p. 135.
38 Ram., p. 310.
39 Shaft., f. 62r.

Like the smiths, some carpenters had service holdings, usually for making ploughs and harrows, and others can be identified only by the occupational name;[40] one specialist at Damerham (Wiltshire) is named as a wheelwright (*factor rotarum*).[41] At Shapwick (Somerset) the mill was held by a carpenter, as also at Stapenhill (Derbyshire), where as part of his rent he had to do all the work in wood or lead needed on the church; at Buckland Newton (Dorset) the miller was required to help in manorial building operations, using his own tools.[42] Milling and carpentry will have combined naturally from the need to maintain the mill machinery.

Other craftsmen are scattered in ones and twos among the tenants of agricultural vills but they are not common; here again we can mostly identify only those who happen to be named by their craft. Some of these names may mislead: they belong to a master craftsman from a town or to a lord's household servant who had acquired a holding by way of investment or endowment but whose craft formed no part of the vill's economy. A likely example is Christian the mason who at South Sherburn (co. Durham) held 40 acres 'which the bishop gave him'; or Geoffrey the marshal who held 2 acres 'at the abbot's will' on Evesham's manor of Hampton (Worcestershire); or Osbert the goldsmith who held a virgate in villeinage at Kempsey but who also had holdings in Worcester, four miles away.[43] Some occupational names themselves pose problems: it is not at all clear just what cook or clerk or merchant (*cocus, clericus, mercator*) meant in terms of everyday life when applied to a village tenant. However, the two thatchers of Millbrook (Bedfordshire) and the three masons of Tysoe (Warwickshire) suggest that some places, without the variety of activities that distinguished the towns, might act as local centres for a particular specialised craft.[44]

Most tenants with craft names held normal agricultural holdings which they (or their families) worked. They cannot have been so very different from their fellow-husbandmen who (or whose families) must often have engaged in other activities as a normal part of their lives. Some of these activities occurred everywhere, such as brewing, often mentioned in the surveys: tenants had to help when ale was brewed for the manorial lord or had to pay toll when they themselves brewed for sale – when they displayed a sign, as the Shaftesbury survey puts it.[45] Others were peculiar to a particular place: tenants of half a hide at Kingston (Dorset) were required to do carrying-service by sea – it seems to have been taken for granted that the

40 E.g. Glast. 1189, pp. 29, 45.
41 Glast. 1189, p. 131.
42 Glast. 1189, p. 57; Burt., p. 239; Glast. 1189, p. 142.
43 Durh., p. 10; Eve., f. 79r,v; Worc., pp. 32, 37, 83.
44 Temp., p. 28–9, 75.
45 Shaft., f. 87r.

tenants would have boats.[46] Rural cloth-working is another problem. In 1965 Dr Miller pointed to such names as weaver and dyer to show that clothmaking was a regular activity in villages as well as towns in the late twelfth and early thirteenth centuries.[47] In fact these names do not occur at all frequently in the surveys, but this could mean that so many people were involved in clothmaking that the occupational name would not serve to distinguish the individual. At Fontmell and Handley (Dorset) the manor's dyers were given corn and an old ewe at Martinmas – but there is no other mention of clothworkers there, showing that this was an activity that could easily pass unnoticed in these records.[48] Perhaps we should not distinguish too sharply between the husbandmen (who were also brewers or shippers or dyers) and the village craftsmen (who were also husbandmen). Yet a distinction there was. It appears in the tenants' obligations at Temple Ewell (Kent): when they repaired the demesne barn the manorial lord had to find the thatcher.[49] There were some jobs anyone could be expected to do; others called for the specialist.

We have not been able to discover change in non-agrarian activities by comparing successive twelfth-century surveys for the same manor. Nor, although nearly a century separates them, can we usefully or safely compare the earliest and the latest of our surveys to find what changes were occurring. For this topic the surveys present an oddly uniform picture. Yet rural society and economy were not static, and the picture they give us is of a particular point in at least two processes of change. The first is the gradual growth of specialised services within the individual agricultural community – or rather, the movement of these services from the towns or other centres into the vills as increased demand produced by rising population or rising living standards enabled the single vill to support them. The surveys show an early stage in this process, a stage where it would be normal for a vill to have a miller and a smith and (though this lies beyond the scope of the surveys and of this paper) a priest. A century before the Norman Conquest it would have been unlikely to have had a priest, it possibly would not have had a miller, and we may wonder whether it would have had a smith: the omission of the smith as well as the miller from the *Rectitudines Singularum Personarum* is interesting and may well be significant. In the thirteenth and fourteenth centuries the process continued, leading in the fifteenth century to the collapse of many of the towns that had existed primarily as centres for providing specialised

46 Shaft., f. 61v.

47 E. Miller, 'The fortunes of the English textile industry in the thirteenth century', *Economic History Review*, 2nd ser. xviii (1965), p. 67.

48 Shaft., ff. 57v, 65r.

49 Temp., p. 23.

services.[50] The village carpenters of the twelfth century can be seen as part of the vanguard of this further progression; so too can the two rural fulling mills recorded in the Templars' survey in 1185.[51]

The second process concerns the vill's internal structure as viewed by the lord of the manor; these surveys are, after all, seignorial records. If we compare them with similar surveys of the late thirteenth century one striking difference is the virtual disappearance of service holdings. It is not just that smiths and carpenters no longer give specialised services in return for their holdings; the twelfth-century surveys often show the demesne's own workers - the ploughmen, the shepherds, the pig-keepers - holding land in return for their work, but a century later they had become simple employees and if they held land this was something separate, for which their obligations were the same as any other tenant's.[52] May we suppose that the smith, who often had a service holding, reached the village earlier than the miller, who hardly ever did? Let us try to see the surveys through the eyes of a twelfth-century manorial lord, without the hindsight even of his thirteenth-century successors, let alone of a twentieth-century historian. He would see no particular distinction between the smith who held land for shoeing the demesne horses, the shepherd who held land for tending the demesne sheep and the villeins who held land for threshing the demesne corn. They, their land and the demesne were bound together in a single nexus that was communal and seignorial rather than simply territorial and economic. But change was going on, and by the end of the thirteenth century the demesne labourer worked for regular payment in cash or corn, labour services were an archaic survival, the village craftsman was an independent trader. Sense of community may well have continued unchanged - of that our records tell us nothing. But the local lord was distancing himself from the vill and its inhabitants: his relations with his tenants were being redefined through the intervention of cash payments and the formalised procedures of the manorial court.[53] The twelfth-century surveys give us our first detailed glimpse of the communities of rural England. But they also give us a final glimpse of a world that was changing.

50 P. D. A. Harvey, 'Non-agrarian activities in the rural communities of late-medieval England', to be published in the proceedings of the 14th Settimana di Studio of the Istituto Internazionale di Storia Economica Francesco Datini, Prato [in fact never published].

51 Temp., pp. 50, 127.

52 M. M. Postan, *The famulus* (Economic History Review Supplement 2; [1954]), p. 33.

53 Cf. P. D. A. Harvey, 'Initiative and authority in settlement change' in *The rural settlements of medieval England*, ed. M. Aston, D. Austin and C. Dyer (Oxford, 1989), pp. 38-43.

NOTE: The *wodecokeres* mentioned on p.106, line 17, were fowlers (woodcockers) not charcoal-burners, and the reference to Burt., p.216, should thus be added to those in n. 27. I am grateful to Dr Jean Birrell for this correction.

Since this article was published in 1990 there have been important editions of twelfth-century surveys by N.E. Stacey in two volumes of the British Academy's Records of Social and Economic History: *Surveys of the Estates of Glastonbury Abbey c.1135–1201* (new series vol. 33; 2001) and *Charters and Custumals of Shaftesbury Abbey 1089–1216* (new series vol. 39; 2006).

V

Initiative and Authority in Settlement Change

INTRODUCTION

Ever since Maurice Beresford first revealed the lost villages of England in 1954, work on medieval settlement has produced more and more evidence of change. Villages decayed and were deserted; but they were also created by simple planting or by bringing together existing villages or dispersed farms, they moved from one site to another, they were entirely rebuilt where they stood. They were endowed with churches – but not just once for all time, for the church might survive or decay, stay still or move, not necessarily in step with the rest of the settlement. The field system too need be neither a surviving relic of dark age colonization nor the product of gradual evolution: the fields, it is now agreed, might be reorganized as suddenly and as drastically as the settlement itself. Where once we saw underlying stability we now look for – and often find – flexibility and sudden change. Some of this change was the piecemeal work of many individual husbandmen, like much of the assarting of waste and reclamation from sea and fen that was achieved in the twelfth and thirteenth centuries.[1] Even a new town of the thirteenth century, Ravenserodd, could be held to have begun its mushroom growth in an enterprising individual's taking over a wrecked ship as a sort of wayside café for people sailing up or down the Humber.[2] But by their very nature many changes now discerned must have involved the cooperation, or the ordering, of whole communities of people. Where should we look for the initiative to plan these changes, the authority to put them into effect? Was it communal agreement that lay behind them or was it seignorial power?

We cannot yet give a definite answer. The sources of local initiative and

[1] For example, E. King, *Peterborough Abbey 1086–1310: A Study in the Land Market* (Cambridge, 1973), pp. 75–80, 83–7.

[2] *Calendar of Inquisitions Miscellaneous (Chancery)* (7 vols, 1916–68), vol. I, no. 1512.

authority varied over time, as we shall see. But they also varied from region to region and from place to place in ways not fully explored. In 1287 agreement was concluded over common pasture rights in Alvechurch and King's Norton, neighbouring places in Worcestershire; for Alvechurch the party to the agreement was the Bishop of Worcester, its manorial lord, but for King's Norton, a manor of royal demesne, it was twenty representatives of the community.[3] More work is needed on such differences in local executive authority. Here two points will be examined: the roles of custom and of lordship on the individual manor and the changing relationship between seignorial power and the local community throughout the middle ages. At best this will provide a background pattern against which we can set particular instances of settlement change.

LORDSHIP AND CUSTOM

The respective roles of manorial custom and manorial lordship have been seen as a problem.[4] Historians working on medieval rural communities have increasingly shown how local custom in practice restrained the lord's powers over his villein tenants. Under the common law, as it came to be defined in the late twelfth and thirteenth centuries, these powers were immense. The villein held his land and all his possessions entirely at the will of his lord, who could deprive him of any or all of them for any reason or for none; he had nothing he could call his own. His rents and services could be increased arbitrarily at any time, but he could not leave the manor. He had no right of inheritance, but having once acquired his holding he could not sell or sublet it and land he got from any other source passed automatically to his manorial lord. This was the law administered in the royal courts and any villein trying to bring a case there in defence of his rights or his holding would be sent away empty-handed: his goods and his lands lay wholly in his lord's jurisdiction, and it was only in his lord's manorial court that he could bring any case concerning them. But once in the manorial court an entirely different set of rules operated. The law administered there was local custom, the custom of the manor, and this allowed the villein all kinds of rights that the common law denied him. His rents and services were defined and fixed, he would inherit his holding by whatever rules applied in the manor, and would continue to hold it unless he voluntarily relinquished it or committed some offence for which custom decreed its forfeiture. There was an implicit assumption that his goods and his money were

[3] R. H. Hilton, *A Medieval Society: The West Midlands at the End of the Thirteenth Century* (London, 1966), p. 152.

[4] For example, E. King in *English Historical Review*, 101 (1986), p. 426 (review of *The Peasant Land Market in Medieval England*, ed. P. D. A. Harvey (Oxford, 1984)).

his own, and would pass to the lord only in accordance with customary rule, like the animal commonly taken as heriot on the villein's death. If he was mulcted ('amerced') in the court for a misdemeanour two of his fellow villagers, the affeerors, had at least a voice, even a decisive voice, in fixing the amount he had to pay. And although custom might place restrictions on his freedom of movement and of action he could usually buy his way out of them for a reasonable price: as John Hatcher has put it, 'Lords were concerned far less with controlling the lives of their villeins than with profiting from their right to do so'.[5]

The detailed customs varied from one manor to the next, but the rule of local custom in manorial courts was universal. This custom was defined and declared not by the manorial lord but by the villagers themselves who, as members of the court, were viewed as 'the guardians of custom, the collective memory of the neighborhood'.[6] Certainly the manorial lord had the right to disregard and override local custom and if he did he could rely on the full backing of the common law and the royal courts. But in practice this did not often happen. In the normal course of events the lord would respect the precedents embodied in local custom, partly, no doubt, because this enabled him to manage his property and his tenants harmoniously, and thus efficiently, but partly too because at his own level he would be as much a respecter of past precedent and declared custom as any of his tenants.[7] Breach of custom could lead to trouble, as at Broughton (Huntingdonshire) in 1291, when villeins doing harvest services went on strike for half a day, claiming that the loaves given them for lunch were smaller than usual.[8] When conflict arose between a manorial lord and his tenants, as from time to time it did, real or alleged breach of custom by one side or the other was often the cause.[9] Sometimes we see lords avoiding difficulties by what seems almost excessive respect for custom; in the early fourteenth century when Titchfield Abbey's tenants at Cadland (Hampshire) refused to elect a reeve, the abbot refrained from the obvious step of nominating someone himself, even though on many other manors such nomination by the lord was customary.[10] And the lord who wanted to alter manorial custom might feel he

[5] The position of the villein, in legal theory and in practice, is fully discussed by J. Hatcher, 'English serfdom and villeinage: towards a reassessment', *Past and Present*, 90 (1981), pp. 3–39; the quotation is from p. 10.

[6] J. S. Beckerman, 'Customary law in English manorial courts in the thirteenth and fourteenth centuries' (unpublished Ph.D. thesis, University of London, 1972), p. 12. I am grateful to Dr Beckerman for permission to cite this important unpublished work.

[7] Cf. Hatcher, 'English serfdom and villeinage', pp. 23–4.

[8] G. C. Homans, *English Villagers of the Thirteenth Century* (Cambridge, Mass., 1941), p. 262.

[9] Many examples are given in the works of R. H. Hilton, such as *The English Peasantry in the Later Middle Ages* (Oxford, 1975), pp. 58–69, or *Class Conflict and the Crisis of Feudalism: Essays in Medieval Social History* (London, 1985), pp. 108–13, 124–35; or, on one estate, by D. G. Watts, 'Peasant discontent on the manors of Titchfield Abbey, 1245–1405', *Proceedings of the Hampshire Field Club and Archaeological Society*, 39 (1983), pp. 121–35.

[10] Watts, 'Peasant discontent', pp. 126–7, cited by S. Reynolds, *Kingdoms and Communities in Western Europe, 900–1300* (Oxford, 1984), p. 133.

had to move with great circumspection. At Povington (Dorset), about 1230, harvest boon-workers, allowed a sheaf of corn at the end of the day's work, had been used to take enormous sheaves, four times larger than normal. However, their manorial lord, the Abbot of Bec, was so nervous of altering established custom that he cited scriptural authority to do so: 'as holy writ says that it is right to do away with evil usage and to change for the better what is evilly attempted', so, henceforth, the harvesters' sheaves should be of normal size.[11] This is not to say that manorial custom was usually static, though lords and tenants alike seem to have thought it ought to be and assumed that it was. D. G. Watts tells how at Titchfield 'The abbot did his utmost to distort custom in his own interest while at the same time regarding himself as bound by it.'[12] Everywhere it was, in R. H. Hilton's phrase, 'a shifting compromise', and it evolved gradually over the years.[13] Sometimes we can see this happening. In 1391 a 5s. fine was imposed when a villein's daughter of Oakington (Cambridgeshire) was unchaste, this being the custom of the manor according to the jurors – but they were wrong, for a hundred years earlier the amount had been 6d. or 1s.[14] This change was to the lord's advantage, but this was not always the case.[15] Despite the strength of their legal position, those thirteenth-century manorial lords who tried to prevent the growth and spread of a peasant land market had little success; where local demand arose among villeins to buy and sell land among themselves their lords seem to have been powerless to prevent them even from breaking up the standard customary holdings, the basis of the manor's tenurial structure.[16]

It is against this background that we must view the evidence for physical changes in the settlements of medieval England. Those who have discovered this evidence have mostly explained these changes as stemming from seignorial power: it was the manorial lords who moved or replanned villages, who reorganized field systems and so on. Often this is the only possible answer. It can only have been the manorial lord who in the twelfth century moved the village of Eaton Socon (Bedfordshire), church and all, so that he could build a

[11] M. Chibnall (ed.), *Select Documents of the English Lands of the Abbey of Bec* (Camden Society, 3rd ser., 73, 1951), p. 62. I owe this reference to the kindness of Dr A. C. Jones, whose articles on 'Harvest customs and labourers' perquisites in southern England, 1150–1350', *Agricultural History Review*, 25 (1977), pp. 14–22, 98–107, covering respectively the corn harvest and the hay harvest, give a valuable conspectus of an item of manorial custom that is often defined in surveys.

[12] Watts, 'Peasant discontent', p. 124.

[13] Hilton, *Class Conflict*, p. 126.

[14] F. M. Page, *The Estates of Crowland Abbey: A Study in Manorial Organisation* (Cambridge, 1934), pp. 334, 339, 416.

[15] Cf. the interesting comparison of a Teddington (Middlesex) virgater's rents and services in the early thirteenth century and the early fourteenth in B. Harvey, *Westminster Abbey and its Estates in the Middle Ages* (Oxford, 1977), pp. 219–22.

[16] P. D. A. Harvey (ed.), *Peasant Land Market*, pp. 344–9.

castle on its old site.[17] And in many more cases it seems, on the face of it, more likely that change came from the lord's initiative rather than from some communally agreed decision. So an odd situation has arisen. Historians, working on documents, often see the lord of the manor as moving gingerly, pussyfooted even, among the intricacies of his tenants' rights and privileges. On the other hand archaeologists and geographers, working in the field, see this same lord sweeping all before him, uprooting entire villages and replanning lands and settlements in the most arbitrary way.

This is less of a problem than it appears at first sight. Our knowledge of manorial custom comes partly from court rolls which record disagreements or statements on particular points, like the Oakington fine for unchastity in 1391, but more from custumals. These were surveys which recorded and defined the obligations between lord and tenant, largely so that the lord could counter with assurance any particularly optimistic recollections of custom that the villagers might produce in the future; thus the Abbot of Ramsey, confronted with his harvest workers' strike over the size of loaves, pointed to the custumal which said that a three-farthing loaf was to be shared between two workers. These records were not meant to be comprehensive codes of local custom, and we may reasonably see them as simply the visible tip of a mass of slowly evolving customs that affected every relationship of village life, not just between lord and tenant but between one villager and another, between bailiff and hired worker and so on. Very occasionally we get glimpses of this: at Wistow (Huntingdonshire) in 1294 the manorial court made Robert Juwel pay 1s. because 'by custom he ought to have given a meal to all the servants of the manor of Wistow on the day when he married his wife, and he did not do so'.[18] More often we get indirect evidence. Thus long after a manor had taken to measuring land by the rod and corn by the struck measure, piecework rates for hired workers might still be based on customary field acres for reaping and on heaped corn measures for threshing, despite the problems this created in accounting and auditing.[19] Within the community, in his dealings with its inhabitants, the lord of the manor was as much involved as anyone else in this close web of custom and precedent; but as its lord he had another role to play. There need be nothing inconsistent in a careful respect for local custom in the internal administration of the manor and the exercise of dictatorial powers in ordering its affairs overall. The lord's position might be summed up in a simple formula: local custom tended to be strongest in his relations with his

[17] P. V. Addyman, 'Late Saxon settlement in the St Neots area: I. The Saxon settlement and Norman castle at Eaton Socon, Bedfordshire', *Proceedings of the Cambridgeshire Antiquarian Society*, 58 (1965), pp. 48–50.

[18] Homans, *English Villagers*, p. 173.

[19] P. D. A. Harvey, *A Medieval Oxfordshire Village: Cuxham 1240–1400* (London, 1965), pp. 44, 55–6. In the case of the threshing, easily identified on manorial accounts, this can be paralleled on many other manors.

individual tenants, officers or employees, while lordship tended to be strongest in matters affecting the community as a whole.

This can only be hypothesis. No instance seems yet to have come to light of the same manorial lord acting in both roles – but given the scantiness of the evidence it would be an extraordinarily lucky chance if it had. All we can say is that it fits the facts as we know them and explains why historians, working on records of manorial administration, are most impressed by the strength of custom; archaeologists and geographers, working on material remains of settlements, by the strength of lordship. Moreover, it fits a much wider pattern. The form of organization that accepts arbitrary direction in external affairs while insisting on careful regulation of internal relationships is, in human terms, a very natural one. We see it in the modern factory. The owner can move it from one side of the town to the other or from one county to the next, changes that may seriously disrupt his employees' pattern of life, yet he is more likely to end up with a strike on his hands if he reduces the tea-break from ten minutes to five. Other analogies, past and present, are not hard to find.[20] But to draw for this purpose a sharp distinction between custom and lordship in the medieval community is misleading. The medieval villein, or the manorial lord, asked about the respective roles of custom and lordship in the community, would not understand the question. Custom and lordship were all one to him. It was custom that gave the villein a loaf of a certain size in return for his harvest work; but equally it was custom that entitled the lord, if he wished, to have the village pulled down and rebuilt on a new site. The exercise of lordship was merely custom as it applied to the lord of the manor. It is with lordship as an aspect of custom that we are concerned.

LORDSHIP AND THE COMMUNITY OF THE VILL

But custom, as we have seen, evolved and changed over the years; and lordship, as an aspect of custom, will have changed too. Nothing said so far has been meant to suggest that it was necessarily the manorial lord who initiated change in settlement, but only that seignorial power was a possible source of initiative: its seeming weakness in the face of local custom is not relevant to its role in settlement change. What was this role? Returning to our original question, was it the manorial lord or the local community that instigated these changes?

The local rural community, the vill – the word comprehends both nucleated village and dispersed settlement – has long been of interest to historians. They have increasingly come to see it as a body which must have had some sort of internal administrative organization, however rudimentary, and which could

[20] It is interesting, for instance, that one thing the British sovereign can do without reference to Parliament is to declare war.

take corporate action. The manor was the local unit of lordship and of estate organization, but royal administration used the vill as its smallest unit of government, never the manor. The vill was the unit of tax assessment and collection, the unit required to produce military levies, the unit represented, if need be, in the county court, the unit responsible for local law and order. It was the four neighbouring vills that testified to a coroner's inquest or, in a forest, to the forester's inquiry when a deer was found dead or wounded. Communal payments by vills are recorded from as early a date as we could expect to know of them. How these payments were divided among the villagers, how the tax collectors, soldiers, constables, representatives of the vill, were chosen is very uncertain; the methods used have left no written records.

But the vill did more than just respond to the demands of royal authority. In 1922 Joan Wake printed two documents, and referred to others, which showed the vill acting corporately in such things as repairing a bridge, holding land for public use, and reorganizing a field system.[21] Since then more evidence of the same sort has been brought together, notably by G. C. Homans in 1941, Helen Cam in 1950 and by W. O. Ault.[22] Ault's account of the vill in 1982 looked particularly at its obligations in royal administration, but earlier he had done much work on the by-laws entered on some manorial court rolls, seeing them as the products not of manorial administration but of the vill itself.[23] In 1985 C. C. Dyer argued cogently that from what we now know, or can deduce, about the organization and effective powers of the medieval vill we need not necessarily see seignorial authority as lying behind change in fields and settlements: it is possible, indeed likely, 'that villagers rather than lords were responsible for the planning of villages and field systems'.[24] Meanwhile, in 1984, Susan Reynolds published an important and penetrating study which sets what we know of the medieval vill in a new, revealing light. Looking at the period 900–1300, and at France, Germany and Italy as well as England, she argued that the vill could act collectively and was accepted as an entity for many purposes long before the idea of a legal corporate body had been worked out. In this it would be no different from other groups of people linked by similar geographical or institutional ties. In other words when we see the vill of the thirteenth, fourteenth or fifteenth century engaging in public works, organizing its fields, holding land, we are not seeing the first abortive fumblings towards legal corporate status that was never achieved, but the

[21] J. Wake, 'Communitas villae', *English Historical Review*, 37 (1922), pp. 406–13.

[22] Homans, *English Villagers*, especially pp. 328–38; H. M. Cam, 'The community of the vill', in *Mediaeval Studies Presented to Rose Graham*, ed. V. Ruffer and A. J. Taylor (Oxford, 1950), pp. 1–14.

[23] W. O. Ault, 'Open-field husbandry and the village community', *Transactions of the American Philosophical Society*, new ser., 55, part 7 (1965), which brings together much work from earlier articles; 'The vill in medieval England', *Proceedings of the American Philosophical Society*, 126 (1982), pp. 188–211.

[24] C. C. Dyer, 'Power and conflict in the medieval English village', in *Medieval Villages: A Review of Current Work*, ed. D. Hooke (Oxford University Committee for Archaeology, monograph no. 5, Oxford, 1985), pp. 27–32; the quotation is from p. 32.

natural, unselfconscious acts of a group long accustomed to acting collectively. This further strengthens Dyer's case: there is nothing inherently wrong in seeing the local community as responsible for planning changes and carrying them out.[25]

Yet some changes by their very nature can only have been the work of manorial lords. We have seen an example already from Eaton Socon; others that come at once to mind are the removals of villagers from properties given to Fountains and some other newly founded Cistercian abbeys in the twelfth century.[26] Mostly we can only guess where any particular change originated. At the same time some broad, long-term trends can be perceived which may at least guide our guessing. Within the vill some inhabitants, some groups, are bound to have carried more weight than others; there will always have been a village elite, even a village aristocracy, who will have had a fair measure of control over the rest.[27] Sometimes we get a glimpse of them, like the eighteen villagers of Pleshey (Essex) who in 1394 agreed to the removal and rebuilding of the parish church on a new site, 'with the assent of all our neighbours, the lesser as well as the greater'.[28] But the vill's most distinguished members would be missing from its counsels, for, as Dyer argues (and this, we shall see, is surely the key to what was happening), manorial lords took little interest in the internal organization of the vill, and were content to leave this to the inhabitants.[29] But if this was true of the absentee lords of great estates in the thirteenth and fourteenth centuries it is much less likely to have been true in the mid-eleventh century, when most local communities included one or more thegns living on the spot, thegns who can hardly have been uninterested in the affairs of the vill and who will presumably have expected to take the lead in ordering them. Taking this as our starting-point we could say, in broad, simple terms, that the period from the eleventh century to the fifteenth saw the gradual withdrawal of the seignorial class from the community of the vill, and that with its withdrawal the other elements in the community gained, perhaps in power and certainly in prominence in the written record.

This can be simply illustrated by four documents concerned with reorganizing lands, in each case lands both of tenants and of manorial demesnes. At Dry Drayton (Cambridgeshire) in the twelfth century lands were redistributed between five manorial lords; the charter recording this mentions the interest or participation of no one but the lords themselves. From Segenhoe (Bedfordshire) we have a record of a similar division of lands between two

25 Reynolds, *Kingdoms and Communities*, esp. pp. 101–54; Dyer's article, originally a paper delivered in 1982, was written before this was published.

26 D. Knowles, *The Monastic Order in England: A History of its Development From the Times of St Dunstan to the Fourth Lateran Council 940–1216*, 2nd edn (Cambridge, 1963), pp. 350–1.

27 Hilton, *English Peasantry*, pp. 54–8; Dyer, 'Power and conflict', pp. 29–30.

28 PRO, DL25/756.

29 Dyer, 'Power and conflict', pp. 27–8.

lords, involving the displacement of some local tenants; later, perhaps in the late twelfth century, all their tenants by general agreement asked to have their lands redefined and redistributed, and after they had surrendered their lands to their two lords for the purpose this was done by six of the vill's older men.[30] At Marton (Yorkshire) in the fourteenth or early fifteenth century the vill's arable was divided into fields for a three-course rotation by nine men, three representing one of the two manorial lords, three representing the other and three representing the vill's freeholders.[31] And at Harlestone (Northamptonshire) in 1410 an adjustment to the vill's fields was recorded in a tripartite indenture of which two parts were held by the manorial lords, one part by the men of the vill, whose six representatives were named as one party to the deed; to supervise the scheme's operation nine men were named, seven for the manorial lords and two for the vill.[32] In these four documents we see the formal role of the manorial lord successively reduced and the formal role of the vill correspondingly enlarged. At the same time the formal roles are not necessarily a guide to the decision-making process or to the realities of power. At Dry Drayton, where the five lords seemingly acted without reference to the vill, are we to take the document at face value and see them simply as five individuals acting alone, or are we to see them as, jointly, the embodiment of the vill, or as its leaders or as its representatives? We cannot tell and at that time the question might well be meaningless; they were simply doing the job of manorial lords as custom allowed and required. But if we see the early medieval lord as an integral member of the community of the vill we need not assume that he (or his official) would join in the village assemblies postulated by historians to discuss the vill's affairs with his tenants, or even to lead their discussion;[33] it is just that without his assent the decisions reached by any group of villagers would be valueless if they in any way affected his interests. What changed in the course of time is what the lord saw as his interests; they seem to have become more and more narrowly conceived in simple economic terms. Just how they were – consciously or unconsciously – defined will of course have differed from lord to lord and from place to place. What the lord thought lay outside his interests at any particular time he was content to leave to the administration, the decision even, of the vill.

Occasionally we get a glimpse of the process of withdrawal, of the lord distancing himself from the vill in one or other detail. In 1424 Merton

[30] Page, *Estates of Crowland Abbey*, p. 162; P. Vinogradoff, *Villainage in England: Essays in English Mediaeval History* (London, 1892), pp. 457–8. Both are discussed by H. S. A. Fox, 'Approaches to the adoption of the Midland system', in *The Origins of Open-field Agriculture*, ed. T. Rowley (London, 1981), pp. 94–8.

[31] Homans, *English Villagers*, pp. 56, 421.

[32] Wake, 'Communitas villae', pp. 409–13.

[33] W. O. Ault, 'Village assemblies in medieval England', in *Album Helen Maud Cam* (Studies Presented to the International Commission for the History of Representative and Parliamentary Institutions, nos 23, 24, 2 vols, Louvain and Paris, 1960–1), vol. I, pp. 13–35.

College, Oxford, which had a manor at Gamlingay (Cambridgeshire), agreed to provide wood to repair a bridge there, but on condition that the vill did not claim in future that the college should contribute to its repair.[34] But it would not be difficult to find contrary instances, for the lords' disinvolvement was not the gradual, steady process that the four records of field reorganization might suggest: it proceeded fitfully, patchily, with backward turns and with much variation between one place and another. We have seen one example of this in the agreement of 1287 between the lord of the manor of Alvechurch and the men of the vill of King's Norton; we can rationalize the difference between the two places by saying that the Bishop of Worcester and the king held different views of their local interests as lords of Alvechurch and King's Norton respectively. Or again when, in 1241, the fen between Whaplode and Holbeach (Lincolnshire) was reclaimed, not just the manorial lords but everyone claiming a share in either vill participated in an agreement for its division; the £40 penalty for a breach of its conditions was to go not to the lords but to all those sharing the fen. Yet in Gloucestershire in 1313, two generations later, the tenants of Coaley and Frocester took no part in an agreement for intercommoning between the two manors; it was drawn up simply between the manorial lords.[35]

But however variable, however piecemeal, the process, we can see how certain broad developments encouraged it. The disappearance of the local thegns at the Norman Conquest and their replacement by mostly absentee lords must have opened a gap between lord and vill; however, subinfeudation will have reduced the effect of this, bringing many vills a resident knightly lord, and so will the farming of manors either by long-term leases that made the farmer little different from a resident lord, or by leases that gave the lord continuing local control. A subtler change was produced by twelfth-century legal developments; by giving new security to the free tenants they produced a class of local tenants with whom the lord would now deal, if not exactly as equals, at least apart from the rest of the vill. This probably divided the lord from the community as a whole far more than it divided the free tenants from their fellow inhabitants. The distinction thus created continued throughout the middle ages – we find, for instance, the lord of Moreton Pinkney (Northamptonshire) making an agreement about enclosure with his free tenants in the 1260s, and in the fifteenth century the free tenants of Lydd (Kent) had their own seal.[36] Lords' withdrawal from demesne farming, first, in the late thirteenth century, by giving increased responsibility to their local officials and latterly by a return to leasing, will again have tended to distance them from the

[34] Merton College muniments 2471.

[35] H. E. Hallam, *Settlement and Society: A Study of the Early Agrarian History of South Lincolnshire* (Cambridge, 1965), pp. 32, 226–7; Hilton, *English Peasantry*, p. 3.

[36] PRO, E326/11479; W. de G. Birch, *Catalogue of Seals in the Department of Manuscripts in the British Museum* (6 vols, London, 1887–1900), vol. II, no. 5145.

vill – as Dyer has said, its internal affairs were now of little interest to them. We can only guess at possible regional differences in the process of disengagement. At the local level it seems likely to have proceeded particularly rapidly on manors of royal demesne (as at King's Norton) and perhaps also at places where the demesne lands were leased to the manors' own tenants, and particularly slowly where the lord of the manor lived on the spot. It has been thought likely that vills of divided lordship had stronger, more developed communal organization than those where vill and manor coincided.[37] Certainly if neither the manorial lords nor their agents were resident in a vill of divided lordship liaison between them may have been difficult, but we should not suppose that lords will have found it harder than the combined body of their tenants to agree among themselves. Thus at Childrey (Berkshire) in the mid-fourteenth century the vill's hayward was appointed by each of the three manorial lords in turn,[38] and it would be interesting to know whether it was communal or seignorial initiative that gave us the terriers covering lands of more than one lordship that we have from some such vills in the fourteenth and fifteenth centuries.[39] The role of the manorial court is of interest here, and also of the parish; these were the two organizations that offered the local community an institutional framework.

The manorial court is first recorded in the early thirteenth century.[40] In the form it had then it is unlikely to have been of great antiquity: it has every appearance of being copied from the courts that twelfth-century lords of honours held for their tenants, adapting – aping even – their procedures to fit the needs of a single manor with its tenantry of peasants. We know of no Anglo-Saxon equivalent to the simple manorial court (the future court baron), though in some places private courts exercised locally certain aspects of royal jurisdiction (the future court leet). There must, though, have been some more or less formal way of settling disputes among the inhabitants of the pre-Conquest vill and this seems to have been taken over or superseded by the manorial court. But even where vill and manor coincided the manorial court practically never became the institutional embodiment of the community of the vill.[41] Certainly it played a part in agrarian regulation and in enforcing

[37] Thus Wake, 'Communitas villae', pp. 407–8; Cam, 'Community of the vill', p. 6; Dyer, 'Power and conflict', p. 29.

[38] *Calendar of Inquisitions Post Mortem* (16 vols, 1904–74), vol. VIII, no. 629.

[39] One such, for Portchester (Hampshire) in 1405, was drawn up by one of the two manorial lords in order to distinguish the two manors' lands: B. Cunliffe and J. Munby, *Excavations at Portchester Castle, vol. IV: Medieval, the Inner Bailey* (Society of Antiquaries of London, Research Report no. 43, London, 1985), p. 280.

[40] W. O. Ault, 'The earliest rolls of manor courts in England', *Studia Gratiana*, 15 (1972), pp. 511–18; P. D. A. Harvey, *Manorial Records* (British Records Association, Archives and the User no. 5, 1984), p. 42.

[41] This did happen at Kingsthorpe (Northamptonshire) but seems to have been most exceptional; interestingly, this was a manor of royal demesne leased to its own tenants (Ault, 'Village assemblies', pp. 26–7).

by-laws. But it did nothing to meet the demands that royal administration laid on the vill: it was not in the manorial court that the vill's constables or foot-soldiers or representatives at other courts were chosen, or that individual contributions to taxation or other communal payments were assessed or collected. Indeed, a manorial court sometimes even hindered the smooth operation of these processes by amercing villagers who in course of these duties were held to have damaged the lord or his tenants.[42] At first sight the manorial court might seem the outward expression both of the power of the lord and of his integration with the village community. In fact it was neither. It was used not as a vehicle for the widest powers of lordship – it is not, for instance, in manorial court records that we look for evidence of fifteenth-century lords' enclosures – but as a means of regulating local custom as between the lord and his individual tenants. In using the court in this way and in applying to local custom procedures borrowed from other courts the manorial lord may well have reduced the authority that custom allowed him, seeking from the court decisions that had once been his alone. And in formalizing these procedures, cutting them off from the vill's collective deliberations and obligations, the lord was taking another step in distancing himself from the community. We can even see the growth of new manors, new manorial courts, in the fifteenth century as evidence that other large freeholders were likewise separating themselves from the vill.

But if the community of the vill did not find its institutional home in the manorial court it had better luck in the parish. The embodiment of the vill in the parish organization was a slow process of great interest. Even its official aspect, the vill's replacement by the parish as the local unit of royal administration, took place gradually; already in 1377 the poll tax was collected by parishes, not vills, in two counties, Cornwall and Middlesex, and some sixteenth-century statutes giving parishes administrative functions provided for some sharing of power with the manorial courts.[43] This reflected still slower change at the local level. A mid-thirteenth-century charter giving land to maintain a light in Castle Donington (Leicestershire) church was placed on the altar 'in the presence of the parish'; but the distinction between vill and parish was not always maintained and already in 1296 we find, exceptionally, land described as lying not in the vill but in the parish of Langham (Essex).[44] The parish church was at first a useful recipient of any financial windfall that came the way of the vill, an object of common concern and obligation,[45] but by

[42] For example, Page, *Estates of Crowland Abbey*, pp. 386–7.

[43] J. C. Russell, *British Medieval Population* (Albuquerque, 1948), p. 123n; Harvey, *Manorial Records*, p. 56.

[44] PRO, DL25/1208, 1712.

[45] See, for example, *Manorial Records of Cuxham, Oxfordshire, circa 1200–1359*, ed. P. D. A. Harvey (Oxfordshire Record Society, 50; Historical Manuscripts Commission, Joint Publication no. 23, 1976), p. 648.

the fifteenth century it was not uncommon for fines for breaches of by-laws to be divided between manorial lord and parish church – an instance of the lord's withdrawal in favour of the vill, here replaced by the parish.[46] By the seventeenth century the parish, its statutory powers still increasing, had outside the northern counties entirely subsumed the vill as a collective body.

But without the manorial lord. In the parish organization that had been the community of the vill there was no place for the lord of the manor; his separation from the vill was now complete. As late as 1395 the manorial lord of Great Ryburgh (Norfolk) was referred to as lord not of the manor but of the vill;[47] two centuries later he would never have been called lord of the parish. But this is not to say that he had forfeited power, only that his perceived field of interest had narrowed. The lord might still in the fifteenth century act in concert with the vill, indeed as an integral part of it; thus the curious parish history of Barholm and Greatford (Lincolnshire), written in the fifteenth century, assumes that land will be bought and sold by the lord and the whole community of the vill acting together.[48] And when we consider the settlement changes of the fifteenth, sixteenth and seventeenth centuries – the evictions, the enclosures, the imparkings – we look to neither vill nor parish as the instigator but to the manorial lord or his local farmer, with or without the collaboration of the other freeholders. The position cannot have been different in earlier centuries.

CONCLUSION

The idea, the initiative, for change may have come from the community of the vill – mostly that is anyone's guess – and increasingly we see the vill, as a collective body, participating in the detailed working, the day-to-day administration, that change may have involved. But if what the manorial lord saw as his interests were affected by change the vill would be powerless to carry it out without his assent. Local custom and the law of property between them left no room for doubt where the authority for settlement change must always have lain.

[46] W. O. Ault, 'Manor court and parish church in fifteenth-century England: a study of village by-laws', *Speculum*, 42 (1967), pp. 53–67. At Shifford (Oxfordshire) in the late fifteenth century we see a halfway stage in the process: half of an amercement for poaching went to the lord, half to the local church, but of an amercement for neglecting to mend a path half went to the lord and half went towards repairing roads in the vill (*VCH Oxfordshire*, vol. II, p. 186n, citing PRO SC2/197/62, m.4r).

[47] BL, Additional Charter 60308.

[48] Edited and discussed by J. A. Cripps in *Local Maps and Plans from Medieval England*, ed. R. A. Skelton and P. D. A. Harvey (Oxford, 1986), pp. 266–88.

Abbreviations used: BL: British Library; PRO: Public Record Office

VI

The Pipe Rolls and the Adoption of Demesne Farming in England

I

It is generally accepted that the management of landed estates in England changed radically between the mid-twelfth and the mid-thirteenth century with the introduction of demesne farming: manors were run by the landlord's local officers, answerable to him for all receipts and expenses, instead of by lessees who paid a predetermined annual rent. But even detailed studies of individual estates have failed to reveal much of the chronology of the change. Where they exist, private estate-accounts provide indisputable evidence; it is through these, for example, that we know that the estates of the bishop of Winchester were mostly under direct management by 1208.[1] But probably very few estates produced such written accounts in the early thirteenth century, and of these fewer still have survived: there is no other estate from which we have complete accounts before 1240 covering all the manors in its possession.[2] A good deal of scattered evidence from individual places certainly survives. A document in the cartulary of Beaulieu Abbey, for instance, provides for an agreement over tithes of Little Faringdon (Oxfordshire) to come into effect when its demesne lands there, leased to men of the vill from Michaelmas 1213, come under the abbey's direct management once more.[3] But evidence of this sort, although valuable, is too haphazard to give us an overall picture. The present article attempts to provide this from the published Pipe Rolls of the Royal Exchequer from 1155 to 1216.[4]

[1] H. Hall, ed. *The Pipe Roll of the Bishopric of Winchester, 1208–9* (1903).

[2] There is, however, an incomplete account for all the manors of the archbishopric of Canterbury in 1236–7.—Lambeth Palace Library, estate documents roll 1193.

[3] B.M. MS Loans 29/330, fo. 26*v* (no. 34 in the forthcoming edition of the cartulary by Dom S. F. Hockey in the Southampton Records Series).

[4] Hereafter referred to as *P.R.* with the regnal year. Those for 2–4 Hen. II and 1 Ric. I were published by the Record Commission, 1844, all the others by the Pipe Roll Society between 1884 and 1964. No Pipe Roll survives for 8 Ric. I (1196) or for 15 John (1213), though for the former year the duplicate Chancellors Roll survives and has been printed in its place. The financial year ran from Michaelmas to Michaelmas. Throughout the article the accounts are referred to by the year of their close; thus "1185" means Michael-

Their evidence for individual estates is summarized in Appendixes I and II and the rest of the article forms a commentary on this evidence, first discussing the problems that arise in using the Pipe Rolls for this purpose and then suggesting some general conclusions.

II

The annual Pipe Roll consists of the accounts, written in Latin, that were rendered by sheriffs and a few others for moneys payable to the Exchequer. Income from lands other than royal demesne was entered only when they came into the king's hands.[1] This might happen by escheat, when a tenant-in-chief forfeited his property or died without heirs; by wardship, when a tenant-in-chief's lands were inherited by a woman or minor; or by vacancy, on the death or translation of a bishop or abbot. These properties were accounted for separately from the normal county revenues even if they remained in the king's hands for a long period. Like the king's own demesnes, however, they were under the general supervision of the itinerant justices in their periodic visitations of the counties. We have the very full instructions given to the justices of 1194 about the stocking and management of these estates, and there is a less detailed reference in the instructions of 1176 (the Assize of Northampton).[2] Other records of the justices' supervision survive in the *Rotuli de Dominabus*, returns made to their inquiries by jurors in 12 counties in 1185,[3] and in the many purchases of livestock recorded on the Pipe Rolls following their visitations.[4] But the justices' visitations came seldom and the continuous administration of these estates was provided in one of two ways. They might be put "in custody", placed in the charge of a keeper (*custos*) who was in effect a bailiff, answering to the Exchequer for all profits and expenditure. Or they might be put "at farm", given to a farmer (*firmarius*) who was virtually a lessee, paying an annual sum fixed in advance from which allowances were sometimes made for losses or unavoidable expenses.[5] The Pipe Rolls seldom say explicitly which method was used on a particular estate and some features of many accounts for estates at farm might at first sight lead one to suppose that they were in custody. Thus the amount of an estate's farm was seldom a round figure, as it was calculated by totalling the annual values of the individual manors and other properties; on the account for the archbishop of Canterbury's estates in 1172, we find 10s. deducted for certain land "that was calculated at 32s. in the farm of the archbishopric and it only rendered 22s."[6] And when some unexpected new source of income accrued to an estate that had been set to farm

mas 1184 to Michaelmas 1185, "1180–5" Michaelmas 1179 to Michaelmas 1185. Sums of money quoted from the Pipe Rolls are normally given to the nearest pound.

1 And not always then, for it might be paid directly into the king's Chamber; revenues from the bishopric of Chichester in 1208–9, for example, were paid thus—*P.R. 11 John*, p. 5.

2 W. Stubbs, ed. *Select Charters* (9th edn, Oxford, 1913), pp. 180, 254–5.

3 J. H. Round, ed. *Rotuli de Dominabus et Pueris et Puellis* (Pipe Roll Soc. 1913), hereafter *Rot. Dom.*

4 E.g. *P.R. 13 Hen. II*, pp. 5, 12, 13, 16, 36–8, etc.

5 The distinction is clearly illustrated by accounts for the bishopric of Hereford, vacant from Feb. 1167 to Oct. 1174; during the first (incomplete) financial year the estate was in custody, and the Pipe Roll gives full details of money received and spent, but for the rest of the vacancy £300 9s. 4d. was entered each year as the farm of the bishopric with deductions for shortages of stock and other allowances.—*P.R. 13 Hen. II*, pp. 76–7; *P.R. 14 Hen. II*, p. 117; *P.R. 15 Hen. II*, pp. 142–3; etc.

6 *P.R. 18 Hen. II*, p. 139.

this would be accounted for as a separate item by the farmer; escheat and wardship of lands held by subtenants in knight service were a frequent source of such additional profits.[1] But it is usually clear from the Pipe Rolls whether a particular estate was at farm or in custody, particularly on comparing accounts for successive years. It is the accounts of estates in custody that tell us most about the way they were managed, and thus about the spread of demesne farming.

Where an estate was in custody, in the charge of a keeper, it might seem very easy to tell from the Pipe Roll whether its manors were being leased to local farmers or run by the estate's own officers. The distinction is obvious on private estate-accounts, which set out detailed receipts and income manor by manor.[2] Occasionally, when the Pipe Rolls set out the accounts of estates in custody in this way it is likewise clear whether lessees or bailiffs are in charge.[3] Often, though, the details given are few and ambiguous, and more often still the accounts on the Pipe Rolls do not proceed manor by manor at all, but enter items of receipt and expenditure for the estate as a whole. The method of management can then be discovered from the accounts only through chance references and hints, and before we can read between the lines in this way we have to resolve certain problems of interpretation. These problems are of three sorts. First are those of vocabulary: what significance to attach to certain words used in the accounts. Secondly, there are problems posed by particular types of entry on the accounts: can they be used as evidence of one or other method of management? Thirdly, there are problems arising from certain contemporary practices in running estates. Most of the Pipe Roll evidence for methods of estate management depends on the answers to these various problems.

In vocabulary the words that present most difficulties are *firma* and *redditus assisus*. When the receipts from an estate in the king's hands are said to come "from the farms of the manors (*de firmis maneriorum*)" we might think that this was an estate where there was no direct management. In fact, these receipts might be simply the agreed annual payment for an entire estate that the crown had put at farm: the plural "farms" merely reminds us that the amount was worked out by adding up the estimated income from the individual manors. Certainly the phrase tells us more when it is used of an estate that was clearly in custody; for instance, when the keepers of the Westminster Abbey estates account in 1175 for a sum "from the farms of the manors for three years"[4] we may reasonably argue that a system of leasing was predominant there. But even here we cannot conclude that no manors at all were under direct management. Receipts entered thus from the bishoprics of Ely (1171) and Winchester (1172) must include income from lands in demesne that are mentioned in the expenses on the same

[1] Examples are the account of the farmer of Albert Gresle's lands in 1188 for income from lands that had hitherto been held by Albert's wife, and the entry for Rotherhithe in the accounts for the honour of Gloucester in 1187.—*P.R. 34 Hen. II*, p. 6; *P.R. 33 Hen. II*, p. 15.

[2] Compare, for example, the brief entries for three properties at farm that open the bishop of Winchester's estate accounts for 1210–11 with the detailed accounts for Bitterne and other manors that were run by the bishop's own officials.—N. R. Holt, ed. *The Pipe Roll of the Bishopric of Winchester, 1210–11* (Manchester, 1964), pp. 5–7, etc.

[3] E.g. the accounts for the estate of Milton Abbey (Dorset), entirely under direct management, in 1212, and for the manors of Henry of Essex (the honour of Rayleigh), all leased to individual farmers, in 1181.—*P.R. 14 John*, p. 122; *P.R. 27 Hen. II*, pp. 107–9.

[4] *P.R. 21 Hen. II*, pp. 79–80.

accounts.[1] More reasonably, perhaps, we find *firma* used in 1184, on an estate in custody, for its entire cash income from fixed sources, not just from leases of entire manors but from urban and peasant rents as well: "the *firma* of the honour of the earl of Gloucester, from the fixed returns (*de redditu assiso*) of the manors that have been placed at farm, from the issues (*de exitu*) of Bristol which is in demesne (*in custodia*) and from the set cash rents (*de redditibus statutis in denariis*) of the other manors of the same honour that are in demesne".[2] Here *firma* includes the rents paid by tenants of manors run in demesne; on the account for the archbishopric of York in 1195 the word sometimes means these local rents by themselves, the rest of the income from each manor in demesne being entered as "*de exitu*", but sometimes too it means the farm paid by the lessee of an entire manor.[3] This variety of usage explains (though still not unambiguously) how the honour of Arundel managed in 1180 to produce £113 "from the farms and perquisites of the manors that were not at farm this year".[4]

Redditus assisus means "set render" or "fixed return", and it is familiar to students of thirteenth- and fourteenth-century manorial history from its use in reeves' accounts and other records; there it refers to the rents paid by the manorial tenants, the "fixed return", unchanging from year to year, in contrast to the sales of corn and livestock and other variable sources of profit. The phrase occurs in the instructions for the itinerant justices in 1176 (the Assize of Northampton),[5] but it does not appear in the Pipe Rolls until 1184,[6] though earlier we find various synonyms: *census statutus, redditus statutus, firme assise*.[7] In 1185 it occurs several times on the Pipe Roll as well as in the *Rotuli de Dominabus*,[8] and from then on it is used frequently, ousting its rival synonyms. Almost certainly it occurred in the lost instructions for the justices of 1185, probably in referring to escheats and wardships.[9] In some instances in 1185 the phrase clearly means the farm paid by the lessee of a manor; after all, if in the context of a single manor the fixed returns were the peasants' rents, in the context of a whole estate the fixed returns were the farms of the lessees. It is in this sense that we have seen the phrase used on the honour of Gloucester's account for 1184.[10] But on the Pipe Roll for 1186 we already find a very clear case of *redditus assisus* being used in its later normal sense of the rents of the manorial tenants: the keepers of the bishopric of Salisbury returned £244 "from the farms (*de firmis*) of five manors of the said bishopric", £63 "from the fixed return (*de redditu assiso*) of Ramsbury and Cannings which were not at farm" and separate amounts from sales of produce and other profits from these two manors.[11] The next year they accounted in the same way (though now a third manor had come under direct management),[12] but in 1188 the lessees'

[1] *P.R. 17 Hen. II*, pp. 115–16; *P.R. 18 Hen. II*, pp. 85–6. [2] *P.R. 30 Hen. II*, p. 109.

[3] *P.R. 7 Ric. I*, pp. 29–32. [4] *P.R. 26 Hen. II*, p. 32. [5] Stubbs, op. cit. p. 181.

[6] In the entry from the account of the honour of Gloucester quoted above.

[7] E.g. *P.R. 13 Hen. II*, p. 114; *P.R. 27 Hen. II*, p. 145; *P.R. 30 Hen. II*, p. 155. Another, *firme statute*, appears later.—*P.R. 33 Hen. II*, p. 21, etc.

[8] *P.R. 31 Hen. II*, pp. 6, 154, etc.; *Rot. Dom.* pp. 10. 19, etc.

[9] It is used in the instructions of 1194, but, as in 1176, only of the king's demesne manors.—Stubbs, op. cit. p. 255.

[10] A later example is the entry of the farms of Bagworth and Desford (Leicestershire), manors of the honour of Leicester, as *firma, redditus*, and *redditus assisus* on successive years' accounts.—*P.R. 11 John*, pp. 24–5; *P.R. 12 John*, p. 96; *P.R. 13 John*, p. 196.

[11] *P.R. 32 Hen. II*, p. 166. [12] *P.R. 33 Hen. II*, p. 187.

farms and the tenants' rents from the manors in demesne were lumped together, and the keepers answered for £277 "from the farms of the manors that are at farm and the fixed return from certain manors of the same bishopric that are not at farm", followed, as before, by separate receipts from sales of produce from the manors under direct management.[1] Elsewhere we find that a similar amalgamation has been made, though not explicitly, so that a single sum "from the fixed return" includes both farms of entire manors and peasant rents from manors run in demesne.[2] *Redditus assisus*, then, like *firma*, can mean lessees' farms, rents of manorial tenants, or a combination of the two. But it is also used to mean all the receipts from manors in demesne, not just tenants' fixed rents but sales of produce and other sources of income as well, or even, more widely still, the whole of an estate's income both from manors at farm and from those in demesne. This appears clearly on accounts for the archbishopric of York. In 1212 and 1214 most of the estate's income came "*de redditu assiso*" (plural in 1212), with another large amount "from demesnes, works and customs farmed out".[3] In 1215 this second item disappears, and we find all the manorial receipts of the estate classed as *redditus assisus*.[4] Even more than *firma*, *redditus assisus* can be interpreted only in the light of its particular context.

Other words too can be misunderstood. One is *perquisitiones*. In later manorial records it means (like *perquisita*) the profits of the manorial court, either all together or the proceeds other than from pleas. It is almost certainly used thus on the bishop of Salisbury's estates in 1186: receipts from the two manors in demesne include £22 "from pleas and perquisites (*de placitis et perquisitionibus*)".[5] But in origin the word simply means receipts, and it was used in the justices' instructions of 1176 to mean all profits other than the fixed returns: the bailiffs should answer both for the *redditus assisi* and for the *perquisitiones* (that is, everything else) that came from their bailiwicks.[6] Certainly it is often used in a very general sense on the Pipe Rolls to mean all the income of an estate or individual manor other than the fixed returns from the lessees' farms or the peasants' rents.[7] *Bladum colligere* is another phrase that demands care. Sometimes it occurs among the expenses of manors that are clearly under direct management, and must mean harvesting corn.[8] But it would be dangerous to assume that it thus provides clear evidence of demesne farming; it might well be used of gathering in tithes or corn-rents from manors that were all leased out. In all these cases we can only interpret

[1] *P.R. 34 Hen. II*, p. 184.

[2] The £341 received "from the fixed returns of the manors" on the bishop of Lincoln's estates in 1186 is probably an example of this.—*P.R. 32 Hen. II*, pp. 83–4.

[3] *P.R. 14 John*, p. 2; *P.R. 16 John*, p. 68. The bishopric of Durham in 1196 provides another clear example of this usage.—*P.R. 8 Ric. I*, p. 253.

[4] *P.R. 17 John*, p. 12.

[5] *P.R. 32 Hen. II*, p. 166. Another example comes from the bishopric of Lincoln in 1183.—*P.R. 29 Hen. II*, p. 33.

[6] Stubbs, op. cit. p. 181. Probably the word came to be applied particularly to the profits of the manorial courts because when estates were entirely leased out the courts were normally kept by the landlord and provided the only receipts apart from the lessees' farms.

[7] Thus on the account of Chertsey Abbey's lands in 1183 the receipts entered are £56 "from the farm of the manors of the abbey", £7 "*de perquisitionibus*" (*P.R. 29 Hen. II*, p. 18); on the lands of William de Curci in 1187 the three manors in demesne produced £9, £2, and £2 "*de redditu assiso*", £7, £3, and £3 "*de perquisicionibus*".—*P.R. 33 Hen. II*, p. 25.

[8] E.g. *P.R. 33 Hen. II*, p. 19, where 13s. was paid "for gathering corn (*pro bladis colligendis*) and for ironwork for the ploughs".

each word from the context and whatever other evidence is available; we must avoid the temptation to give it a precise technical meaning that it patently does not bear.

Passing from difficulties of vocabulary, we meet other problems in using the contents of the accounts as evidence of the method of management. It might be thought that certain items could serve as a touchstone: if corn or livestock was bought or sold, or payments made to manorial officers, then some part of the estate must be under direct management. In fact these can be misleading guides. If a tenant-in-chief fell foul of the crown so that his estate was declared forfeit, the royal officers would deliberately sell off the corn and livestock of his manors. There are many such cases on the Pipe Rolls for years of rebellion, such as 1173 or 1194.[1] These sales tell us nothing of how the estate was normally run; they merely mean that when the owner regained the crown's favour, and with it his manors, he would find they would need extensive restocking before they could be made to pay again, whether they were farmed out or run in demesne. A similar policy was followed, though less drastically, on other estates that came for a time into the king's hands.[2] Other sales might be of produce stored for domestic use, or, on monastic estates, of corn that came from tithes, though neither of these possibilities can be formally demonstrated from the printed Pipe Rolls. The existence of food farms must also be borne in mind: although leasing manors for a rent in produce was uncommon by the late twelfth century the practice was far from dead.[3] Sometimes the Pipe Rolls say that items sold came from rents in kind, but probably many other sales of corn and livestock came from this source; for instance the 289 *summe* of wheat "from the issue of the manors" of the bishopric of Hereford in 1167 look suspiciously like the product of food farms rather than of demesne farming, for there are no entries for other types of corn.[4] Purchases of livestock and corn are equally doubtful evidence of demesne farming. In most years livestock was bought for manors in the king's hands, especially following the justices' visitations, but this could perfectly well be to restock manors that were leased out in order to raise the rent, as is clear from many entries in the *Rotuli de Dominabus*.[5] Sometimes the amount of stock bought for an estate is astonishing,[6] and what the Pipe Roll called purchase of stock may often have been simply a change in the form of a lease: the lessee would be paid

[1] *P.R. 19 Hen. II*, pp. 20–1, 37–8, 49–50, etc.; *P.R. 6 Ric. I*, pp. 1–4, 14, etc. An obvious later example is on the lands of William Braose in 1209.—*P.R. 11 John*, p. 91.

[2] E.g. the sale of stock worth £225 from the archbishopric of York in 1194 (*P.R. 6 Ric. I*, p. 165) and the large sales of stock in 1203 and 1205 from the lands of William de Stutville that were in the king's hands during the heir's minority.—*P.R. 5 John*, p. 222; *P.R. 7 John*, p. 39. Wastage from lands in wardship was condemned by Magna Carta, capp. 4, 5; cf. J. C. Holt, *The Northerners* (Oxford, 1961), pp. 230–1.

[3] Cf. R. Lennard, *Rural England, 1086–1135* (Oxford, 1959), pp. 135–9. Chalfont St Giles (Buckinghamshire) and Saham Toney (Norfolk) provide examples of rents in oats, malt, salt, and rams in 1185.—*Rot. Dom.* pp. 40, 95.

[4] *P.R. 13 Hen. II*, pp. 76–7. The 60s. each year "from the corn of North Walsham sold" on the accounts of St Benet of Hulme Abbey in 1169–70 may well be another example.—*P.R. 15 Hen. II*, p. 106; *P.R. 16 Hen. II*, p. 13.

[5] E.g. the orders given, on jurors' advice, to buy specified stock for property at Little Wigborough (Essex) so that its farm could be raised from £11 to £16 a year.—*Rot. Dom.* p. 80. Restocking for leasing might include buying seed-corn, as at Great Tew (Oxfordshire) in 1168.—*P.R. 14 Hen. II*, p. 205.

[6] E.g. when the lands of Jordan de Buisseio, worth £103 a year, came into the crown's hands in 1163, 184 oxen, 21 plough-horses, 70 cows, 1,600 sheep, 60 sows, and 6 boars were bought for £67 "for restocking the same land".—*P.R. 9 Hen. II*, p. 73.

a lump sum for the animals that he was already maintaining on the manor and his rent would be raised accordingly. Purchases of livestock for these lands might, then, be a first timid step towards agricultural entrepreneurship, but certainly need not imply demesne farming. Sometimes a manor was restocked before being given away by the crown, the converse of the deliberate wastage of forfeited lands.[1] Payment of manorial officers is better evidence of demesne farming than sales and purchases of livestock and corn but must still be regarded with caution. The *servientes* whose food or wages are charged may be household officers, not estate workers, and even when they are "on the manors (*per maneria*)" they may still be employed in residences maintained on manors that have been leased out. Sometimes the earlier Pipe Rolls mention individual employees of this sort, and among them we often find semi-agricultural workers of the precincts, such as the gardeners at Banbury and other residences of the bishop of Lincoln, or the vineyard workers of William de Mohun.[2] And even where all the manors were leased out, some estate officers might still be needed: rent-collectors, men to hold the manorial courts, or the *servientes* of the bishop of Ely who had to summon the bishop's knights.[3] Once, in 1185, the keeper of an estate was specifically allowed board and wages for his seven *servientes* "appointed to look after the said land"; this payment was distinguished from the "wages of the *servientes* on all the manors".[4] In none of these cases could we use the payment of officers as evidence of demesne farming.

If purchases and sales of livestock and corn and payments to manorial officials do not prove that manors were under direct management can we take their absence as evidence of leasing? Unfortunately we cannot. For one thing the produce of an estate might not be sold at all, but delivered for the king's use. Very occasionally the accounts mention this,[5] but usually they deal with cash only and we need not assume that an estate had little land in demesne because little corn was sold. This can be well illustrated from the accounts for Canterbury Cathedral Priory in 1211–12: sales of corn from the manors brought in no more than £3 altogether in the two years, yet we know from the detailed account of the estate's issues in 1207–8 that it then had many manors in demesne.[6] Or again, even if the Pipe Roll gives a single entry of receipts from one or more manors with no sales, purchases, or manorial expenses at all it still does not follow that the properties were leased out, for the amount may simply represent the net profits

[1] E.g. the purchases "for stocking the lands of Offham and Preston for the use of Jordan, the duke of Saxony's seneschal" on the honour of Arundel in 1185.—*P.R. 31 Hen. II*, p. 171–2.

[2] *P.R. 14 Hen. II*, p. 77; *P.R. 23 Hen. II*, p. 25. If anything, entries such as these point to the leasing of manors, for where demesne farming was practised such payments would be included with other wages in the manorial expenses. The *bubalarius* with his *bubali* on the archbishop of Canterbury's estates in 1171–2 may well have been another of these semi-agricultural workers, a deer-park keeper rather than an oxherd (*P.R. 17 Hen. II*, p. 142; *P.R. 18 Hen. II*, pp. 139–40); Prof. F. R. H. Du Boulay takes him to be an oxherd (*The Lordship of Canterbury* (1966), p. 204) and so too, though with some doubt, does Mr R. E. Latham (*Revised Medieval Latin Word-List* (1965), p. 58), but there is no other hint of demesne arable that the oxen could have ploughed, nor of fellow manorial *famuli*.

[3] *P.R. 16 Hen. II*, p. 96. [4] *P.R. 31 Hen. II*, p. 239.

[5] E.g. the deliveries of corn from the bishop of Chichester's estate in 1210–12.—*P.R. 12 John*, p. 4; *P.R. 14 John*, p. 9.

[6] *P.R. 13 John*, p. 244; *P.R. 14 John*, p. 40; Patricia M. Barnes and W. R. Powell, eds. *Interdict Documents* (Pipe Roll Soc. 1960), pp. 69–80. But failure to cultivate may have been partly responsible for the small amounts of corn sold.—Ibid. pp. 48–9, 81–5.

of demesne farming. Probably the clearest evidence of leasing of manors comes from allowances made on the account for shortages of livestock, provided it is clear that the estate as a whole was run in custody and not at farm.[1]

The third type of difficulty in assessing the evidence of the Pipe Rolls arises from certain features of contemporary estate management. We are investigating the adoption of demesne farming as a regular practice on substantial landed estates. It is hardly relevant to this inquiry that someone who held a single manor managed his property himself. The late Anglo-Saxon thegn who owned 5 hides of land probably directed the tilling of his own inland, just as—to take a specific example—Henry de Bray ran his own 250 acres or so of demesne at Harlestone (Northamptonshire) in the early fourteenth century.[2] There are many such properties mentioned in the *Rotuli de Dominabus*, and accounts for them appear on the Pipe Rolls, especially—and it is here that they may be misleading—as escheats or wardships that fell to larger estates in the king's hands. Thus the account for Ralph Gaugy's lands in 1187 includes £2 4s. 2d. "from the proceeds (*de exitu*) of the land of Adam Ribald", against which is set 10s. 7d. spent "in the cost of cultivating the said land"; clearly Adam Ribald had managed his holding himself, but this tells us nothing of how Ralph Gaugy ran his own demesne lands, for Adam had been a subtenant.[3] Another practice, too, might lead us to suppose that certain manors were under direct management when in fact they were normally leased out. It is very likely that after a long lease a manor might be run by the landlord's own officials for perhaps a year or more, to replace missing stock, repair buildings, and so on, while a new lessee was being found and terms negotiated. Indeed, where a lease was for terms of lives, some interval of direct management must have often followed the death of the last lessee. On the Pipe Rolls we find several cases where this seems to have happened. The most likely example comes from the bishopric of Ely during the vacancy of 1169–73: wages and other manorial expenses at Hartest (Suffolk) and Rettendon (Essex) are entered in 1171, but not in the other years.[4] Thus where accounts point to demesne cultivation on some small part of an estate, what we see may not be the first step towards direct management but a normal feature of a system of leasing; this creates a particular risk of misinterpreting the practice of estates that appear on the Pipe Rolls for a single year only.

These are the main problems that arise in drawing evidence of methods of estate management from the Pipe Rolls. Certainly we cannot proceed by simple rule of thumb: the evidence from each estate must be considered as a whole and its assessment must often be tentative, even subjective. The conclusions set out here should not be taken as the final verdict for any particular estate.

[1] When the honour of Lancaster returned to the crown in 1194 allowance was made for 15 plough-teams, 15 vaccaries, 80 mares, and 100 sheep that had disappeared from its stock while it was in the hands of the future King John, but the deduction came from a fixed sum due for the entire honour, and it thus tells us nothing of how either John or the crown's farmer actually managed the honour's properties.—*P.R. 6 Ric. I*, p. 123; *P.R. 7 Ric. I*, p. 148.

[2] Dorothy Willis, ed. *The Estate Book of Henry de Bray* (Camden, 3rd ser. 1916), pp. xv, 48–54, etc.

[3] *P.R. 33 Hen. II*, p. 14. Another clear example in the same year is the property at Wickford that had been held by Durand de Bolonia of the honour of Rayleigh.—Ibid. p. 19. The holding of Ralph son of Mainus de Foresta that fell to the crown in 1185 is an example of a small property held in chief of the crown by a tenant who ran his own demesne.—*P.R. 31 Hen. II*, p. 10.

[4] *P.R. 17 Hen. II*, p. 116; *P.R. 16 Hen. II*, pp. 95–6; *P.R. 18 Hen. II*, pp. 115–17.

III

In Appendix 1 are listed 42 estates for which the Pipe Rolls seem to give definite information about the method of management. In ownership and location they range widely: 12 are episcopal, 13 monastic, and 17 lay-owned, and they come from all parts of the country. In size they are less representative, being the whole or part of large and medium-sized estates. The list makes no claim to be definitive, and the evidence is too imprecise to permit exact statistical analysis.[1] Overall, however, a consistent pattern emerges, and although its details may be questioned and revised it provides a basis for some provisional general conclusions.

Before turning to these, it is important to note two unavoidable risks involved in using the Pipe Rolls as evidence for the chronology of demesne farming. One is that a system of leasing might contain within it a system of demesne farming: a man might lease several manors from one or more landlords and run the estate he thus built up by direct management. Jocelin of Brakelond refers to several lessees who held more than one manor of St Edmund's Abbey;[2] on the other hand, the only list of lessees on the Pipe Rolls, for the lands of Henry of Essex in 1181, seems to name different farmers for each of the manors.[3] The other risk arises from the evident prevalence of leases on a very short-term, even annual, basis:[4] in such cases an estate's system of management may have been changed rapidly and radically when it came into the king's hands. However, this probably happened seldom:[5] on estates that were in the king's hands on more than one occasion we find a pattern that accords with uninterrupted development. And although we should be cautious on both these counts, the picture presented by the evidence drawn from the Pipe Rolls is clear and consistent and is unlikely to be distorted for either of these reasons.

What conclusions, then, can we draw from the evidence of the Pipe Rolls? In the first place, while there was a little demesne farming in the 1170's it is only from 1184 that any general trend becomes visible. Thirty years later it was still far from universal: the last case of an estate (or portion of an estate) entirely at farm is in 1196, the first of one entirely in demesne in 1194, but in the five years 1210–14 nearly all still had some manors at farm, even though on some two-thirds direct management preponderated. Although the change seems to have begun at about the same time on all types of estate, it was probably accomplished fastest on estates in lay ownership. Of the six estates (or portions of estates) recorded on the Pipe Rolls as entirely given over to demesne farming, five were lay-owned. A more surprising result emerges from a geographical comparison:

[1] My interpretation of the Pipe Roll evidence differs for certain estates from Mr E. Miller's in 'England in the Twelfth and Thirteenth Centuries: An Economic Contrast?', *Economic History Review*, 2nd ser. XXIV (1971), 3–4. This demonstrates the ambiguity of some of the accounts and the impracticability of using them for detailed statistics.

[2] H. E. Butler, ed. *The Chronicle of Jocelin of Brakelond* (1949), pp. 32, 59, and (possibly) 6; cf. Lennard, op. cit. pp. 150–3.

[3] *P.R. 27 Hen. II*, pp. 107–9.

[4] Thus Abbot Samson of Bury St Edmunds was able to take his manors into his own hands the Michaelmas after his installation without waiting for long-term agreements to expire.—Butler, op. cit. p. 32. Similar arrangements are implied by many returns in *Rot. Dom.*, where we see a manor's farm being altered frequently, even from year to year.

[5] However, the wording of the accounts for the honour of Arundel in 1179–80 (*P.R. 25 Hen. II*, pp. 38–9; *P.R. 26 Hen. II*, pp. 32–3) and the lands of the count of Perche in 1207 (*P.R. 9 John*, pp. 112–13) suggests that here changes were made when they came into the king's hands.

eastern England was markedly more conservative than the rest of the country. All the estates (or portions of estates) that were apparently wholly at farm in the 1180's lay in the east, between the Thames and the Humber, and so, too, were all those that seem to have been more than half at farm in the reign of John. Between the north, the south, and the west, however, no variation is perceptible and the size of an estate does not seem to have been a significant factor.

These conclusions are necessarily provisional and there is room for much further research. Appendix II is one possible starting-point. This is a list of estates for which the Pipe Rolls provide information that by itself is ambiguous; where they can be supplemented from other sources these accounts may well show how a particular estate was managed. Again, accounts for some considerable estates were entered piecemeal among the sheriffs' and escheators' accounts together with those for small estates; these have not been investigated, and although a thorough analysis of them would be complicated it would probably yield further information. Work on the estate accounts on the Pipe Rolls of Henry III and on the records of individual estates could tell us still more. The present study provides merely a framework for future work. The subject deserves further investigation. The movement towards demesne farming was important not only in itself but also as a sign of other economic change. More knowledge of its chronology may well shed new light on its causes as well as on its course.[1]

[1] For recent discussions of its possible causes see Miller, loc. cit. 8–14; C. G. Reed and T. L. Anderson, 'An Economic Explanation of English Agrarian Organization in the Twelfth and Thirteenth Centuries', *Econ. Hist. Rev.* 2nd ser. XXVI (1973), 134–7; E. Miller, 'Farming of Manors and Direct Management', *Econ. Hist. Rev.* 2nd ser. XXVI (1973), 138–40; and P. D. A. Harvey, 'The English Inflation of 1180–1220', *Past and Present*, LXI (Nov. 1973), 4–9.

APPENDIX I

The Management of Private Estates Accounted for on the Pipe Rolls, 1155–1216

Notes:

1. Unless otherwise stated the source of information is the printed Pipe Roll for the year in question; the relevant pages can easily be found from the list of contents or the index.
2. The years mentioned are those for which the accounts provide information on the actual running of the estate, and are not necessarily all the years for which its accounts are entered on the Pipe Rolls.
3. In a few cases a brief note is given to draw attention to significant detail that might be overlooked in reading the account on the Pipe Roll.
4. Sometimes an estate in the crown's hands was divided so that a particular account on the Pipe Roll may be for only part of it; thus the accounts for the archbishopric of York in 1214–15 exclude its Gloucestershire properties, that for the Braose estate in 1210 is for the Sussex lands only. Often it is not clear whether an account covers the whole estate or not. This complication has been ignored in the list: in establishing the general pattern it makes little difference whether a particular account is for an entire estate or only a sample portion.
5. Although the entry of the main income of an estate in custody as "*de firmis maneriorum*" suggests the predominance of leasing, it is a hazardous guide and accounts where it occurs are mentioned only if there is other evidence of the method of management. Glastonbury Abbey in 1181–2 is an example of an estate thus omitted from the list.

Bishopric of Lincoln

1167–72: all or nearly all at farm.

1181–3: Dorchester in demesne, remainder probably at farm, Tinghurst (1181) and Empingham (1183) being in hand between leases.

1185–6: part in demesne, but mostly at farm, Willoughby (1185) perhaps being in hand between leases.

1206–7, 1212: a substantial part, but probably less than half, in demesne.

Bishopric of Ely

1170–2: all or nearly all at farm, Hartest and Rettendon probably being in hand between leases in 1171.

Honour of Haughley (honour of the Constable; lands of the count and countess of Perche)

1170: part at least at farm.

1184–5: probably all or nearly all normally at farm.

1207: probably all in demesne.

1211–12: probably at least a substantial part in demesne.

Bishopric of Winchester

1172: probably mostly at farm, but some land just outside Winchester in demesne (costs of cultivating "*terram dominii extra Wint'*", harvesting "*bladum dominii de Wint'*").

Hyde Abbey

1173–5: a small part possibly in demesne, perhaps a home farm (repairs of barns ("*curie Wint'*" 1173), ironwork for ploughs, horses, etc., 1173–4; but the corresponding entry in 1175 mentions carts, not ploughs, perhaps a mistake in 1173–4).

1206: Candover, Headbourne Worthy, Laverstoke, and Stoneham at farm, other six manors in demesne.

Peterborough Abbey

1177: probably mostly at farm, but some land in demesne (sale of wool "*dominicarum ovium*").

1210–11: a substantial part, but probably less than half, in demesne.

Honour of Rayleigh (lands of Henry of Essex)
1181: all at farm.

Honour of Chester
1182–3: all at farm except probably Macclesfield and possibly some other property, as under Earl Hugh (d. 1181).

Archbishopric of Canterbury
1184: probably mostly at farm, but with at least some land in demesne (lands that the archbishop "*excoli fecit in dominio*").
1206, 1211–12: probably mostly in demesne.

Honour of Alnwick (lands of William de Vesci)
1184–5: probably part, but much less than half, in demesne (manorial wages, etc.; but "*porcariorum*" should perhaps be "*portariorum*").

Honour of Gloucester
1184–5: up to half in demesne, as under Earl William (d. 1183).
1199–1208: entirely in demesne, except Brasted 1199–1200, Petersfield 1200, Castle Hanley 1201–3, Bisley 1206–8, Cheltenham 1206–8, Fairford 1206–8; of these the leases of Brasted, Petersfield, and Castle Hanley preceded their alienation or grant in fee, those of Bisley and Fairford were to other landlords, not local farmers, and Cheltenham is not accounted for before 1206, so it seems that farming of manors was resorted to only in special circumstances.

Bishopric of Rochester
1184–5: probably part, possibly a substantial part, in demesne.

Lands of Herbert son of Herbert
1185: probably part in demesne.

Lands of Thomas Muschamp
1185: probably part in demesne.

Bishopric of Salisbury
1185: part in demesne.
1186: Ramsbury and Bishops Cannings in demesne, other five manors at farm.
1187–8: three manors in demesne, other four manors at farm.
1211–12: probably more of the estate in demesne, but part at farm.

Abingdon Abbey
1185–6: probably part in demesne.

York Abbey
1185–6: probably part in demesne.

Archbishopric of York
1186–9: Wilton at least at farm (allowance for defects of stock).
1194–7: probably a substantial part, possibly mostly, in demesne, but Cawood at least at farm in 1195 (Cawood *firma* a round sum, few other receipts there, no expenses).
1212, 1214–15: probably over half in demesne, but Ripon and some manors at farm.

Cirencester Abbey
1187: probably part, possibly a substantial part, in demesne.

Lands of William de Curci
1187: four manors at farm, Chalton and two smaller properties in demesne, Farlington uncertain.
1188–9: five manors, including Farlington, at farm, Chalton and appurtenances in demesne.

Malmesbury Abbey
1187: probably part in demesne.

Honour of Boulogne (lands of the count of Flanders)
1187–8, 1194–6: all at farm.

Bishopric of London
1187–9: probably part in demesne.

Leicester Abbey
1188: probably part in demesne.

Sherborne Abbey
1189: probably part in demesne.
1212: 14 manors in demesne, Abbotskerswell and Horton at farm (*redditus assisus, firma*, from Abbotskerswell, Horton, round sums without other receipts).

Lands of the count of Brittany
1190–1: all at farm.

Lands of Ralph of Cornhill
1194: all in demesne.

Honour of Eye
1194–6: three manors at farm, Laxfield and possibly Thorndon in demesne (receipts "*de exitibus*", etc., from Laxfield, Thorndon, only).

Bishopric of Chester and Coventry
1195–6: probably all at farm.

Bishopric of Durham
1196: probably mostly in demesne, but some manors at farm (*carruce* farmed "*pro denariis*", "*pro blado*").
1208–12 (1208–11 on *P.R. 13 John*): probably mostly in demesne, but some manors at farm in 1210–12 only.

Ramsey Abbey
1202: Barton, Broughton, Cranfield, Gretton, and Shillington in demesne, rest at farm.
1207: Barton, Brancaster, Gretton, Hemingford Abbots, and Ringstead in demesne, other 16 manors (including Shillington) at farm.
1210–12: probably about the same proportion of the estate in demesne.

Lands of William de Longchamp
1204: probably all in demesne.

Bishopric of Exeter
1207, 1211–12, 1214: probably a substantial part in demesne.

Lands of William Braose
1210: probably mostly in demesne, but Shoreham (borough) and Warnham at least at farm.

VI

358

Abbotsbury Abbey
1210–11: probably mostly in demesne, but Graston and Portisham at farm in 1210, Wootton Abbas in 1211.

Lands of William de Beauchamp
1210–11: probably a substantial part in demesne.

Eynsham Abbey
1210–11: probably mostly in demesne, but South Stoke at least at farm in 1210 (receipts from South Stoke, some smaller properties, "*de firma*", "*de redditu*", or "*de redditu assiso*" alone).

Bishopric of Chichester
1210–12 (1211–12 on *P.R. 14 John*): a substantial part, possibly all, in demesne.

Honour of St Valéry (lands of Thomas de St Valéry)
1211: probably about half in demesne.

Canterbury Cathedral Priory
1211–12: probably mostly in demesne.

Milton Abbey
1212: probably all in demesne.

Honour of Berkhamsted
1214: probably all in demesne.

APPENDIX II

Pipe Roll Accounts Giving Ambiguous Evidence of Estate Management, 1155–1216

For the following estates the Pipe Rolls give information which by itself is ambiguous but which might be used as evidence of the method of management where it can be supplemented by other records:

Bishopric of London, 1162
Bishopric of Hereford, 1167
Honour of Belvoir, 1168–9
St Benet of Hulme Abbey, 1169–73, 1210–11
Lands of Geoffrey Tresgoz, 1171, 1175
Honour of Eye, 1175
Lands of William de Mohun, 1177
Honour of Petworth, 1180
Bury St Edmunds Abbey, 1181–2
Archbishopric of York, 1182
Honour of Leicester, 1183–4,[1] 1209–12
Bishopric of Chester, 1183–5
Bardney Abbey, 1185
Lands of Robert, son of Bernard, 1185
Lands of Guthlev, 1185
Lands of William de Luvetot, 1185
Selby Abbey, 1185
Bishopric of Worcester, 1185, 1212
Bishopric of Exeter, 1185–6
Honour of Striguil, 1185–8
Tavistock Abbey, 1186
Fécamp Abbey, 1188
Bishopric of Winchester, 1189
Honour of St Valéry, 1192–6
Honour of Gloucester, 1194
Coventry Priory, 1195
Lands of the count of Brittany, 1200, 1204
Lands of Walter de Dunstanville, 1200
Lands of William de Stutville, 1203, 1205
Lands of Henry de la Pomeraie, 1208
Lands of William Braose, 1209
Lands of Roger de Tony, 1209–12, 1214[2]
Kenilworth Priory, 1210–12
Stone Priory, 1210–12
Bishopric of Bath, 1212, 1214

[1] In 1184 there are accounts for two portions of the estate (*P.R. 30 Hen. II*, pp. 50, 94).

[2] These are entered with the accounts of Robert Peverell in 1211 and 1212, with the honour of Ongar in 1214.

VII

THE ENGLISH INFLATION OF 1180-1220 *

HISTORIANS HAVE FOR LONG BEEN AWARE OF THE RISE IN PRICES THAT occurred in thirteenth-century England. In 1908 Sir James Ramsay gave examples to show that "prices were rising steadily and continuously" from 1200 to the death of Henry III.[1] In 1914 S. K. Mitchell, citing Ramsay, considered that "a rise in prices was going on, probably slowly, from about 1190 to 1250".[2] N. S. B. Gras in 1915, writing particularly of the price of corn, discussed the general increases that he found throughout the century for all agricultural commodities and suggested that they were caused by an "increase in the volume of trade due in part to a growing foreign demand and more especially to the growth of the town population, a subject bound up with the evolution of the local market".[3] Three years later Gras also commented on the increase in the price of wine between the reigns of Henry II and Edward I.[4] But it was not until the work of the future Lord Beveridge in 1927, A. L. Poole in 1940 and Dr. D. L. Farmer in 1956-8 that the chronology and scale of these price rises were defined with real precision.[5] Their tables, drawn from the royal Pipe Rolls and from the estate accounts of the bishops of Winchester, show that the prices of corn, livestock and the few other goods for

* I am grateful to Mr. T. H. Aston for some very valuable comments and suggestions which I have taken into account in revising this article for the press and for letting me read the unpublished paper mentioned in note 72. Throughout the notes I have referred to the editions of the Pipe Rolls from 5 Hen. II to 2 Hen. III (1159-1218) published by the Pipe Roll Society (old ser., vols. i-xxxviii, 1884-1925; new ser., vols. i-xxxiv, 1925-72) as *P.R. 5 Hen. II*, etc.

[1] *The Dawn of the Constitution* (London, 1908), p. 301.

[2] *Studies in Taxation under John and Henry III* (Yale Historical Publications, Studies, ii, New Haven, 1914), p. 2.

[3] *The Evolution of the English Corn Market* (Harvard Economic Studies, xiii, Cambridge, Mass., 1915, repr. 1926), pp. 15-16.

[4] *The Early English Customs System* (Harvard Economic Studies, xviii, Cambridge, Mass., 1918), p. 42.

[5] W. H. Beveridge, "The Yield and Price of Corn in the Middle Ages", *Economic History*, i (1926-9), pp. 162-6; A. L. Poole, "Livestock Prices in the Twelfth Century", *Eng. Hist. Rev.*, lv (1940), pp. 284-95; D. L. Farmer, "Some Price Fluctuations in Angevin England", *Econ. Hist. Rev.*, 2nd ser., ix (1956-7), pp. 34-43, and "Some Grain Price Movements in Thirteenth-Century England", *ibid.*, x (1957-8), pp. 207-20. These can now be supplemented by T. H. Lloyd, *The Movement of Wool Prices in Medieval England* (Econ. Hist. Rev. Supplement no. 6, 1973), pp. 38-9, 45, 50, whose series start in 1209. A convenient table of prices of wheat, oxen and wool is given by J. Z. Titow, *English Rural Society, 1200-1350* (London, 1969), pp. 100-1.

which we have evidence doubled or trebled between 1180 and 1220; they continued to rise slightly until about 1260 but then levelled off, just at the date when manorial accounts start to provide real profusion of evidence on prices, evidence that had been tabulated and analysed by Thorold Rogers as long ago as 1866.[6] The enormous local, seasonal and annual variations make it very difficult to discover long-term price trends from medieval records, but the conclusions that Poole and Dr. Farmer reached have not been questioned. Rapidly rising prices are now well established as a feature of late twelfth- and early thirteenth-century England and have been mentioned among the causes of various phenomena of the time: the increasing legal disabilities of the peasantry, the change from leasing manors to demesne farming on large estates, and the political troubles of the reign of John. It is not now too early to attempt to survey the overall consequences of these price changes, to see what effect they had on the society of the time, what was their economic significance, and to take a fresh look at their possible cause.

In the twelfth century large estate owners drew profits from their lands by leasing them to farmers who would pay an agreed rent in cash or (decreasingly in the course of the century) in produce. In the thirteenth century the lessees had been replaced by the landlords' own local agents — reeves and bailiffs — who tilled the demesne lands on their behalf and were answerable to them for all profits. The change may seem a small one, but, as Mr. E. Miller has recently pointed out, the shift in emphasis from local to seigneurial entrepreneurship in agriculture lay at the root of the "economic contrast" between the twelfth and thirteenth centuries in England.[7] Our only source of general evidence for the chronology of the change is the accounts for estates temporarily in the king's hands that are entered on the Pipe Rolls, and the information these give us is sparse and difficult to interpret. The earliest clear case they provide of part of a substantial estate being run by a local agent as regular practice is in 1171-2, when some small part of the bishop of Winchester's lands were under direct management; the same is found at Peterborough Abbey in 1176-7 and on the honour of Chester in 1181-2, and thenceforward the estates for which there is evidence were turning more and

[6] J. E. T. Rogers, *A History of Agriculture and Prices in England,* 7 vols. (Oxford, 1866-1902), i and ii.

[7] "England in the Twelfth and Thirteenth Centuries: an Economic Contrast?", *Econ. Hist. Rev.*, 2nd ser., xxiv (1971), pp. 1-14.

more to demesne farming.[8] By the end of the reign of John there were still some estates of which substantial portions were leased out — probably three-quarters of the manors of Ramsey Abbey were farmed out in 1211-12 — but most were all or nearly all under direct management, like those of Sherborne and Milton Abbeys and Canterbury Cathedral Priory in the same year.[9] The coincidence between the period when estate owners were turning to demesne farming and the period of rapidly rising prices is very striking. At first sight, indeed, it might seem altogether too striking: some delay would be necessary if we are to see them as cause and effect. But this is probably due to the nature of our records; new price levels would have to be well established before the Exchequer would allow them to sheriffs and others, and thus bring them into recorded history on the Pipe Rolls. Given hindsight and an elementary knowledge of economics it is clear that estate owners were taking exactly the right steps to meet the situation; even if rents could be raised to new levels there would always be a time-lag in which the lessees would benefit unduly from the increased prices that their landlords and others would have to pay for the produce they sold and direct management was the only way to avoid this. The failure of the landlords' successors in the sixteenth century to adopt a similar policy in like circumstances meant that some of their lessees made considerable gains at their expense.[10] But it is equally clear that twelfth-century landlords had neither the theoretical knowledge nor the information about contemporary price trends that would enable them to see their position in these terms. If they adopted the correct policy it was for immediate and empirical reasons. Ultimately these reasons all stemmed from the rise in prices,[11] but in practice the form in which the landlords saw the problem probably depended on whether the particular manor was let for a short or a long term.

It is not difficult to find examples from the late twelfth century of manors held on long leases. One is Benfleet (Essex), which the abbot of Westminster leased out at £24 a year for the lifetime of the lessee,

[8] *P.R. 18 Hen. II*, pp. 85-6; *P.R. 23 Hen. II*, pp. 104-5; *P.R. 28 Hen. II*, pp. 148-9; P. D. A. Harvey, "The Pipe Rolls and the Adoption of Demesne Farming in England", *Econ. Hist. Rev.* (to be published).

[9] *P.R. 14 John*, pp. 6, 40, 121-2; Harvey, *op. cit.*

[10] P. Bowden in *The Agrarian History of England and Wales*, vol. iv, *1550-1640*, ed. Joan Thirsk (Cambridge, 1967), pp. 687-9.

[11] I thus find myself in full agreement both with the conclusions of C. G. Reed and T. L. Anderson, "An Economic Explanation of English Agrarian Organization in the Twelfth and Thirteenth Centuries", *Econ. Hist. Rev.*, 2nd ser., xxvi (1973), pp. 134-7, and with Mr. E. Miller's refutation of the arguments used to reach them, "Farming of Manors and Direct Management", *ibid.*, pp. 138-40.

an agreement that is enrolled on the first Pipe Roll of Richard I's reign.[12] Another is the Bury St. Edmunds manor of Runcton (Norfolk) which the wife of Herlewin likewise held for life: "Yesterday I would have given sixty marks to free that manor; but now the Lord hath freed it", said Abbot Samson on hearing of her death.[13] In a valuable discussion of the abandonment of leasing in the late twelfth century Mr. Miller shows the drawbacks to contemporary landlords of leases of this type.[14] It was not only that rising prices would before long make the annual rent worth far less in real terms than when the agreement had been made; an increase in the number of peasant sub-tenants, assarting and other improvements on the manor could bring the lessee an income appreciably larger than the landlord had anticipated or bargained for. Moreover there was a real risk that the lease might become hereditary, effectively alienating the manor from the estate: Mr. Miller cites the interesting case of the Ramsey Abbey manor of Over (Cambridgeshire) which, after four generations in the hands of a single family of lessees, was recovered by the abbey only by long litigation.[15] Just what tenures and offices could be inherited was of course much in question at that time, and a landlord may well have been afraid of binding his heirs for ever to an agreement made originally for a limited term. But leasing out manors for cash rents differed from demising them for military or other services that changing social conditions would make nugatory; given an unchanging economy, both locally and nationally, the landlord had little to lose from a lease that was inherited. The only exceptions would be where, for a large initial payment or for some other consideration, a lease had been granted for a particularly low annual rent, but such cases were probably a minority. Otherwise it was only by extensive local development or by a general rise in prices that the lessee would start to make profits so conspicuous (for it is thus that the situation would come to his landlord's notice) that the landlord would see the perpetuation of such leases as a real risk. Certainly assarting, peasant population increase and more intensive husbandry must have brought enhanced profits to some lessees in

[12] *The Great Roll of the Pipe for 1 Richard I*, ed. J. Hunter (Record Commission, London, 1844), p. 230.

[13] Jocelin of Brakelond, *Chronicle*, ed. H. E. Butler (London, 1949), pp. 32-3.

[14] Miller, "England in the Twelfth and Thirteenth Centuries", pp. 8-10.

[15] *Ibid.*, p. 9, citing *Cartularium Monasterii de Rameseia*, ed. W. H. Hart and P. A. Lyons (Rolls Series, 1884-93), i, pp. 120-7. R. V. Lennard, *Rural England, 1086-1135* (Oxford, 1959), pp. 107-11, gives earlier evidence of the same trend; it was not a new phenomenon.

the late twelfth century.[16] But local developments of this sort had for many generations been part of the background of the leasing of manors,[17] and we have no reason to suppose that at this point they became especially widespread. The rise in prices, on the other hand, must everywhere have brought immediate benefits to the long-term lessees, benefits that were clearly visible in their effects even if their cause was obscure. It was probably this that brought home to landlords the disadvantages of so inflexible a method of running their properties, whether or not there was the risk that a long lease might become hereditary. It is worth noting that in other parts of Europe too there was a tendency at this time for leases to become hereditary, but that only in England was this tendency halted by the introduction of direct management of estates; as we shall see, the rise in prices was probably a purely English phenomenon, and it would satisfactorily explain why the adoption of demesne farming was a development peculiar to England.[18]

But if some manors were let out on long leases by fixed agreement, others, perhaps the majority, were farmed under much more flexible arrangements. When Abbot Samson of Bury St. Edmunds decided on a policy of direct management of his estates he did not have to wait for the expiry of leases or buy out lessees' rights on most of his manors: he simply waited until the Michaelmas after his installation and then took them into his own hands.[19] Leases on a short-term, even annual, basis are implied by the pattern of re-stocking manors that were in the king's hands by wardship or escheat: for the most part new stock was bought for them (and, thus, the rents raised) not piecemeal as one would expect if leases had to expire before their conditions were changed, but *en bloc* following the

[16] An example of a manor leased for a farm that was well below its true value is provided by Waddington (Lincs.), which in 1185 rendered £22 a year but was said to be worth £40: *Rotuli de Dominabus*, ed. J. H. Round (Pipe Roll Soc., old ser., xxxv, 1913), p. 15.

[17] Lennard, *Rural England*, pp. 182-3, describes arrangements for increasing the amounts of the farm within the term of a lease.

[18] Dorothea Oschinsky, *Walter of Henley and other Treatises on Estate Management and Accounting* (Oxford, 1971), pp. 72-3, 213-14, suggests that the existence in England of well-developed accounting techniques was one cause of the adoption of demesne farming there but not in Germany; but this argument implies that written accounts were in general use by estate owners from the very beginning of the period of demesne farming, an assumption that is open to question.

[19] Jocelin of Brakelond, *Chronicle*, ed. Butler, p. 32; M. M. Postan in *The Cambridge Economic History of Europe*, vol. i, *The Agrarian Life of the Middle Ages*, 2nd edn., ed. M. M. Postan (Cambridge, 1966), p. 586, provides some other examples.

periodic inquiries made by the itinerant justices.[20] Tunstead (Norfolk) was in the king's hands in the early 1180s; for two years its custodians received £19 farm, but they then added a hundred sheep to the manorial stock and it thenceforth produced an extra pound a year.[21] The lessees in such a case would probably be one or several of the sub-tenants of the manor or, more vaguely, men of the vill (*homines ville*) acting jointly or even communally.[22] Certainly where leases were as flexible as this it cannot have been fears of lessees' hereditary rights nor even of rents fixed irrevocably at unrealistic levels that led landlords to turn to direct management. Jocelin of Brakelond speaks of the dilapidations that had occurred on the Bury St. Edmunds manors at the hands of their farmers;[23] this points to the laxity and negligence of Abbot Samson's predecessor, but it was something that had always had to be guarded against in a system of leasing and it can hardly have suddenly become a universal problem. In fact we can again see rising prices as bringing about the general change to demesne farming, though not in quite the same way as where manors were let for long terms. Clearly a rise in prices would benefit any lessee who had a marketable surplus of produce, whether he held by a long lease or a short one. But we must remember that it would not be a question of gradual, steady improvement in his position so that each year he did a little better than the year before. There would be good harvests and bad ones; there would be years when local demand was strong and prices high, but there would be other years when the market was depressed. There would increasingly be years when the lessee did very well indeed — outrageously well from his landlord's point of view. But there would still be years when he might hardly be able to find the cash to pay the stipulated farm. And once rising prices had introduced a measure of long-term change into the local economy, making the established farms palpably unrealistic, it was very difficult to discover what new levels of rent would be fair to landlord and lessee

[20] Thus there are many cases of restocking on *P.R. 13 Hen. II* (pp. 36-8, 102-4, etc.) following the judicial visitation of 1166, but hardly any on the Pipe Rolls for the three following years.

[21] *Rotuli de Dominabus*, ed. Round, p. 48.

[22] *P.R. 27 Hen. II*, pp. 107-9, gives a list (unique on the Pipe Rolls) of the farmers of the manors of the honour of Rayleigh (lands of Henry of Essex) who seem in each case to be one or two of the local inhabitants; cf. Lennard, *Rural England*, pp. 153-5.

[23] Jocelin of Brakelond, *Chronicle*, ed. Butler, pp. 1, 32. We find that deterioration or improvement of the manor at the hands of a lessee was one of the matters on which the jurors of the St. Paul's manors were questioned in the inquiries of 1222: *Domesday of St. Paul's*, ed. W. H. Hale (Camden Soc., old ser., lxix, 1858), pp. 1-2, 8, 13-14, etc.

alike. In fact we see just this difficulty arising at this time. The *Rotuli de Dominabus* of 1185 record that when Billingford (Norfolk) came into the hands of the Crown it was sworn to be worth £15 a year, but its custodian for the king was able to let it out at £20.[24] More dramatically, when the archbishop of York acquired Billinghay (Lincolnshire) through a wardship he substantially increased its farm "and as a result, so the jurors say, the vill has been wasted and its men harassed".[25] We can easily see how the landlord might turn to direct management when confronted with recurrent problems of this sort and confronted too with the local resistance that would be aroused by any proposal to increase the annual farm.

The rise in prices, then, provides a simple and straightforward explanation of landlords' replacement of lessees by their own manorial officers in the late twelfth and early thirteenth centuries.[26] We may even conjecture that had a system of food-farms instead of money-rents prevailed until after prices stabilized in the mid-thirteenth century there would have been no period of demesne farming in England, for landlords as well as their lessees would then have benefited directly from the rise in prices. But there was one landlord in England to whom rising prices presented as much of a problem as to any other magnate but who could not turn to direct management as a way out of the difficulty; this was the king, the greatest landowner of all. For the most part the royal estates were run by the sheriff of each county who was in a sense a lessee: he would keep for himself the income from certain of the king's resources in the shire — not only the royal demesne but the dues from certain boroughs and the profits of the hundred and county courts — and he paid to the king simply a fixed sum, the farm of the county. Not only was the sheriff a lessee; he might even become an hereditary lessee, for from John's reign onwards the shrievalties of a few counties were granted or claimed as hereditary, a custom which had been known under William I and his sons but which had lapsed since.[27] But even where, as in most counties, the sheriff held office for only a few years

[24] *Rotuli de Dominabus*, ed. Round, p. 57.

[25] "et ideo, ut aiunt juratores, est villa destracta et homines destracti": *ibid.*, p. 3. At Ospringe (Kent) in 1240 we see a later case of difficulty in fixing a fair farm in view of annual variation in profits: R. S. Hoyt, *The Royal Demesne in English Constitutional History, 1066-1272* (Ithaca, 1950), pp. 158-9.

[26] Interestingly, the return to leasing has been associated with the decline of prices in real terms in the fourteenth and fifteenth centuries: Postan in *Cambridge Economic History*, i, 2nd edn., p. 588.

[27] W. A. Morris, *The Medieval English Sheriff to 1300* (Manchester, 1927), pp. 50-2, 76-7, 84, 145, 179-82; S. Painter, *The Reign of King John* (Baltimore, 1949), pp. 40, 59.

at a time the farm of the county could not be increased to accord with changing economic circumstances, for its amount was fixed by custom; with few exceptions the amount of the farm was the same at the end of the thirteenth century as it had been in 1170.[28] In effect, however, the farms were augmented by making the sheriff answer for a further amount known as the *crementum* or *incrementum*, increase, a practice that first appears in the 1160s; but by 1201-2 this added less than one-tenth, by 1211-12 less than a quarter, to the amounts of the traditional farms, and even this modest supplement was forbidden by Magna Carta.[29] Another device that Richard I and John used to tap the wealth of their own estates and other resources was to exact a premium from an incoming sheriff for the privilege of holding office; this practice had existed since the eleventh century, but it was now exploited so that a shire could be practically sold to the highest bidder for a sum that might be as much as £2,000.[30] At best this was a hit-and-miss method of bringing the profits of the shires into the Exchequer, but it points to the value of the office to the sheriff and the amount of revenue that the Crown was losing through having to adhere to custom. Finally in 1204 John took the step that so many English magnates had already taken on their estates: he appointed sheriffs who, designated keepers (*custodes*), were required to answer not only for the traditional farms of their counties but for for additional profits (*proficua*) as well, so that, in effect, they became bailiffs instead of lessees. The system was applied to about half the counties, and in its first year it increased their revenues by about one-third, but after four years it was rapidly abandoned and the sheriffs became simple lessees once more.[31] Both political and administrative reasons have been suggested for its failure. In fact it

[28] Painter, *op. cit.*, p. 243; G. J. Turner, "The Sheriff's Farm", *Trans. Roy. Hist. Soc.*, new ser., xii (1898), p. 122.

[29] Mitchell, *Studies in Taxation*, pp. 15-16; Mabel H. Mills, "Experiments in Exchequer Procedure (1200-1232)", *Trans. Roy. Hist. Soc.*, 4th ser., viii (1925), pp. 159-60; Morris, *op. cit.*, pp. 128, 144; Painter, *op. cit.*, p. 115. Both Mitchell and Mabel Mills seem to have misinterpreted the Buckinghamshire increment of £10 and three hawks on the list of county farms given by Turner, *op. cit.*, p. 143; as Morris shows, other increments, some both earlier and larger, are recorded on the Pipe Rolls of Henry II. The reissues of Magna Carta omitted the prohibition of these increments (*cap.* 25 in Magna Carta of 1215: W. Stubbs, *Select Charters*, 9th edn., ed. H. W. C. Davis [Oxford, 1913], p. 296), and they reappeared after a few years' lapse (Mills, *op. cit.*, pp. 166-7).

[30] Morris, *op. cit.*, pp. 64, 83-4; introduction by Doris M. Stenton to *P.R. 6 Ric. I*, pp. xvii-xviii.

[31] Morris, *op. cit.*, p. 144; Painter, *Reign of King John*, pp. 118-23; J. C. Holt, *The Northerners* (Oxford, 1961), pp. 152-3; introduction by Patricia M. Barnes to *P.R. 16 John*, pp. xx-xxiii.

seems unlikely that it could have been made to work successfully. When magnates turned to direct management of their properties, they appointed bailiffs to have oversight of several manors and stewards who would supervise agriculture over an entire estate, so that the reeve on the individual manor was carefully and closely watched in all his activities; only in the late thirteenth century were accounting techniques developed to the point where it became possible to relax this continual supervision and give greater freedom to the manorial officer who would answer for his actions simply in the searching scrutiny of the annual audit.[32] Royal Exchequer procedure had certainly not reached this level of sophistication by the reign of John; indeed, compared even with private estate accounts of the same period, its written records, the Pipe Rolls, are clumsy, obscure and riddled with archaic conventions. The sheriffs were too dispersed and their duties too multifarious for the alternative of close oversight to be practicable, even if the traditions of the shrievalty had permitted it, but it may be that the appointment of a second official to some counties represents an attempt at supervision and that his function was to act as a check on the honesty of the first; the account for Warwickshire and Leicestershire in 1204-5, for instance, is rendered by "Hugh de Chaucumb as keeper and Master Hilary with him".[33] We do not know how the recorded profits of the keeper-sheriffs were determined,[34] but if they really represent an attempt to set down the total profits from the counties they must have given rise to much bullying and extortion on one side of the Exchequer board, much concealment and fraud on the other; politically, if not financially, the system would soon be self-defeating.

The Crown then, unable to apply contemporary methods of direct management to its own estates and other regular sources of revenue, was powerless to draw from them the greatly increased income that the rise in prices would warrant. Even on the small but growing number of royal manors that were run directly from the Exchequer, by-passing the sheriffs, conservatism or administrative difficulties seem to have prevented the introduction of demesne farming, though here the annual farms could be and were raised substantially;[35] it is interesting that these properties were specifically excluded from

[32] See chaps. iii-v (especially the last two paragraphs of ch. iii) of the introduction to *Manorial Records of Cuxham*, ed. P. D. A. Harvey (Hist. MSS. Commission and Oxfordshire Rec. Soc.; in the press).

[33] *P.R. 7 John*, p. 28.

[34] S. Smith in introduction to *P.R. 7 John*, p. xxvi.

[35] Painter, *Reign of King John*, pp. 123-4.

Magna Carta's prohibition of increments.[36] Eventually the Crown solved the problem for its demesne manors by the reforms of 1236-40 that were probably the work of Peter des Rivaux: the royal estates were all taken from the sheriffs' control and organized centrally by two keepers who in principle let each manor to its own tenants (*hominibus singulorum maneriorum*) under a lease reviewable every five years.[37] The date is significant, for by the 1230s prices in the long term were becoming steadier so that a policy of leasing was once more compatible with efficient management; the Crown thus passed through the period of massive price increases without gaining direct control of its estates. Even if the experiment of 1204 had proved workable, turning the sheriffs into bailiffs of the royal estates, it is unlikely that the Crown's profits from this source would have been commensurate with the rise in prices, though, as we have seen, the premiums that men were prepared to pay for appointment show that the office was a profitable one. We know little of how the sheriffs ran the royal manors in their care, but they probably adhered to a system of leasing. With frequent changes of sheriff demesne farming would be very difficult to organize and from its opportunities for abuse it is unlikely to have been viewed favourably by the itinerant justices who exercised ultimate control of the royal estates; certainly the instructions for the judicial visitation of 1194 imply a system of leasing.[38] This meant that the sheriffs suffered the same disadvantages as landlords who leased out their estates: the case of the Ramsey Abbey manor of Over, which its lessees claimed as an inheritance, can be paralleled by that of the royal manor of Matlock (Derbyshire) which its lessee claimed to hold in fee-farm after it had been let to his family for five generations.[39] The other resources of the Crown for which the sheriff answered in his farm were probably even less amenable than the royal demesne to increased profits corresponding to the rise in prices: the revenues from the boroughs were fixed by custom and so too, very probably, were the fines and amercements of the local courts so that any increase, let alone one of two or three hundred per cent, would appear grossly extortionate.

[36] Presumably because no powerful interest was concerned with them.

[37] Mabel H. Mills, "The Reforms at the Exchequer (1232-1242)", *Trans. Roy. Hist. Soc.*, 4th ser., x (1927), pp. 121-3; F. M. Powicke, *Henry III and the Lord Edward*, 2 vols. (Oxford, 1947), i, pp. 102-3; Hoyt, *Royal Demesne*, pp. 156-61.

[38] *Select Charters*, 9th edn., p. 255; *assisus redditus* here almost certainly means the fixed return or farm from the entire manor (Harvey, "The Pipe Rolls"). Cf. B. P. Wolffe, *The Royal Demesne in English History* (London, 1971), p. 25.

[39] Hoyt, *Royal Demesne*, pp. 150-1.

As we shall see, the boroughs were probably as capable as the royal demesne lands of paying increased dues exactly related to the rise in prices, and it is in this context that we should see the increasingly heavy tallages that were laid on boroughs and royal demesne in the reigns of Richard I and John:[40] they were attempts not merely to tap wealth that was immune from most other forms of taxation, but to tap wealth that was inaccessible to the king as landlord because his administrative system was unable to take account of economic change. Professor R. S. Hoyt has argued that these tallages had an important consequence: they forced the barons and, in Magna Carta, the king to accept a distinction between the boroughs and the king's manors which amounted in effect to the earliest clear definition of the royal demesne.[41] If this is so, it is one constitutional development that can be closely linked with the rise in prices.

It is by no means the only one. Dr. B. P. Wolffe has recently suggested that for most of the middle ages the royal estates were regarded as a source of patronage rather than of cash, and that the kings lived on the proceeds of taxation and other extraordinary income rather than from their so-called ordinary revenues.[42] The rise in prices certainly ensured that this was so in the reigns of Richard I and John, and it may have tilted the scales in this direction for the next two hundred years. Giraldus Cambrensis says of Henry II and his sons that "what they lacked in fixed returns (*in redditibus*) they made up in casual profits, trusting more to accessories than to principals".[43] Knowledge of price trends gives the comment more significance than its author intended. The kings had no alternative. Their failure to raise the income that reached them through the sheriffs' farms to anything like the level demanded by the rise in prices threw the whole burden of their increased expenditure on other forms of revenue. It was not just that John in 1215 had to pay about three times as much as his father fifty years before when he bought food for his household or livestock and corn for his manors; he also had to pay his troops at three times the earlier rates. Henry II had paid his knights 8d. a day, his foot-soldiers 1d.; by about 1200 the rates were 2s. or 3s. for the

[40] Painter, *Reign of King John*, pp. 125-6; S. K. Mitchell, *Taxation in Medieval England* (Yale Historical Publications, Studies, xv, New Haven, 1951), pp. 313-5.

[41] Hoyt, *Royal Demesne*, pp. 144-6.

[42] Wolffe, *Royal Demesne*, p. 10.

[43] Giraldus Cambrensis, "De Principis Instructione Liber" in *Opera*, ed. J. S. Brewer, J. F. Dimock and G. F. Warner (Rolls Series, 1861-91), viii, p. 316, cited by F. M. Powicke, *The Loss of Normandy, 1189-1204*, 2nd edn. (Manchester, 1961), p. 232 n.

knights, 2d. for the foot-soldiers.[44] There is evidence too of increasing standards of luxury or comfort at the royal court, which must also have increased its expenses.[45] How Richard I and John exploited every variety of taxation known to them, every available source of casual profits, is well known and there is no need to rehearse the details here.[46] Despite expensive commitments abroad both kings managed to remain solvent. But in the end the defiance of custom that knew nothing of price trends was too great. Whatever other personal and political factors were involved, it was the king's continual financial exactions of one sort or another that lay at the root of the rebellion of 1215. Certainly, if John's expenditure had been reduced by one-half or two-thirds, enabling him to keep his financial demands within the limits of feudal custom, most of his political difficulties would have vanished. No landmark in English constitutional history was more clearly brought about by economic change than Magna Carta.

Historians' assessments of John's ability as king have exactly kept pace with growing knowledge of the rise in prices and the financial administration of his reign. No one would now say, as F. M. Powicke did in 1913, that "By careful management Henry II had doubled his income; by mismanagement John brought chaos".[47] Even in 1914 S. K. Mitchell remarked that it was likely that the rise in prices "increased the difficulties of the government somewhat".[48] To the late Lady Stenton in 1951, "No king of England was ever so unlucky as John".[49] Dr. W. L. Warren in 1961 described John's government as "enterprising and intelligent" in its attempts to bring more money to the Crown.[50] In a way John's predicament was very like Henry

[44] Sir J. H. Ramsay, *The Angevin Empire* (London, 1903), p. 373, and *A History of the Revenues of the Kings of England, 1066-1399*, 2 vols. (Oxford, 1925), i, pp. 211, 221; Mitchell, *Studies in Taxation*, p. 309. But at the beginning of John's reign a knight's wage could clearly vary a good deal: Doris M. Stenton in introduction to *P.R. 1 John*, p. xiv; cf. A. L. Poole, *From Domesday Book to Magna Carta*, 2nd edn. (Oxford, 1955), pp. 371-2.

[45] Painter, *Reign of King John*, p. 112.

[46] The details are most fully set out in the introductions to the Pipe Roll Society's editions of the Pipe Rolls of their reigns; a general description is given by W. L. Warren, *King John* (London, 1961), pp. 148-52.

[47] *The Loss of Normandy, 1189-1204* (Manchester, 1913), p. 350 (p. 237 in the 2nd edn. [1961], where the author acknowledged that subsequent investigation had shown that John was in fact more business-like than he had supposed).

[48] *Studies in Taxation*, p. 2.

[49] *English Society in the Early Middle Ages* (Pelican History of England, iii, 2nd edn., Harmondsworth, 1952), p. 44.

[50] Warren, *King John*, p. 152. Of modern writers on John and his reign it is Dr. Warren who attaches most weight to the price rises in considering his financial and political difficulties: *op. cit.*, pp. 145-53.

IV's two hundred years later: in a changing economy fiscal custom denied him access to the real wealth of the country. Certainly when we recognize the extraordinary difficulties he had to meet we must admire his ingenuity and tenacity in managing the royal finances. The Interdict was, of course, financially a piece of great good fortune, enabling the king to take major ecclesiastical revenues into his own hands for up to five years, apparently without serious protest from the laity;[51] this almost certainly staved off the final debacle. But when we find that for the last year of his reign, when the Exchequer had ceased to function, John was able to subsist largely on the money that he had saved and stored in scattered treasuries,[52] we can only see his financial administration as an astonishing achievement in the face of economic pressures that he was probably unaware of and certainly could not control. Conversely it is easy for us to condemn the baronial opposition to John. Having secured their own financial position by direct management of their estates, and enjoying the enlarged incomes that rising prices were bringing them,[53] the magnates combatted every attempt of the king to bring his revenues into line with theirs. But they could hardly see the position in these terms. Even if they were aware of the problems posed by rising prices it is unlikely that they realized how ruthlessly they were exploiting the Crown's declared respect for custom. There was probably nothing cynical in their appeals to irrelevant precedent. And if they were simply unaware of the economic weapon that fate had placed in their hands, this very ignorance would give the weapon added power.

We have seen that soldiers' pay in this period increased in the same proportions as the prices of corn and livestock. This raises a funda-

[51] Painter, *Reign of King John*, pp. 174-6, 183-4; Poole, *Domesday Book to Magna Carta*, pp. 446-9; Warren, *King John*, pp. 166-8, 173. This is not to say, however, that the Crown's tenure of these revenues caused no resentment among laymen; cf. Holt, *Northerners*, pp. 235-6.

[52] J. E. A. Jolliffe, "The Chamber and the Castle Treasuries under King John", in R. W. Hunt, W. A. Pantin and R. W. Southern (eds.), *Studies in Medieval History presented to F. M. Powicke* (Oxford, 1948), pp. 125-38; C. F. Slade, introduction to *P.R. 12 John*, pp. xxxi-xxxii. The king did, however, get some revenue at this period from the confiscated estates of rebels: Painter, *Reign of King John*, pp. 350, 371.

[53] Calculations showing increases in manorial revenues have been made by S. Painter, *Studies in the History of the English Feudal Barony* (Baltimore, 1943), p. 160, who finds an average increase of 60% from 1086 to 1220 and a further 51-64% from 1220 to 1250. There is some reason to suppose that estates in eastern England were slower to adopt demesne farming than those elsewhere (Harvey, "The Pipe Rolls"); this may help to explain the opposition to John in this area (Painter, *Reign of King John*, pp. 298, 360; Poole, *Domesday Book to Magna Carta*, p. 469 n.), as landlords there would be less able to meet the king's financial demands.

mental question. So far we have been considering the direct consequences of a rise in prices that is well attested. But what was the significance of this rise in prices? Are we dealing with a simple change in price structure or with a general inflation of the currency? Unfortunately we just have not got the evidence to give an indisputable answer to this question. Virtually all we know of prices at this period concerns agricultural products, and their prices depended not on costs of production but on market levels regulated by supply and demand.[54] Thus they might change quite independently of wage-rates or the prices of other goods and, indeed, we find just this happening in the mid-fourteenth century: the loss of population in the Black Death led to marked rises in wages and in prices that depended on wage-rates, while the prices of agricultural products were unaffected.[55] Was it the reverse process that occurred in the late twelfth century, a sudden vast rise in the value of foodstuffs that might be attributed to population pressure and growing shortage of land? Or are these increases in prices the only clear traces to survive of a general decline in the value of money that affected all prices, wages and other payments? Historians differ in their views. Dr. Warren takes for granted that it was an inflation of the currency: "It was John's misfortune to reign in a period of monetary inflation .... There was more silver available, trade was flourishing, and the standard of living was rising; but the productivity of the land was increasing only slowly, so the prices of agricultural produce went up, and labour was dear".[56] Professor Postan in 1944 was prepared to countenance the same view, but by 1952 he had concluded that the price increases affected foodstuffs alone.[57] At first sight, of course, the rise in soldiers' pay might seem conclusive evidence of a general rise in wages. But this could possibly be a special case: the increases may reflect the more elaborate (and thus more expensive) equipment that was becoming necessary,[58] while the military were in a peculiarly strong position to bargain for rates of pay that took into account the

[54] It is a great pity that we have no more evidence of the price of salt in this period; the few examples on the Pipe Rolls from 1171 to 1210 can be interpreted either way: Farmer, "Price Fluctuations", p. 37; cf. Rogers, *History of Agriculture and Prices*, ii, pp. 409-18. The increase in the price of wine, as noted by Gras (above, note 4), is more significant, though there is here the complication that we are dealing with an imported product.

[55] A. R. Bridbury, *Economic Growth: England in the Later Middle Ages* (London, 1962), p. 92.

[56] Warren, *King John*, p. 145.

[57] "The Rise of a Money Economy", *Econ. Hist. Rev.*, 1st ser., xiv (1944), p. 128; *Cambridge Economic History*, vol. ii, *Trade and Industry in the Middle Ages*, ed. M. Postan and E. E. Rich (Cambridge, 1952), p. 167.

[58] As suggested by Mitchell, *Studies in Taxation*, p. 309.

increased cost of living. Other evidence of wages, however, points in the same direction. The presumably skilled workers in charge of the king's vineyards in Herefordshire and at Windsor were paid 1d. a day at the beginning of Henry II's reign;[59] in 1210-11 carpenters working on the bishop of Winchester's residence at Marwell (Hampshire) were paid from 1½d. to 2d. a day, and by the mid-thirteenth century 2½d.-3d. a day seems to have been the normal rate for work of this sort.[60] But twelfth-century examples of civilian wage-rates are very few indeed, and in view of later medieval evidence of wide seasonal and regional variations in wages it would be wrong to attach much importance to them. There are two more cogent arguments that there was a real inflation of the currency. The first is the apparent prosperity of towns at this period. If prices of corn and livestock rose while wages and other prices remained static we would expect the prosperity of towns to diminish. Pressure of population might indeed bring about expansion of old towns and the creation of new ones, which would bring more rent into the pockets of the king or borough-owning magnate, but the individual trader or craftsman, paid the same as before for his own work yet paying two or three times as much for the food he ate, would be much worse off and would have little cash to spare for schemes of municipal improvement or aggrandizement. Yet the evidence suggests that the towns — and that is the burgesses, not the borough-owners — were prospering and were ready to buy the privileges that Richard I and John were offering to sell them in royal charters.[61] Admittedly it was in John's reign that we find the first signs of decline in the urban cloth industry, but this can be traced to causes other than a general decline in the towns' wealth.[62] Their continuing prosperity is a serious difficulty in the way of seeing the rise in prices as some-

[59] *The Great Rolls of the Pipe for 2, 3 and 4 Henry II*, ed. J. Hunter (Record Commission, London, 1844), pp. 18, 51; payments at this rate continued until 1204-5 and at least 1205-6 respectively (*P.R. 7 John*, p. 271; *P.R. 8 John*, p. 226) but this attests the Exchequer's conservatism rather than stability of wage rates.

[60] *The Pipe Roll of the Bishopric of Winchester, 1210-1211*, ed. N. R. Holt (Manchester, 1964), p. 13; W. Beveridge, "Wages in the Winchester Manors", *Econ. Hist. Rev.*, 1st ser., vii (1936-7), pp. 40-1, and "Westminster Wages in the Manorial Era", *ibid.*, 2nd ser., viii (1955-6), p. 27.

[61] J. Tait, *The Medieval English Borough* (Manchester, 1936), pp. 178-9. Admittedly there were delays and difficulties over payment in some cases.

[62] E. M. Carus-Wilson, "An Industrial Revolution of the Thirteenth Century", *Econ. Hist. Rev.*, 1st ser., xi (1941), pp. 39-60; E. Miller, "The Fortunes of the English Textile Industry in the Thirteenth Century", *Econ. Hist. Rev.*, 2nd ser., xviii (1965), pp. 71-4. In the article cited below (note 93) I suggest a way in which the change might, indirectly, be linked with the cause of the inflation.

thing other than an aspect of monetary inflation. The second piece of evidence comes from the countryside. On manors where villeins did labour services in the thirteenth and early fourteenth centuries custom often gave the lord the option of taking in any year a stipulated payment in cash instead of the basic week-work. Sometimes this is specified in a custumal, sometimes it appears only when works were thus commuted and entered as "sold" in the annual account. But when this was done we very often find that the amount paid was much less than the value of the work. At Combe (Hampshire) in the mid-thirteenth century the tenant of a half-virgate was required either to pay 4s. rent or to do some 150 days' labour service in the course of a year, work that at contemporary rates was worth about 12s. 6d.[63] Again, at Cuxham (Oxfordshire) in the early 1330s the week-works of a villein tenant were valued at just over 10s. a year, yet if they were remitted the charge levied was only 5s.[64] These instances can be matched from many other places. The simplest explanation is that these customary charges were fixed at a time when labour was worth rather less than half what it cost by the mid-thirteenth century,[65] and it seems most likely that they date from the twelfth century, when so many landlords were taking money rents as an alternative to the traditional labour services. It

[63] *Select Documents of the English Lands of the Abbey of Bec*, ed. Marjorie Chibnall (Camden Soc., 3rd ser., lxxiii, 1951), p. 43; theoretically his obligation is for 184 days' week-work (three days a week for 38 weeks, five days for 14 weeks) but 30 days seems a reasonable allowance for holidays together with the harvest boon-works which he had to perform even when paying rent and which must have been deducted from his week-work.

[64] P. D. A. Harvey, *A Medieval Oxfordshire Village* (London, 1965), pp. 82-3.

[65] It might be argued, following M. M. Postan, "The Chronology of Labour Services", *Trans. Roy. Hist. Soc.*, 4th ser., xx (1937), pp. 186-9, that the disparity was due to heavy increases in labour services in the thirteenth century. The return to labour services is discussed below, pp. 21-2. I do not question that in the thirteenth century they were sometimes increased above the former customary level. I doubt, however, for three reasons, whether these increases accounted for more than a small part of the disparity between labour services and the alternative money rents: (i) the disparity seems too large and too consistent to be explained thus; (ii) it is difficult to see why an authoritarian doubling or trebling of labour services should not be accompanied by an equivalent increase in the alternative money rent; (iii) there are anomalous cases that will not fit this explanation. Combe (Hants.) in fact provides an example of these last: as already shown, the half-virgaters in the mid-thirteenth century did week-work worth about 12s. 6d. a year or paid 4s. rent, but the full virgaters simply paid 10s. rent; if the need for labour led to doubling or trebling the half-virgaters' works, why were the full virgaters allowed to continue to pay a money rent? More likely the imposition of the half-virgaters' week-work at the customary level produced as much labour as the demesne could use from this source.

follows that agricultural wages rose in this period in very much the same proportion as agricultural prices.[66]

Once we accept that the rise in prices was only one aspect of a general inflation of the currency, other developments of the time start to fall into place as part of a connected pattern of events. In an inflationary period those who would suffer would be those who, like the king, drew their income from fixed charges. Those who paid these fixed charges, of course, stood to gain. In the countryside the most obvious class of beneficiaries was the group of manorial tenants who held land permanently for money rents, with or without other light services, on what was acknowledged to be a contractual basis or even explicitly by written charter. These men, the sokemen of eastern England and the forerunners of the thirteenth-century free tenants found elsewhere, must have greatly strengthened their economic position at this time as the real value of the rents they paid fell by half or two-thirds. This may well have contributed to the growth of the peasant land-market in the thirteenth century:[67] rising prices and static rents gave free tenants both the opportunity and the incentive to improve their holdings by purchase wherever possible. But it was not so much in economic terms that the position of these tenants was visibly improving at this time but rather in legal and social status *vis-à-vis* their fellow manorial tenants; these were the villeins whose tenure was held to be not contractual but customary. They were affected by the inflation in much more complicated ways.

The tenure of the villeins was regulated by immemorial custom of the manor. Our earliest descriptions of their services date from before the Norman conquest and correspond closely to the custumals

[66] I venture here to disagree with the conclusions that Professor M. Postan illustrates from an analysis of wages and labour services at Cottenham (Cambs.) in *The Famulus* (Econ. Hist. Rev. Supplement no. 2, 1955), pp. 41-2. What seems to me particularly significant about the Cottenham evidence is that the allowances of rent made to the tenants for doing their harvest works in the thirteenth century (half-virgaters 12d., cottagers 6d.) were less than half the values attached to their harvest works in 1322 (half-virgaters 2s. 1d., cottagers 18d.) after the reorganization that made it the norm for these works to be performed; I would not, however, attach particular importance to this example in default of evidence that this same reorganization did not increase the number of works demanded: Frances M. Page, *The Estates of Crowland Abbey* (Cambridge Studies in Economic History, ii, Cambridge, 1934), pp. 86-7, 180, 220, 244-6.

[67] G. C. Homans, *English Villagers of the Thirteenth Century* (Cambridge, Mass., 1941), pp. 201-4; Postan in *Cambridge Economic History*, i, 2nd edn., pp. 625-8; P. R. Hyams, "The Origins of a Peasant Land Market in England", *Econ. Hist. Rev.*, 2nd ser., xxiii (1970), pp. 18-31, especially p. 20 on the sources of finance.

of the thirteenth and fourteenth centuries. These custumals sometimes go a little beyond the services of the tenants and mention other customs as well, and we may reasonably suspect that villein tenure was only a part of a vast mass of unwritten, slowly evolving custom that controlled many aspects of life in the medieval village and laid obligations on landlord and lessee, bailiff and villagers alike. In one notable respect the customs of villein tenure were changing in the twelfth century. As already mentioned, the period saw money rents replacing labour services on many manors, either altogether or as an alternative option at the landlord's choice from year to year.[68] This was the equivalent of the contemporary change from food-farms to the money rents paid by the lessees of entire manors; both may well have been made possible by increasing cash sales of surplus produce to the growing towns and other markets. This spread of commutation is often spoken of as if it were wholly to the advantage of the tenants, but in fact the immediate beneficiary may well have been the landlord. We know little of the day-to-day work of a manor in the twelfth century, but from the time detailed records become abundant in the late thirteenth century landlords clearly found it difficult to exact efficient labour services, and they were used only for the most unskilled work of the manorial demesne: the threshing, the ditching, the muck-carting, rather than the ploughing, the sowing or the keeping of animals.[69] Money rents, on the other hand, were easy to collect and difficult to evade, and Mr. Miller has suggested that the lessees of manorial demesnes may well have encouraged their spread;[70] certainly they must have made it much easier for both landlord and lessee to evaluate the tenants' services when negotiating a new lease. In some cases, of course, the change may have come about through a reduction in the demesne lands, which would make labour services redundant, or through a lease of the demesne to the villagers themselves.[71] Nevertheless, this process of commutation does not

[68] Postan, "Chronology of Labour Services", pp. 174-85.

[69] A. Elizabeth Levett, *The Black Death on the Estates of the See of Winchester*, in P. Vinogradoff (ed.), *Oxford Studies in Social and Legal History*, 9 vols. (Oxford, 1909-27), v (1916), pp. 35-6, 60, 66-7, 89-94; J. A. Raftis, *The Estates of Ramsey Abbey* (Pontifical Institute of Medieval Studies, Studies and Texts, iii, Toronto, 1957), pp. 190-201; Harvey, *Medieval Oxfordshire Village*, pp. 75-86.

[70] E. Miller, *The Abbey and Bishopric of Ely* (Cambridge Studies in Medieval Life and Thought, new ser., i, Cambridge, 1951), p. 112; Lennard, *Rural England*, p. 198, gives examples of tenants installed in their holdings by the farmer of the manor.

[71] Postan, "Chronology of Labour Services", pp. 173-85, sees a close connection between the reduction of demesnes and the swing to money rents in the twelfth century.

seem to have continued into the thirteenth century. At the very least the swing towards commutation stopped or was drastically reduced. There may even have been a general return to labour services, as Professor Postan suggested in 1937; he argued that where the option of a money rent existed it was often dropped or, as we have seen, allowed exceptionally, while elsewhere money rents were often simply replaced by the labour services that had been commuted in the previous century.[72] Professor Postan accounted for this marked reaction by the increasing areas of manorial demesne lands: larger demesne lands needed more labour to till them, and labour services provided the answer.[73] But there are difficulties in the way of this explanation, which implies a very sensitive connection between labour services and demesne area. By this date labour services can seldom have played a very important part in demesne agriculture. At Cuxham (Oxfordshire), where heavy week-work was normally exacted in full, they provided about one-third (the least skilled third) of all the work done on the demesne at the end of the thirteenth century, and this is probably typical.[74] There was no shortage of labour in the thirteenth century, a period when population was growing fast, and it seems improbable that landlords had to use labour services as the only way — or even the most convenient way — of tilling their lands. More likely the inflation forced labour services upon them. This does not credit the landlords with a precocious knowledge of economics: it would come about very naturally. On manors where an option existed it would become increasingly clear that the value of the services was greater than the alternative money rent, which would gradually fall into near disuse or be dropped altogether.[75] This would serve as an example

[72] *Ibid.*, pp. 186-9. It is likely, though, that the restoration of labour services was much less widespread than is there suggested (cf. Titow, *English Rural Society*, p. 60). The question is a complicated one that badly needs further investigation and re-assessment. I have learned much from Mr. T. H. Aston's unpublished paper of 1953, which deals with this and other aspects of manorial rents and labour.

[73] Postan, "Chronology of Labour Services", p. 186.

[74] Harvey, *Medieval Oxfordshire Village*, pp. 84-5.

[75] A graphic case of the opposite process occurs in the 1366 custumal of Wood Eaton (Oxon.): under a new agreement after the Black Death, the virgater was to do certain services or else pay 13s. 4d. money rent if the lord preferred, to which the writer of the custumal has added the sardonic note "and indeed the lord will prefer this for ever, for the said services are not worth so much": *Eynsham Cartulary*, ed. H. E. Salter, 2 vols. (Oxford Hist. Soc., old ser., xlix, li, 1907-8), ii, p. 20. The alternatives must have been fixed at the time of very high wages immediately after the Black Death. Interestingly, the same custumal gives the pre-plague alternatives (*ibid.*, p. 19): 5s. money rent, plus boon-works totalling 16 man-days, or else about 180 days' week-work (worth about 15s. by the mid-thirteenth century); need we doubt that given this choice it was the latter that the lord would go on preferring "for ever"?

for the restoration of labour services on manors where they had been entirely abandoned. In this way the inflation may have acted directly in bringing about a return to labour services.

This does not explain, however, why the process of commutation did not advance as it did in the twelfth century, replacing existing labour services with money rents set at new, realistic levels. In fact commutation now brought a positive disadvantage to the landlord. That it did so was again a consequence of inflation, but indirectly, as part of more far-reaching legal and social change. In restoring labour services we see the landlords acting with remarkable respect for manorial custom. They did not invent new burdens for their tenants: they brought back old ones. It is very likely that in a period of economic change the landlords found themselves as much hampered by custom in their financial relations with their tenants as the king was in his relations with the landlords. Historians have long recognized that the legal position of the villein changed for the worse in the course of the twelfth century — certainly in relation to other classes of rural society and probably absolutely. R. V. Lennard, following A. Ballard, referred to these tenants at the beginning of the century as *villans*, so as to make it plain that their Domesday predecessors were legally quite different from the thirteenth-century *villeins*.[76] The Domesday villan was classed among the free men of society; his landlord could not eject him from his holding if he did the services due for it, and his oath was accepted in the king's court. None of this was true of the thirteenth-century villein, whom the common lawyers were enmeshing in an elaborate net of rulings that in some ways equated him with the slave of Roman law. In practice the villein was, more often than not, sheltered by manorial custom from the new insecurity of his position, but this was only by the grace of the lord of the manor, who now had the unquestionable right to override this custom in almost every detail if he chose. Recently Professor R. H. Hilton has argued that the villein's legal status did not decline steadily throughout the twelfth century but changed catastrophically in its last two decades: 1180-1200 was "the crucial period for the depression of the majority of the rural population".[77] This certainly accords well with the suggestion that a fall in the value of money would tend to bring landlords into conflict with manorial custom. Legal changes in the twelfth century did not come about by formulating rules in the abstract, but from actual decisions reached in

[76] *Rural England*, p. vi.

[77] "Freedom and Villeinage in England", *Past and Present*, no. 31 (July 1965), p. 13.

the courts. If lords of manors acquired new powers over their tenants, enlarging, defining or wholly setting aside manorial custom, this was because disputes were arising and were being brought to the royal courts for settlement. Professor Hilton connects this with "the rising prices, the intense fiscal pressure" of the time, especially in John's reign, and suggests that it was only natural that "rising costs should be passed on to the basic producer" and that "this must involve a worsening of his legal and social status".[78] But if we accept that there was a monetary inflation we see the decline in villein status as the landlords' response to a situation where manorial custom combined with the decreasing value of money was endangering their income from their tenants. Overall, indeed, the landlords may even have lost financially: in many places labour services were not reimposed and the money rents continued unchanged. But potentially they gained enormously in the clear backing of royal authority for their powers over the lands, goods and persons of their newly unfree tenants, whether these powers were entirely new or just defined and extended. Certainly throughout the next two hundred years we see these powers being exercised, intermittently and with varying severity, but bringing to the landlords advantages that probably lay beyond the reach of their twelfth-century predecessors. In the process, however, their free tenants had undoubtedly gained, and not only economically. For the first time a sharp line of legal definition was drawn between them and the customary tenants, and as this depressed the one group so, correspondingly, it elevated the other, not only legally but socially as well. It was through this new demarcation of villeinage that the inflation served to maintain labour services. Of the various tests that the courts applied to decide whether a man was a villein and unfree, one of the commonest was his obligation to do unspecified work on his lord's demesne.[79] It was thus to the landlord's long-term advantage to have his villeins render labour services rather than money rents since this helped to identify them as unfree tenants. Directly or indirectly the inflation delayed the effective disappearance of labour services from England by some two hundred years. At the same time it brought the class that paid its rents in this way into a legal degradation from which it took even longer to emerge.

To the people of the countryside, then, the inflation brought changes that were profound and long-lasting. In the towns its

[78] *Ibid.*, p. 14.

[79] F. Pollock and F. W. Maitland, *The History of English Law before the Time of Edward I*, 2nd edn., 2 vols. (Cambridge, 1898), i, p. 372.

direct effects were less dramatic. At the same time it must have contributed much to two long-term developments: the rise of one class, the decline of another. The burgesses certainly did not suffer from the inflation. Whether their town was at farm to them or whether they paid individual rents to the borough-owner their annual payments, similarly fixed by custom, did not rise with the fall in the value of money. Not only did existing rents not increase: the rents charged in new towns and suburbs seem to have kept the same general level. When the bishop of Salisbury founded the new Salisbury in the 1220s he let out the house-plots at 12d. a year each, the same rent as the bishop of Lincoln had charged in his new town of Banbury eighty or ninety years before; when a new suburb of Banbury was laid out in the mid-thirteenth century an annual rent of 6d. was charged for its cottages (the plots were clearly smaller than in the old town).[80] Rents, of course, were only one of the charges on the burgess, who had to meet increases in royal taxation and perhaps in local dues as well. Even so, it was probably cheaper in real terms to be a townsman in the thirteenth century than in the twelfth. But whether or not the burgesses gained financially over this period they certainly gained by the new legal definitions that created an unfree peasantry. The burgesses were set firmly among the classes of free men recognized by the common law. Increasingly the difference between *burgensis* and *villanus* came to mean more than merely the difference between borough-dweller and village-dweller: it implied a great difference in legal status and privilege. It was not merely through the king's need for ready money that the period of inflation began the steady growth of privilege and self-government in the towns; it probably contributed a little to their economic prosperity and it certainly brought a new status to their inhabitants and with it a new self-confidence and self-respect.

Equally it is no coincidence that the same period saw a sharp decline in the position of the Jews in England, a decline that continued down to their expulsion in 1290. There can have been no group in the country that would be worse hit by a fall in the value of money. The wealth of the Jewish community was based almost entirely on money-lending. The dislocation that the inflation caused may at first have brought the money-lenders some extra business from those whose income came from estates farmed out to lessees; we perhaps see an example of this at Bury St. Edmunds, where the abbot and

[80] *Historic Towns*, ed. M. D. Lobel (London, 1969-), i, Salisbury, p. 4, Banbury, pp. 2, 3.

obedientiaries were heavily in debt to Jews by the 1180s.[81] But over some forty years the real value of the Jews' working capital must have fallen by half or two-thirds. Clearly John's tallages on the Jews were aimed at entirely the wrong target: like the king himself they were victims of the economic trends of the time. But neither king nor magnates could be expected to see this, and the Jews seemed an obvious source of revenue to help the Crown in the difficulties that inflation had brought. A very heavy tallage was levied on them in 1210, and the cruelties with which it was exacted may not be entirely the invention of later chroniclers; moreover, as C. Roth put it, this tallage "set a new fashion in the manner of exploiting the Jewish wealth", initiating the crippling series of levies inflicted on the Jews by Henry III.[82] The expulsion of the Jews can reasonably be seen as a natural conclusion to the steady attrition of their wealth throughout the thirteenth century.[83] It was the inflation that began the Jews' financial weakness, and indirectly it contributed to their impoverishment over a much longer period through the new severity of royal taxation and the wealth and growing political power of the landed gentry. There were, of course, many other factors involved. But it is significant that the decline of the Jewish community in England began in the period of inflation and it is not difficult to see why this was so.

The fact of a sharp rise in prices from 1180 to 1220 is generally accepted, and the outlines of its economic and political effects are clear. It is less certain that there was a monetary inflation, but what evidence there is points to this and, as we have seen, it accords well with various social and economic developments of the period and helps to explain them. But in considering not the effects but the cause of this inflation we are straying much further into an area of speculation and hypothesis. However, one possible explanation can be ruled out completely. There was no deterioration in the

[81] Jocelin of Brakelond, *Chronicle*, ed. Butler, pp. 1-3, 5-6, 10, 30-1, 32; cf. D. Knowles, *The Monastic Order in England*, 2nd edn. (Cambridge, 1963), pp. 304-5, 353-4. But many monasteries were chronically in debt throughout the thirteenth century: J. R. H. Moorman, *Church Life in England in the Thirteenth Century* (Cambridge, 1945), pp. 294, 303-4.

[82] C. Roth, *A History of the Jews in England* (Oxford, 1941), pp. 35, 43-50, 270-1 (but cf. Powicke, *Henry III and the Lord Edward*, i, p. 311 n.); Painter, *Reign of King John*, p. 144; Doris M. Stenton in introduction to *P.R. 13 John*, pp. xix-xxi; H. G. Richardson, *The English Jewry under Angevin Kings* (London, 1960), pp. 166-72.

[83] Cf. Roth, *op. cit.*, pp. 84-5; Powicke, *loc. cit.*

standard of the English coinage that would lead to its depreciation. The silver content of the penny, the only coin in normal use, barely altered from A.D. 800 to 1250.[84] If, then, the real value of money fell, this is to say that silver was worth less than before, and we must see the inflation of 1180-1220 as a catastrophic fall in the value of silver to about one-third of what it had been. Was this a local phenomenon peculiar to England, or did t occur throughout Europe so that we should look for an explanation in the development of new silver mines and over-production of the metal? Unfortunately, sparse though our evidence is for twelfth-century prices and wages in England, it is plentiful compared with what we find elsewhere. However in Normandy, where one would most expect a pattern similar to England, at least until 1204, there seems to have been a slight upward movement of prices between the mid-twelfth century and the mid-fourteenth, but there is no trace of changes of the scale or rapidity of those in England.[85] Nor does there seem to have been any similar rise in prices elsewhere in Europe.[86] On other grounds too it is unlikely that Europe generally was flooded with silver at this time. The silver content of the currencies of France and Italy in the early thirteenth century was very low: the *denarii* of Tours, Genoa and Milan contained less than a quarter, those of Venice less than a sixteenth, of the amount of silver in the English penny.[87] It seems to have been shortage of silver as well as new imports of gold from West Africa that led to the introduction of gold coinage alongside the silver from 1252 on.[88] In fact England, in the thirteenth century and throughout the middle ages, was quite exceptional in her wealth in silver. It is attested by the high standard of her silver currency and by the fact that it was not permanently supplemented with gold until 1344. It is attested too by the comments of contemporaries. The Venetian diplomat who wrote the *relazione* on England about 1500 was notably impressed by the quantity of silver plate he saw in churches, in private houses, in shops and even in inns.[89] Even in the

[84] C. M. Cipolla, "Currency Depreciation in Medieval Europe", *Econ. Hist. Rev.*, 2nd ser., xv (1962-3), p. 422.

[85] L. Delisle, *Études sur la Condition de la Classe Agricole et l'État de l'Agriculture en Normandie au Moyen Âge* (Évreux, 1851), pp. 572-626. The most useful evidence is the prices of livestock (pp. 610-19), as these avoid the variations in local measures though not the variations in the currencies used for payment.

[86] Postan in *Cambridge Economic History*, ii, pp. 165-7.

[87] Cipolla, "Currency Depreciation", p. 422.

[88] R. S. Lopez, "Back to Gold, 1252", *Econ. Hist. Rev.*, 2nd ser., ix (1956-7), pp. 233-4.

[89] *A Relation, or rather a True Account, of the Island of England,* ed. Charlotte A. Sneyd (Camden Soc., old ser., xxxvii, 1847), pp. 28-9, 42-3.

early twelfth century Henry of Huntingdon, explaining that England acquired silver from her exports to Germany, added that there was consequently as much silver in England as there was in Germany itself.[90] Historians agree with what both Henry of Huntingdon and the Venetian diplomat say of the origin of medieval England's silver: it came from the mines of Germany and was acquired through a favourable balance of trade with northern Europe, latterly in particular through supplying wool to the cloth industry of the Low Countries.[91] Relatively little silver was mined in England itself, and in the late twelfth century production was declining.[92] It follows that if England was suddenly flooded with silver in the late twelfth and early thirteenth centuries we must look to a massive increase in her exports, especially of wool to Flanders.

There is nothing to make this impossible in what we know of the history of the English wool trade or the Flemish cloth industry.[93] On the contrary, the period of the inflation seems a likely enough time for the development of the close interdependence that is so well recorded in the late thirteenth and fourteenth centuries. We know that in the mid-twelfth century monasteries at least were developing large-scale sheep-farming on the moors of northern England.[94] The end of the disturbances of the Anarchy and the more settled conditions of Henry II's reign were a time when a dramatic increase in trade might be expected. It would take some years for the new imports of silver to have any noticeable effect on prices and, as we have seen, probably still longer for the price rises to appear on the Pipe Rolls. This is not to say, though, that exports to the Low Countries and Germany now first became important. Henry of Huntingdon, after all, was writing before the reign of Henry II; he mentions "very valuable wool" among England's products,[95] and we know of Flemish merchants who came to England to buy wool in

[90] Henry of Huntingdon, *Historia Anglorum*, ed. T. Arnold (Rolls Series, 1879), pp. 5-6.

[91] As Cipolla, "Currrency Depreciation", p. 420.

[92] Poole, *Domesday Book to Magna Carta*, p. 82.

[93] I explore some of the implications of this suggestion for the history of the English wool and cloth trade in "The English Trade in Wool and Cloth, 1150-1250: Some Problems and Suggestions", to be published by the Istituto Internazionale di Storia Economica "Francesco Datini", Prato, in the proceedings of its second *Settimana di Studio* held in April 1970.

[94] D. Knowles, *Monastic Order*, pp. 352-3, and *The Religious Orders in England*, 3 vols. (Cambridge, 1948-59), i, pp. 65-73; R. A. Donkin, "Cattle on the Estates of Medieval Cistercian Monasteries in England and Wales", *Econ. Hist. Rev.*, 2nd ser., xv (1962-3), p. 45.

[95] *Loc. cit.*

1113.[96] Recently, indeed, Professor P. H. Sawyer has convincingly argued that England's wealth in silver can be carried back to the end of the tenth century and was what attracted successive generations of raiders and invaders; he suggests that it may already have been exports of wool that brought the silver to England.[97] None of this contradicts the hypothesis proposed here: it is suggested only that this export trade vastly expanded in the second half of the twelfth century, not that it began then.[98] Nor need we look only to exports of wool. There is substantial evidence of exports of finished cloth from the towns of eastern England.[99] Production and export of tin were growing rapidly at just this period.[100] There is also evidence of export of corn on a large scale to the Low Countries, a sign of their growing urban markets and urban industries.[101] We may bear in mind the possibility that it was a decline in exports of cloth and corn (while the export of wool continued) that first slowed down and then ended the inflation in the course of the thirteenth century. Movements in the price of gold are in keeping with a connection between the inflation and the export trade. Gold was not in everyday currency at this period, though gold coins (*besants*) were used for certain formal or ceremonial payments and gold was in use along with silver as a medium for large-scale trade: in 1257 we find London merchants warning Henry III against the general circulation of gold in coinage, as this might reduce its value.[102] During the inflation the value of gold seems to have fallen, but not quite so much as the value of silver: in 1159 the Exchequer accepted twenty silver pence for one *besant* (a gold: silver ratio of 1:10), in 1186 twenty-four (a

[96] Eileen Power, *The Wool Trade in English Medieval History* (London, 1941), p. 52; P. H. Sawyer, "The Wealth of England in the Eleventh Century", *Trans. Roy. Hist. Soc.*, 5th ser., xv (1965), p. 162.

[97] *Op. cit.*, pp. 145-64.

[98] This hypothesis does, however, call in question Professor Sawyer's suggestion that there were at least as many sheep in England in the late eleventh century as in the late thirteenth (*op. cit.*, pp. 162-3); cf. the paper cited above, note 93.

[99] E. M. Carus-Wilson, "The English Cloth Industry in the late Twelfth and early Thirteenth Centuries", *Econ. Hist. Rev.*, 1st ser., xiv (1944-5), pp. 32-4, and in *Cambridge Economic History*, ii, pp. 374-5; Miller, "Fortunes of the English Textile Industry", p. 68.

[100] Poole, *Domesday Book to Magna Carta*, p. 83.

[101] Doris M. Stenton in introduction to *P.R. 10 Ric. I*, p. xiv; there is evidence also of export of corn to Norway (*P.R. 32 Hen. II*, p. xxi) and, probably, western France (*P.R. 29 Hen. II*, p. xxviii; *P.R. 30 Hen. II*, p. xxx).

[102] *De Antiquis Legibus Liber*, ed. T. Stapleton (Camden Soc., old ser., xxxiv, 1846), p. 30.

ratio of 1:12).[103] By 1265 the workable ratio seems again to have been 1:12, but there may well have been fluctuations meanwhile.[104] Conservative tradition and international custom would tend to keep the ratio stable; the change recorded would be consistent with the import of both metals, but especially silver, in exchange for exported goods.

If excessive imports of silver really were at the root of the inflation there is a strange irony in the financial difficulties of Richard I and John. The policies of both kings led to great quantities of silver being sent abroad, especially in the decade between the payment of Richard's ransom and the final hopeless defence of Normandy. These exports of coin were on a scale to serve as a strong check to the rise in prices.[105] Yet nothing did more than this deflationary policy, which the kings pursued unwittingly, to bring about the Crown's serious financial embarrassments in John's reign. But whether or not this explanation of its origins is accepted, it is clear that 1180-1220 was a period of economic crisis or change in England. Its most spectacular manifestations were political, but its effects on the society and economy of the country were no less profound and long-lasting. It was the more devastating in that people at the time were probably completely unaware of what was happening. In 1927 Lord Beveridge suggested that the rise in prices from the mid-twelfth century to the late thirteenth "when fully examined, may prove to have been one of the most violent price revolutions in English history, comparable in speed to the revolution of the sixteenth century though not having equal social and economic consequences".[106] Professor Postan in 1944 wrote that "Historians and economists will easily find instances of social transformations brought about by additions to the mere volume of the circulating medium .... There was the well-known and much publicized increase in bullion in the sixteenth century, and a similar influx may also have occurred in the late twelfth and the

[103] *P.R. 5 Hen. II*, p. 42; *P.R. 32 Hen. II*, p. 149. The ratios assume that these *besants* were twice the weight of a silver penny, like the gold pennies of Aethelraed II (approximately) and Henry III (exactly): G. C. Brooke, *English Coins*, 3rd edn. (London, 1950), pp. 68, 107, 110.

[104] *Ibid.*, p. 110. In 1257 the London merchants told Henry III that the value of gold had recently been falling by 10 or 20%: *De Antiquis Legibus Liber*, p. 30.

[105] Powicke, *Loss of Normandy*, 2nd edn., pp. 233-5, 298; Doris M. Stenton, introduction to *P.R. 5 John*, p. xi; Poole, *Domesday Book to Magna Carta*, pp. 364-6, 373; Warren, *King John*, pp. 63, 91-2.

[106] Beveridge, "Yield and Price of Corn", p. 164.

thirteenth centuries".[107] Work over the last forty years has increasingly confirmed these predictions. It is now time for 1180-1220 to take its place beside the sixteenth and the twentieth centuries as one of the three great inflationary periods of recorded English history. Historians and others attribute profound and dire consequences to the two later inflations of the currency. Need we still hesitate to attach a like importance to the first?

[107] Postan, "Rise of a Money Economy", p. 128. If we compare the suggested size of the English coinage at the end of the tenth century, from 12 to 15 million pennies, with its known size of 168 million pennies, 20 million halfpennies and 20 million farthings at the end of the thirteenth century (D. M. Metcalf, "How Large was the Anglo-Saxon Currency?", *Econ. Hist. Rev.*, 2nd ser., xviii [1965], pp. 476, 481-2), and take into account both the reduction of its real value by two-thirds in the inflation and an estimated three-fold growth of population (J. C. Russell, *British Medieval Population* [Albuquerque, 1948], p. 280), we find that the purchasing power of the money owned by the average Englishman rose by not more than 69% between 1000 and 1300. This is a curious result, though not quite an impossible one; of the figures involved in the calculation the amount of depreciation by inflation is by no means the most vulnerable.

VIII

Boldon Book and the Wards between Tyne and Tees

BOLDON BOOK is a survey of the estates of the bishop of Durham. Its opening words tell how Bishop Hugh of le Puiset had it drawn up in his presence on the feast of St Cuthbert in Lent – 20 March – in the year 1183; but as we do not know whether the year began on 1 January or 25 March this may have been either 1183 or 1184 by our reckoning. The estates of the cathedral priory had been separated from those of the bishop nearly a century before,[1] and we have no contemporary survey of the estates of the Durham monks. Almost certainly none was made. Two of the four surviving medieval texts of Boldon Book come from the records of the cathedral priory, and it would be an odd mischance if the monks, in their remarkably well-managed archives, had kept copies of the early survey of the bishop's estates but had preserved not even a reference to their own.

In fact there seem to be only twelve other estates from which we have such manorial surveys dating from the twelfth century;[2] no doubt others have failed to survive, but it is unlikely that it was yet the normal practice for large estate-owners to compile records of this sort. Among these surviving surveys Boldon Book has long been considered especially important. Partly this is because it is the only one to come from a wholly northern estate, though the Templars' survey of 1185 has a substantial section on their Yorkshire properties. But even more it is because it covers an area excluded from Domesday Book, so that it is our earliest systematic record of any part of Britain north of Lancashire or Yorkshire. The particular significance that historians have attached to Boldon Book is reflected in the publication of these twelfth-century surveys. Of one of the others, from Burton Abbey, there have been two editions, but all the rest have been printed only once or not at all. In contrast, the Latin text of Boldon Book has been printed three times and there are three published translations.[3] There has been more historical research on

1 *FPD*, pp. xxvii–xxxii (the possible division of estates with the pre-monastic convent is discussed on pp. xiv–xxv); VCH *Durham*, ed. W. Page (3 vols.; 1905–28), II, 86.

2 To those listed in P.D.A. Harvey, 'Non-Agrarian Activities in Twelfth-Century English Estate Surveys', in *England in the Twelfth Century*, ed. D. Williams (Woodbridge, 1990), p. 101n, should be added the survey of Abingdon Abbey's properties in the time of Henry II: *Chronicon monasterii de Abingdon*, ed. J. Stevenson (2 vols.; RS 2, London, 1858), II, 296–334.

3 *Libri censualis, vocati Domesday-Book, additamenta*, [ed. H. Ellis] (Record Commission; 1816), pp. 565–87, text; *Boldon Buke*, ed. W. Greenwell (SS 25; 1852), text and translation;

Boldon Book than on any of the other surveys, but it probably remains the least understood and it certainly still presents a formidable range of unsolved problems. This is partly for want of comparable documents from this region; in some ways, being a craggier, less polished record, it offers more footholds for research than most of the rest.

One serious difficulty is that none of the four surviving texts of Boldon Book is contemporary with the survey itself. The earliest may have been copied 150 years later, and all four incorporate changes made at some time or another to bring parts of it up to date. Some of these additions and alterations can be plainly identified and others we can guess at, but there are probably others too that are impossible to detect. The late Professor H.S. Offler has placed all future workers on Boldon Book greatly in his debt by the article, to be posthumously published, in which he definitively sets out the relationship of the four copies to each other and to the survey as first written.[4] The discussion that follows accepts his arguments and simply builds a little on the foundations that he has laid.

In three of the manuscripts of Boldon Book the text is badly disarranged: its last sections have been moved to an earlier part of the survey and the second quarter has been moved to the end. One or other of these three manuscripts is the basis of all editions and translations except Greenwell's in 1852, though the latest editor, Mr David Austin in 1982, sets out the correct arrangement in a note. Offler convincingly suggests that these three copies all derive from one written on four small quires that were put together in the wrong order, perhaps after having been dropped. The fourth manuscript, printed by Greenwell, presents the text in a different order; that this was the original and correct arrangement is clear from its sequence of places.

Most of the correctly ordered text forms a more-or-less straightforward itinerary through the bishop's estates between the Tyne and the Tees, the historic County Durham. It begins at Durham itself, and passes first to the nearby property at Newton, then north to Plawsworth and Gateshead, returning to Chester-le-Street before moving decisively north-east to Washington and Boldon. It was from Boldon that the survey took its name – it was already known as Boldon Book in the late fourteenth century[5] – because nineteen later entries refer back to the customs and services there set out in full: 'the holdings, rents and works are as at Boldon'. From Boldon, the survey proceeds clockwise round the county – Wearmouth, Houghton-le-Spring, Easington, Sedgefield, Stockton, Darlington, Bishop Auckland, Stanhope, and Lanchester are among successive places entered – and the itinerary concludes with the bishop's properties on the Tyne above Newcastle: Ryton, Crawcrook, Winlaton, and Barlow along with (as Offler shows) four properties elsewhere that were administratively connected with Winlaton.

VCH *Durham*, I, 327–41, translation; *Boldon Book: Northumberland and Durham*, ed. D. Austin (Chichester, 1982), text and translation.

4 'Re-reading Boldon Book', to be published in a forthcoming collection of essays by H.S. Offler. I am grateful to the late Professor Offler for letting me read a draft of this article.

5 *Bishop Hatfield's Survey*, ed. W. Greenwell (SS 32; 1857), pp. 8, 11, 20, etc.

There follow entries for eight places held by the form of tenure called drengage; and the survey ends with an account of the bishop's properties in Northumberland that were centred on Bedlington and Norham.

Two hundred years after Boldon Book another survey of the bishop's estates was compiled, this time for Bishop Thomas Hatfield. The Hatfield Survey was probably drawn up between 1377 and 1380, but in its surviving form it incorporates revisions down to 1382, the year after Hatfield's death.[6] It is much fuller than Boldon Book – it is about five times as long – and is in four sections, one for each of the four wards into which the county palatine was now divided for administrative purposes: Darlington, Chester-le-Street, Easington, and Stockton. Boldon Book does not name places in at all the same order as the Hatfield Survey. All the same, if we assign the places named in Boldon Book to the wards of the Hatfield Survey, Boldon Book likewise divides neatly into blocks corresponding to these wards. Setting aside Durham City as a special case (the Hatfield Survey places it in Easington Ward), we find that every place entered in Boldon Book from Newton-by-Durham to Whitburn is in Chester Ward, every place from Wearmouth to Tursdale is in Easington Ward, every place from Sedgefield to Grindon is in Stockton Ward, every place from Ricknall to Frosterley is in Darlington Ward and every place from Lanchester to Winlaton and Barlow is again in Chester Ward. At first sight this might seem no more than one would expect: the wards, closely defined by the Hatfield Survey, formed compact quarterings of the palatinate and a systematic itinerary around the bishop's estates would naturally pass through each in turn. But in fact at the local level the itinerary represented by the succession of places in Boldon Book follows a quite intricate route, passing to and fro, doubling back on itself, and so on. It cannot be through mere chance that at no point does it move back and forth across the ward boundaries that existed two centuries later.

Why, however, does the survey divide Chester Ward in two, describing part of it at the start of the itinerary, part at the end? A likely explanation is that this itinerary was the route followed by the palatinate's steward or sheriff in making a tour of inspection. He would start from Durham itself and, though for much of the tour Boldon Book's itinerary would not be the quickest way to cover the ground, the route could well have been determined by historical precedent or administrative needs and we can certainly see that it might well be practical to divide the area that formed Chester Ward between the east parts visited at the beginning and the west parts visited at the end. Whether this itinerary was recorded by tradition or in writing, using it as the basis of Boldon Book would mean that no place was omitted, something that would otherwise be difficult to ensure if there was no existing survey either of the estate as a whole or of its component parts.

Whatever the origin of the route followed in Boldon Book, it is clearly related to the wards as recorded about 1382. Either the wards were created after the time of Boldon Book, but on the basis of the survey itself or of the itinerary that it follows, or else the wards were already in existence by 1184. That they did exist already and

6 Ibid., p. vii.

that they played a part in the bishop's administration of his estate is confirmed by the content of the entries in Boldon Book. A pattern of differences between the wards emerges if we take the wards of 1183–4 to be identical with those of about 1382. It emerges even more clearly, however, if a small adjustment is made to the ward boundaries in the north-east, placing Boldon, Newton-by-Boldon, Cleadon, and Whitburn not in Chester Ward as in the Hatfield Survey, but in Easington. In describing below the differences between the wards that appear in Boldon Book, all references to Chester and Easington Wards take this alteration into account. It is not unreasonable to assume that such a change was made in the course of the intervening two hundred years; it will have involved simply moving the northern limit of Easington Ward southward from the mouth of the Tyne to the mouth of the Wear (fig. 23).

It is the more reasonable to postulate this adjustment of the boundary in that some changes seem to have been made to the overall structure of the wards before the late fourteenth century. The register of Bishop Richard de Bury in 1344 lists four wards under the same names as the Hatfield Survey, but earlier in the fourteenth century there are references to Lanchester Ward as a forest area distinct from Chester Ward.[7] In 1293 there seem to have been neither four nor five wards but three: there was a coroner for each of the three wards in the liberty of Durham.[8] We shall see that in Boldon Book, though there are some features peculiar to a single ward, there are others that are common either to Easington and Stockton Wards on the one hand or to Darlington and Chester Wards on the other, perhaps pointing to division at some time into two wards only. It looks very much as if the wards, while fixed in broad outline, were subdivided or recombined from time to time in the course of the twelfth, thirteenth, and fourteenth centuries.

That the information in Boldon Book was collected ward by ward is suggested by some tiny verbal differences in the entries. These are not strikingly obvious; however it was assembled, the text of the survey must, as Offler points out, have undergone some degree of editing to iron out inconsistencies. However, he himself notes that one (perhaps the original) version of the text uses the phrase *e contra* in defining obligations of week-work at different times of year, but uses it only in Darlington Ward. Other differences can be found. In Easington and Stockton Wards entries of places where there are tenants of standard holdings always begin with the number of tenants: 'In Carlton there are 23 rent-payers who hold 46 bovates'. In Darlington and Chester Wards on the other hand they often begin with the number of holdings: 'In Blackwell there are 47 bovates which villeins hold'.[9] Again, in Chester Ward hamlets that return only a money rent mostly do

7 *Registrum palatinum Dunelmense*, ed. T.D. Hardy (4 vols.; RS 62, 1873–8), II, 1134, 1161, 1162, 1216; IV, 271, 273–7; *Richard d'Aungerville, of Bury: Fragments of his Register, and other Documents*, [ed. G.W. Kitchin] (SS 119; 1910), p. 131.

8 *Placita de quo warranto*, [ed. W. Illingworth] (Record Commission; 1818), p. 604.

9 *Boldon Buke*, pp. 15, 17.

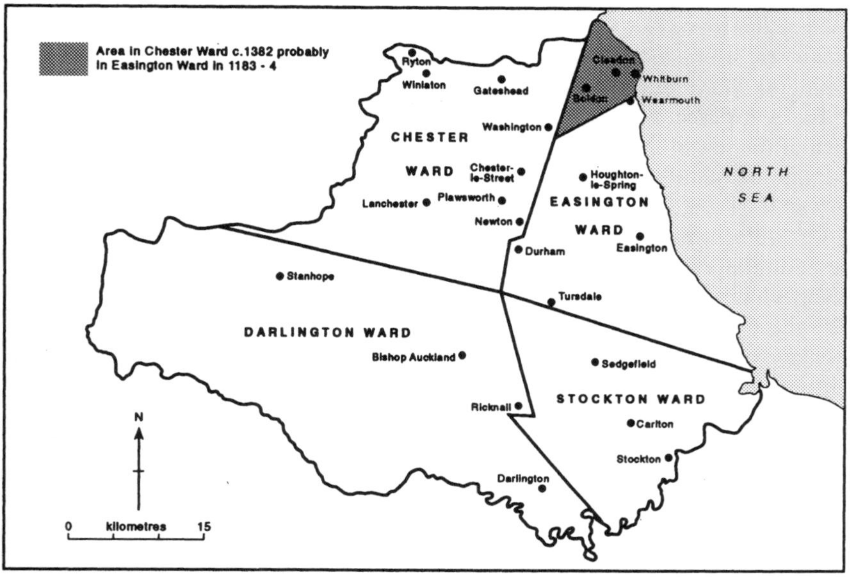

Fig. 23. Boldon Book and the wards.

not name the tenant: 'Medomsley renders 22s.' In the other wards the tenant is usually named: 'William holds Hardwick and renders 10s.'[10]

This last suggests that the information supplied for the survey differed from ward to ward. We may suspect that some apparent differences in custom from one area to another reflect inconsistencies in the collection of data rather than real differences in fact. Thus all the places where it is stated that tenants are required to work on the mill or mill-pond are in Darlington and Chester Wards;[11] there is no obvious reason for this, and it could well be that the same service was required in Easington and Stockton Wards too, but was either overlooked or thought too obvious to be worth mentioning.

Alternatively it may point to different practice. Some other differences between the wards certainly result from actual differences in management. This appears clearly in the entries of demesnes that have been farmed out. Thus, in Chester Ward, but nowhere else, all demesnes at farm are let out along with the entire vill, its services, rents, and mill.[12] Again, in Darlington Ward four of the five demesnes at farm have been let out by charter (*in cyrografo*).[13] This does not occur elsewhere, and it is unlikely that it was the general practice but recorded only in this one ward, as it would be such an obvious inconsistency to remove on editing. Leasing by

[10] Ibid., pp. 11, 32.
[11] E.g. ibid., pp. 26, 31.
[12] E.g. ibid., pp. 3 (Chester-le-Street), 34 (Whickham, Ryton).
[13] Ibid., pp. 18 (Darlington), 19 (Great Haughton), 20 (Ketton), 29 (Wolsingham with Rogerley).

written charter was anyway unusual in the twelfth century, and the survey's phraseology here seems to echo the wording of the charters themselves.[14] Then again, in Easington Ward four demesnes are farmed for 8 chalders of wheat, 8 of oats, and 4 of barley for each of some of the ploughs, plus (in three cases) 5 marks in cash for each of the rest of the ploughs. Elsewhere this pattern occurs only at Heighington, the one demesne at farm in Darlington Ward that is not said to be leased by charter.[15]

But differences between the wards appear in other aspects of management too. In Easington Ward there are seven entries of smiths who have service holdings, doing whatever smith's work was needed on the demesne in return for their land. Elsewhere smiths with service holdings are mentioned only twice, once each in Stockton and Darlington Wards.[16] By the 1180s this form of tenure was becoming a little old-fashioned; more often a smith would hold his land by normal rents and services and would be paid for whatever work he did.[17] Perhaps these service holdings had been more or less phased out in three of the wards but not in the fourth. The tenants of Boldon and the other manors with the same services all render each year two hens and ten eggs if they are villeins, twelve hens and sixty eggs if they are cotters.[18] Renders of hens and eggs are stipulated elsewhere too, but this ratio of one to five otherwise occurs only in Easington Ward, and it is only at some of the places in Darlington Ward that there is an obligation to render hens but no eggs.[19]

Some of these differences may well arise from older local custom rather than from managerial decisions in one or other ward. But it is a different matter when we come to the Boldon-type vills themselves – Boldon and the nineteen places where tenants held by the same customs and services 'as those of Boldon'. For the villeins, these services included heavy works, three days a week on the bishop's demesne lands. Other twelfth-century estate surveys offer no parallel to a long series of manors sharing exactly the same set of customs and services and, as Dr Brian Roberts points out,[20] such total uniformity can only have been imposed from above, though not necessarily on a single occasion.

In 1979 Dr William E. Kapelle argued that the Boldon-type vills, all in the east of County Durham, were places devastated by William the Conqueror's punitive expeditions of 1069–70: the harrying of the north. When he reconstructed them and reinstated the peasantry the bishop was able to dictate his own terms, and the

14 In the references to enclosed yards, barns, and other buildings: ibid., pp. 18–20.

15 Ibid., pp. 4 (Boldon), 5 (Cleadon with Whitburn), 6 (Ryhope with Burdon: 3 ploughs plus ½ ploughland), 9 (Shotton), 21 (Heighington).

16 Ibid., pp. 5 (Tunstall), 7 (Newbottle, Houghton), 8 (Easington with Thorpe, Shotton), 10 (South Sherburn, Quarrington), 11 (Sedgefield), 16 (Darlington).

17 Harvey, 'Non-Agrarian Activities', pp. 108, 111.

18 *Boldon Buke*, pp. 3–4.

19 E.g. ibid., pp. 6 (Newbottle), 22 (Middridge, Thickley).

20 In an important forthcoming article: B.K. Roberts, ' "And They Work as They of Boldon": Tenure and Settlement on the Cuthbertine Estates in County Durham in the Eleventh and Twelfth Centuries'. I am grateful to Dr Roberts for allowing me to read this before publication.

heavy labour-services of the Boldon-type vills were the result.[21] This is an interesting suggestion. However, it raises some difficulties. The bishop was not the only property-holder in the area and it seems unlikely that his vills would have been particularly singled out for destruction; we thus have to assume the total devastation of half the lowland area of the modern county – a possible but not very probable hypothesis. It is noteworthy that all twenty of the Boldon-type vills are in Easington and Stockton Wards – and are all the places in those two wards where there were tenants that Boldon Book calls villeins (*villani*). This does not explain why the Boldon-type services were imposed, but at least it suggests strongly that it was a simple decision of the estate – or ward – management, unrelated to outside events. It also suggests that they do not pre-date the creation of the wards between Tyne and Tees; but in any case their complete uniformity in Boldon Book implies that they did not originate in the distant past.

What conclusions can be drawn from this discussion? The most obvious are that the administrative wards of the palatinate were already in existence by 1183–4 and that it was through them that the bishop's estates were managed.[22] But we may suppose that it was not only the management of the bishop's own lands that was based on them, and that the official in charge of each ward served as the representative of the bishop – or of the palatinate's steward or its sheriff – for all kinds of business there. Indeed, it would be anachronistic to draw a distinction in the late twelfth century between the management of landed property and the exercise of judicial rights over a wider area. In the other counties of England the sheriff served as the king's representative for all kinds of business, including managing the royal demesnes. The differences between the wards revealed by Boldon Book suggest that the bishop's officials were allowed some degree of initiative and independence. We know so little of twelfth-century estate management in general that this is of interest beyond the local history of Durham.

Many problems remain. Why does the survey omit parts of the estates outside County Durham? What, indeed, was the purpose of Boldon Book? Was it meant primarily to bring together information on the whole estate between Tyne and Tees with a view to making practice more uniform? We can only guess. Why were the entries of the eight places with drengage tenure placed together at the end of the lands between Tyne and Tees?[23] This is one of the most puzzling features of Boldon Book. They must have been extracted from the survey and placed together at quite a late stage of compilation, for their sequence is that of the main itinerary. Offler suggests that changes to these archaic tenures may have been contemplated, indeed, perhaps as part of a new policy of uniformity; but if so, as the Hatfield Survey shows and as he himself points out, it was not implemented. In all this, however, Boldon Book does not stand alone: we do not know precisely why any surviving estate survey of the twelfth century was compiled.

21 W.E. Kapelle, *The Norman Conquest of the North: the Region and its Transformation, 1000–1135* (London, 1979), pp. 181–8.
22 For the possible connection with episcopal churches, see p. 147 n. 22.
23 *Boldon Buke*, pp. 35–8.

Abbreviations used:
FPD: *Feodarium Prioratus Dunelmensis*, ed. W. Greenwell (Surtees Society, vol.58; 1871)
VCH: Victoria History of the Counties of England

Note on Chapter IX

These three pieces are from *The Peasant Land Market in Medieval England*, ed. P.D.A. Harvey (Oxford, 1984). The first, on 'The Standard Holding' is from its Introduction, the others, on 'Forms of Tenure' and 'Land Market, Lord, and Tenant', form its Conclusion. The book defines its subject as 'the lands held by the small-scale landholders, whom we call peasants, and the way these lands moved, by mutual agreement, from one living landholder to another, which we call the peasant land market' (p. 1), and it contains accounts of what happened in particular places in four limited regions. Janet Williamson wrote on Gressenhall, Martham, and Sedgeford in Norfolk; Rosamond Faith on Brightwalton, Coleshill, Englefield, Letcombe Regis, Mackney, Sotwell Stonor, South Moreton, Speenhamland, and Woolstone, all in Berkshire, and Eastrop in Wiltshire; Andrew Jones on Arlesey, Blunham, Leighton Buzzard, Shillington, and Willington in Bedfordshire; and Tim Lomas on Billingham, Carlton, Cowpen Bewley, Hardwick, Hartburn, Newton Bewley, Norton, Seaton Carew, Stillington, Stockton, and Wolviston in south-east County Durham. The pieces printed here make many references to these places – often as 'we shall see', 'we have seen' – with relevant page numbers introduced as 'below', 'above'; these references are too numerous to be altered here to a different form.

On p. 8, line 18, an author's error has been corrected: the eruings at Martham were of 12 acres, not the 24 acres originally printed here.

IX

Aspects of the Peasant Land Market in England, Thirteenth to Fifteenth Centuries

THE STANDARD HOLDING

In most manorial records of the thirteenth and early fourteenth centuries, and in most writings on the medieval English countryside, we find the standard peasant holding: on a manor there would be one or more groups of holdings whose tenants owed customs and services uniform for everyone in the group and which thus were presumably more or less alike in size or value. Each holding would usually consist of a dwelling, areas of arable and meadow, pasture and other rights, but its exact composition and its form—whether the arable was a single block for instance, or dispersed among common fields—varied from one place to another. The names of these holdings varied too, but on a regional basis. In the present four studies we find them called tenements in Norfolk, virgates (or yardlands) and half-virgates in Berkshire and Bedfordshire, bovates (or oxgangs) in Durham; if one of the studies had covered Kent we should have been dealing there with yokes. It was very unusual for all the tenant lands of a manor to be held in standard holdings of this sort; generally they would comprise the larger unfree holdings—those of the villeins rather than the cottagers—but many kinds of variation occurred.

In these studies we are particularly concerned with these standard holdings in their decay: their fragmentation and often their complete disappearance. The chronology of this varied from region to region, even from place to place. On our three

Norfolk manors the standard tenements had all but vanished from sight by the end of the thirteenth century, whereas throughout south-east Durham the bovate was still a basis of tenure two hundred years later. In fifteenth-century Bedfordshire virgates and half-virgates disappeared from Leighton Buzzard but remained the normal holdings at Shillington. How, why, and when these standard holdings broke up in each area is discussed below. But behind these discussions lies the assumption—often demonstrably true—that at an earlier date, let us say at the end of the twelfth century, it would be normal for a high proportion of a manor's tenant lands to be held in standard holdings with uniform obligations and apparently of uniform size. It is here that the problem of these standard holdings arises. Let us take two specific examples. We shall see how at Martham (Norfolk) there were, about 1220, some twenty holdings called, unusually, eruings, nearly all held as single units or in halves or quarters. These eruings were very precisely defined: each consisted of exactly 12 acres (measured by a perch of 18½ feet) made up of as many as thirty or more separate pieces which might be as small as ½ rood. By 1292, though manorial administration retained a recollection of these eruings, they were no longer effective tenurial units: they had been split into as many as ten or more portions, each held by a different tenant.[1] Our other example comes from Combe, on the borders of Berkshire and Hampshire, a manor of Bec Abbey in Normandy. A custumal of about 1230 lists thirteen tenants who held a virgate of land each for more or less identical rents and services; later versions of the same custumal, made during the next twenty years, show two or three acres of assart, newly cleared land, added to a number of these hitherto identical holdings in return for extra rents.[2] Both these cases, Martham and Combe, present the same difficulty. Either these standard holdings, these eruings and virgates, were quite recent creations in the early thirteenth century. Or a long tradition of unchanging stability was in the thirteenth century suddenly broken, at Martham by the rapid fragmentation of age-old

[1] Below, p.65; a good example of fragmentation is the 12-acre holding that had been Thomas Knight's, its 33 separate parcels being held in 1292 by 11 tenants (BL Stowe MS 936, fos.39v-40v.

[2] *Bec Documents*, pp. 41-5.

holdings, at Combe by the extension of the cultivated area for the first time in centuries. Both explanations invite disbelief.

And indeed, though we must always reckon with possible local peculiarities, neither explanation fits what we know about standard holdings of this sort before the thirteenth century. Certainly we cannot doubt their antiquity; we find ample evidence for them in the late, even the mid-, eleventh century. Several of their names correspond to assessment units of the Anglo-Saxon taxation system. The bovate and the virgate were regular subdivisions of, respectively, the carucate (used in eastern and northern England) and the hide (used in most other parts of the country). But even before this taxation system fell into disuse in the twelfth century[3] the virgates of the standard holdings were no longer the same as the virgates of tax assessment—if, indeed, they had ever been. It is only for Middlesex that Domesday Book regularly defines the holdings of individual manorial tenants; there it was common in 1086 for groups of villeins on a particular manor to have holdings called virgates or half-virgates, but on only about half the manors with full details does the sum of these holdings, when added to the assessment of the demesne and other lands, equal the assessment for the entire manor.[4] Domesday Book's smallest tenant holdings in Middlesex—those of the bordars and cottars—were defined not as fractions of a virgate but in acres; but these again were measures not of area but of assessment, of liability to tax, for they were fiscal acres, the smallest unit of the taxation system, of which there were thirty to the virgate, 120 to the hide.[5] Just as the villeins on any one manor often had holdings all with the same description so too did the bordars and cottars: five acres was a standard holding on many of these Middlesex manors, and it is interesting that the *Rectitudines Singularum Personarum*, a mid-eleventh-century text, remarks

[3] The last Danegeld was in 1162, but the carucage of 1194 seems to have been based on the same assessment.

[4] Maitland, *Domesday*, pp. 477-8; Lennard, *Rural Eng.*, p. 341; T.G. Pinder in *VCH Middx.*, i. 86-8, 130-2. A similar problem on the Ramsey Abbey estates is discussed by Raftis, *Ramsey*, pp. 69-71.

[5] The classic account of the three medieval meanings of the word acre—the customary (or field) acre, the measured acre, the fiscal acre—is in Maitland, *Domesday*, pp. 373-93; a summary is in *Man. Records of Cuxham*, pp. 77-8. The use of the three sorts of acre is discussed, with many examples, by A. Jones, 'Land Measurement in England, 1150-1350', *AgHR*, xxvii (1979), 10-18.

that the cottar (*kotsetla*) often holds five acres.[6] It is presumably these same fiscal acres that lie behind the standard holdings, the tenements (*tenementa*), of thirteenth-century East Anglia (at Martham, we shall see, it was not the fiscal acres of tax assessment but acres measured on the ground that were in question). There the holdings on any one manor tended to fall into groups each with the same number of acres: in Suffolk ten-acre units were frequent, in Norfolk units of twelve acres or multiples of twelve acres.[7] East Anglia had been assessed for taxation in carucates and bovates, not hides and virgates, but the fiscal acre was still the lowest unit of the system, one-120th of the carucate as well as of the hide. In Kent there was a different terminology, but again we find that units of land, much split up by the thirteenth century, had apparently been, a century earlier, standard holdings in the hands of individuals;[8] these units were called *iuga*, yokes, a name again harking back to eleventh-century tax assessment, which in Kent was returned in sulungs of four yokes each.

But if these holdings took their names from a tax-assessment system that flickered out in the twelfth century it does not follow that they then became fixed and stabilized. On the contrary, there is much evidence of flexibility and change in their structure in the course of the twelfth century. New ones might be created. At Brancaster (Norfolk) the number of twelve-acre villein units was increased in the mid-twelfth century by letting out lands from the manorial demesne.[9] At Barnetby-le-Wold (Lincolnshire) four new bovates were formed about 1160 by combining some demesne with other adjacent lands.[10] In these and other cases surviving records show us the process as it actually occurred; in many more we can infer the creation of new standard holdings from the increasing number of tenants and holdings in successive surveys. Nor were the existing holdings immutably fixed. At Theddingworth (Northamptonshire) and Keddington (Lincolnshire) we find one acre being subtracted

[6] Lennard, *Rural Eng.*, pp. 340-4; *Die Gesetze der Angelsachsen*, ed. F. Liebermann (3 vols., Halle, 1903-16), i. 446.

[7] Dodwell in *EcHR*, 2nd ser., xx (1967), 55-8.

[8] A.R.H. Baker in Baker & Butlin, pp. 398, 407-8.

[9] Dodwell in *EcHR*, 2nd ser., xx (1967), 58.

[10] *Danelaw Ch.*, pp. 180-2 (cf. Baker & Butlin, p. 267).

from each virgate or bovate to endow local churches.[11] Conversely where new land was brought under the plough each standard holding might be enlarged by taking a share of it; this is specifically envisaged, for instance, in a charter granting a bovate at Spaldington (Yorkshire).[12] In these cases the reduction or enlargement would apply equally to all the standard holdings in a particular manor; but this need not be so. At Kirton (Nottinghamshire) the meadow belonging to a single bovate was taken from it and granted away separately.[13] This is a useful reminder that none of the names given to these holdings—not even those defined as so many fiscal acres—carried any implication of specific size. Those of ancient origin may all at one time have owed the same amount of tax. Generally all those in a single group on any one manor owed the same rents and services to the lord. But it need not follow that even on the same manor they were more than very roughly alike in area or resources. At Gillingham (Kent) the yokes demonstrably varied in size, apparently reflecting the varying quality of soil in different parts of the manor.[14] Here the yokes can be identified on the ground, as Kentish holdings were generally compact blocks of land. But elsewhere, bovates and virgates might well be made up of scattered strips and pieces of arable and meadow, together with shared rights over wide areas of pasture. The form and composition of the standard holding varied enormously, depending on the region or the particular vill where it lay.

Given this variety of form, given their different origins, given the changes that we find occurring in these standard holdings in the course of the twelfth century, we may well wonder whether by the thirteenth century we can properly regard them as a single type of unit, a single phenomenon. Can we see the decay of twelve-acre holdings in thirteenth-century Norfolk and of virgates and half-virgates in fifteenth-century Bedfordshire as different aspects of a single process? We have seen that standard holdings were far from being new creations at the end of the twelfth century, and that they had by then already

[11] *Danelaw Ch.*, pp. lxx-lxxi, 342-3 (cf. Baker & Butlin, pp. 271-2).

[12] Homans, *Eng. Villagers*, p. 84.

[13] *Danelaw Ch.*, pp. xl, 274.

[14] Du Boulay, *Canterbury*, pp. 121-2; A.R.H. Baker, 'Open Fields and Partible Inheritance on a Kent Manor', *EcHR*, 2nd ser., xvii (1964-5), 5.

undergone much local change. We might reasonably ask whether the standard holdings of late-medieval England had any discoverable shared baseline, any common starting-point that would make it meaningful to compare or contrast their varying fortunes in different places. The answer is that they had, but that it is not to be sought in the ultimate origin of these holdings. It is to be sought rather in a process which occurred in the twelfth century and which might be called a process of territorializing the holdings of manorial tenants.

Put simply, we might say that in 1100 the lord of a manor was the lord of men who held lands of him; in 1200 he was the lord of lands that were occupied by tenants. The change is slight but significant. In 1100 the tenant's holding could be viewed simply as a standard share in the vill's resources; by 1200 it was far more likely to be viewed as precisely defined in its area of land and other rights. At most, of course, it is no more than a change of emphasis. It would be absurd to suggest that the eleventh-century manorial lord cared nothing for precisely defined lands and boundaries, or that his thirteenth-century successor was uninterested in his seignorial rights over his tenants. All the same a clear shift can be seen. Its precise manifestations varied from region to region and from place to place. As we shall see, tenurial forms differed a good deal from one manor to another in the late Middle Ages and there is no reason to assume any greater uniformity in earlier periods—indeed, Domesday Book points quite the other way, despite its bureaucratic tendency to iron out variations. Just as local conditions and forms of tenure varied, so the effects of this territorializing process were bound to vary too, but everywhere the trend was the same. It was a trend towards precise definition on the ground of the lands and territorial rights of the individual tenant, and it is this process of definition that forms the common starting-point, the basis for the comparisons and contrasts we draw in later developments.

This same trend may be compared with the changes in relationship between lord and free tenant in the twelfth century that S.F.C. Milsom has described.[15] It lies behind the view of William Rees, in 1924, on developments in south Wales and

[15] Milsom, *Legal Framework*, pp. 34-5, 65-6, 90.

the March[16] as well as the views of two recent writers on the development of English field systems. R.A. Dodgshon, taking up a phrase of Vinogradoff, has argued that subdivided fields originated in a tenurial structure based on shareholding by the vill's tenants rather than on specifically delimited properties.[17] And H.S.A. Fox, writing of twelfth- or early-thirteenth-century charters that describe arable holdings simply as lying in the vill's open fields (sometimes these fields are named, but without further topographical detail), asks 'might the idea behind this terminology have been to reinforce the grantee's claim to a stake in a *system* of husbandry involving the two or three great fields?'[18] It would thus not be necessary to define the holding more precisely—and indeed it might not have been possible to do so. There are indications that in some places the trend towards precise definition of holdings took the form of changing from a system of regular re-allotment of strips in the arable fields to one where all the strips were attached permanently to particular holdings. We get, for example, a possible hint of this (and of other agrarian changes as well) in an extent of Woodeaton (Oxfordshire) in 1366: the manor's twelfth-century owner had given the nuns of Littlemore one acre in an unmanured furlong, a grant since replaced by the gift of the wheat from two selions of the manorial demesne.[19]

Better documented, and probably much more widespread, was the growing tendency to define assarts, newly cultivated lands, in precisely measured acres, roods, and perches,[20] and to regard them as additions to the standard holding, not merely as realizations of potential assets that had always belonged to it. The holding was now defined and fixed, and any additions to it must be paid for in the form of extra rent. The distinction was made quite explicit in certain places in Cheshire, where

16 Rees, *S. Wales and March*, pp. 150-1.

17 R.A. Dodgshon, 'The Landholding Foundations of the Open-field System', *P&P*, lxvii (1975), 14-26, and Dodgshon, *Origin*, pp. 34-41, 52-3.

18 Fox in Rowley, *Origins*, p. 78 (the reference on p. 107 to *Danelaw Ch.* should be to pp. xxxi-xxxii, 232-3).

19 *Eynsham Cart.*, ii. 17-18 (cf. p. lxxii). The same 12th-century grantor's gift of an acre of meadow, specifically 'per assignacionem custodis manerii' but in 1366 a permanently identifiable strip (ibid., p. 17), points to the same process in the manor's meadows.

20 This was just one aspect of increasingly precise measurement of lands in general (Jones in *AgHR*, xxvii (1979), 12-13, 15-16).

lands were referred to as either *oxgang lands*, that is the ancient holdings, or *rodelands*, the assarted lands measured in roods or by the rod.[21] In south Lincolnshire the change is very clear in the first half of the thirteenth century: until then newly reclaimed lands were held to belong to existing bovate holdings as of right, but they then start to be seen as additions to the holdings, carefully measured and defined, usually in terms of acres but on one occasion as seven-twelfths of a bovate.[22] In Durham the bovates of the bishop's local tenants scarcely changed between the Boldon Book survey of 1183 and Bishop Hatfield's survey of about 1382, but a new class of holding, exchequer lands, had appeared; these were small areas defined in acres and held by money rents and seem to have been lands newly brought under cultivation.[23] In East Anglia, where the standard holdings were themselves defined in units called acres, the distinction is less obvious, but some thirteenth-century surveys show tenants holding extra lands with very irregular acreages, suggesting that here we have assarts defined in acres of actual measurement, not the fiscal acres used for the basic holdings.[24] This does not mean that we cannot still find, as in parts in Yorkshire in the thirteenth and fourteenth centuries, new oxgangs being created from waste land, and in the fifteenth century assarts might still be divided in strict proportion between the tenants of standard holdings.[25] But at least it became a very general corollary to the stricter territorial definition of standard holdings that any additional lands should be no less strictly measured and paid for. Thus where rents for small assarts suddenly start to appear, as in the early-thirteenth-century surveys from Combe, it need not follow that cultivation was being expanded for the first time since the standard holdings had come into existence, nor that an age-old total equality of tenure was now being disrupted. The provision of the Statute of Merton (1236)

[21] G. Elliott in Baker & Butlin, p. 45.

[22] H.E. Hallam, *Settlement and Society: a Study of the Early Agrarian History of South Lincolnshire* (Cambridge, 1965), pp. 9-10, 11, 13, 15, 20, 27-8, 44, 51, 55, 61-2, 68-9, 99, 104, 110, 158-61.

[23] B. Roberts, 'Townfield Origins: the Case of Cockfield, County Durham', in Rowley, *Origins*, pp. 150, 153.

[24] Dodwell in *EcHR*, 2nd ser., xx (1967), 58-9.

[25] J.A. Sheppard in Baker & Butlin, pp. 174-5; Homans, *Eng. Villagers*, p. 425 n4; Raftis, *Tenure*, p. 29.

that asserts the manorial lord's rights over waste is almost an explicit statement of the same change as affecting free tenants' claims over common lands.

We find too that landlords were starting to apply exact measurement to the standard holdings themselves. F.M. Stenton drew attention to a grant of a bovate at Dunholme (Lincolnshire) which lists and measures each of its individual strips of arable; he dates this probably to the reign of Henry II, certainly not later than 1200, and comments that 'it would be difficult to find the record of any earlier measurement of the component parts of an open field bovate'.[26] But from about this time some landlords began to record at least the approximate acreage of the virgate or bovate on each of their manors (whether in measured acres or in field acres is seldom clear). This is information that we are given in a list of Ramsey Abbey's manors; here the virgate varied from 15½ to 44 acres, and J.A. Raftis has shown that in manors on the clay uplands it was larger than on the gravels of the valleys.[27] A register of about 1240 shows that the number of acres in a virgate was a question regularly asked when manorial surveys were drawn up on the estate of Worcester Cathedral Priory; the answers would be given in field acres or measured acres, whereas the acreages of the 1182 surveys for the bishop of Worcester's estate are simply the figures of fiscal assessment.[28]

We have seen enough of the earlier history of standard holdings to realize that on a single manor they need not all be the same size, even if measured in the somewhat loose units of field acres; and indeed we find that the 1244 survey of Ramsey Abbey's manor of Cranfield (Bedfordshire) says that some virgates there were 48 acres, some were smaller.[29] But we can see how easy it was (and still is), once the way of thinking in strictly territorial terms was adopted, to assume that, because the holdings all had the same description—a bovate, half-virgate, and

[26] *Danelaw Ch.*, p. xlvi n; Dodgshon, *Origin*, p. 40, slightly mistakes Stenton's dating of this charter and the significance he attaches to it.

[27] *Cart.Mon.Ram.*, iii. 208-15, discussed by Maitland, *Domesday*, pp. 393-5, and by Raftis, *Ramsey*, p. 68.

[28] Hilton, *Medieval Society*, pp. 115, 122; Dyer, *Lords and Peasants*, pp. 87-9. Another list of acres per virgate for various manors on an estate is supplied by Harvey, *Westminster*, pp. 208, 434.

[29] *Cart.Mon.Ram.*, i. 438, discussed, with other examples, by P. Vinogradoff, *Villainage in England* (London, 1892), p. 240.

so on—and all owed the same services to the lord of the manor, they must therefore be alike in every respect and in particular must comprise exactly the same area of land. This may well lie behind the planning of villages investigated by B.K. Roberts in Durham and June A. Sheppard in the East Riding of Yorkshire, where the length of each house-plot's street frontage seems directly related to the definition of the associated holding in terms of the bovate. It is not certain when these villages were laid out but it may well have been in the twelfth or thirteenth century in Yorkshire, though Roberts argues for a late-eleventh- or early-twelfth-century date in Durham.[30] Also in the East Riding, in Holderness, Mary Harvey has found fields set out with a regularity that seems to stem from translating standard holdings into egalitarian divisions on the ground; again, a date in the twelfth or early thirteenth century is possible.[31]

In this context let us return to the Martham survey and its group of holdings each consisting of small pieces that add up to exactly 12 acres. Historians of medieval English fields and measures have often been confused by the bewildering use of the word acre with three different meanings and have failed to distinguish clearly between fiscal acres, field acres, and measured acres. Medieval estate administrators were just as likely to fall into the same trap as we see from notes and tables which make a single system of units of measurement and of ancient tax assessment.[32] What more natural than that some official at the beginning of the thirteenth century, confronted with what were called twelve-acre holdings, should assume that they were, or had once been, or ought to be, 12 acres each? The fiscal acre could well have been entirely forgotten. In these circumstances it would seem only proper, in any reorganization of the vill's fields, to assign to each of these holdings exactly 12 acres of arable; this would do no more than restore their pristine integrity. It is very likely that some such line of argument underlies the 12-acre holdings of the Martham survey and that

[30] B.K. Roberts, 'Village Plans in County Durham: a Preliminary Statement', *Medieval Archaeology*, xvi (1972), 33-56; J.A. Sheppard, 'Metrological Analysis of Regular Village Plans in Yorkshire', *AgHR*, xxii (1974), 118-35; J.A. Sheppard in Baker & Butlin, pp. 183-6.

[31] Mary Harvey, 'The Origin of Planned Field Systems in Holderness', in Rowley, *Origins*, pp. 184-201.

[32] e.g. *Walter of Henley and Other Treatises*, ed. D. Oschinsky (Oxford, 1971), p. 475.

they date from some more or less thoroughgoing reallocation of the vill's arable that occurred not long before 1220, the time of the 'former' tenants referred to in the survey of 1292.[33] It is most improbable that these holdings of exactly 12 acres were already ancient in the early thirteenth century. Although they must have been based on measurement and calculation this measurement was not very precise; the areas of the pieces of land making up each holding seem to have been entered only to the nearest ½ rood (20 perches), suggesting that their basis was, at best, a rather roughly measured acre.[34] This meant the process was unrepeatable: remeasurement would not produce the same result, the same exact 12 acres in every holding. Once they had been laid out, the detailed composition of each holding and the definition of each of its component parts must have been preserved either in the memory of each tenant, in which case it is hardly possible that there would be no divergence from the exact 12-acre totals, or in the form of a written survey—but a survey of such detail and complexity that it would be unique among estate records before the thirteenth century. The reallocation of lands in exactly equal units and the construction of detailed surveys based on careful description or measurement can be seen as a part of the general move towards the precise definition of tenants' holdings; it is in this light that we should see the 12-acre holdings of thirteenth-century Martham and the early-thirteenth-century survey in which their composition was probably first recorded.

In the studies that follow we get further glimpses of this move towards precise territorial definition of holdings. In thirteenth-century Gressenhall (Norfolk) it may well have underlain the new practice of dividing holdings among heirs; if this marked a change from joint tenure by a group of heirs then the connection is clear, but if (as seems likely) it marked a change in local

33 Although only six of the 'former' tenants held an entire eruing the subsequent fission of the eruings proceeded so fast that this need not stop us from seeing them as still fairly new in 1220.

34 Interestingly, though the ½-rood is the smallest unit used in measuring the 12-acre holdings, the survey elsewhere defines parcels of land much more precisely: 3 perches, 1 rood 21 perches, 13 perches 6 feet, etc. This reflects the exact measurement of arable that would have become normal by 1292 (Jones in *AgHR*, xxvii (1979), 12-13, 15-16); applied to the 12-acre holdings it would almost certainly produce the wrong totals, so the older less precise computation was retained for them.

custom from impartible to partible inheritance the connection may still be there, for the former sole heir may effectively have supported his brothers, whose claims were simply made specific, were defined territorially, by the change in custom.[35] In thirteenth-century Kent the division of the standard holdings, the yokes, may well have come about through changing interpretation of the custom of partible inheritance, replacing joint tenure by individual holdings.[36] Pursuing this thought a little further, we may wonder whether the new emphasis on territorial definition may not have brought about in many places the first definition of inheritance custom, the change away from a vague claim of the family corporately to inherit and towards an explicit, specified claim of one or more of its members—but this is a point we shall return to later.

With more confidence we can see a connection between the new approach to manorial tenants' lands and the emergence of a distinction that brought many problems to the lawyers of the thirteenth century and later, the distinction between the status of the tenant and the status of the lands he held.[37] While the lord of the manor was lord primarily of the tenant and only secondarily of his holding there was little scope for differentiation between the status of the land and the status of the tenant. Whatever was held by a villein was self-evidently villein land. As P.R. Hyams has put it, in the twelfth century 'lordship over men and over land were still much the same thing'.[38] But once the lord started to look more to the land than to the tenant, started to define precisely the holding and the rights on the ground, it became easier to see the land apart from the tenant, to see the land as property that might pass to new and desirable tenants who were prepared to take over all its traditional rents and obligations but who were not at all prepared to accept the disabilities (themselves ever more closely defined) that were attendant on the personal status of villeinage.

And behind it all, behind the whole process of closer definition of tenants' lands, lies the growth of population and the looming land shortage which may have been felt in some areas

35 Below, Ch. II, section 2.

36 Du Boulay, *Canterbury*, pp. 146-8.

37 Hyams, *King, Lords, and Peasants*, pp. 107-19.

38 Ibid., p. 228.

by the late twelfth century and which a hundred years later was widespread and critical. It is against the same background of rising population and increasing pressure on land that we should see the development of the local land market in the thirteenth century.

FORMS OF TENURE

An important book by C.M. Gray on the origin of the legal protection offered to copyholders was published in 1963. In it he looked critically at the conclusions of A. Savine and I.S. Leadam, who had argued that copyholders could successfully bring cases concerning their holdings to the Court of Chancery before (probably well before) 1467 and to the Court of Common Pleas by 1482. Gray showed that the evidence for these dates can be differently interpreted and argued that both were too early. Instead it was only in Henry VII's reign that these cases could be brought to Chancery with any hope of success and not until the 1550s that the common-law courts began to take cognizance of cases over copyhold.[1]

This revised chronology at once leads us to ask why the royal courts were so slow in bringing these copyhold cases within their scope.[2] The last three chapters help to provide an answer. We have seen how John Broughton, esquire, the manorial lord of Toddington (Bedfordshire) held some copyhold land at Leighton Buzzard in 1466; R.H. Hilton has shown that copyhold tenants at Cheltenham included John Throckmorton, esquire, in 1430 and William Grevile, lawyer and judge, later in the fifteenth century.[3] But on the whole it was clearly still unusual for lands to be held in copyhold by members of the gentry, or even by the moderately well-to-do; the speculative market in copyhold land was in its infancy, though we have seen it operating in or beside small towns at Speenhamland and

[1] Gray. *Copyhold*, pp. 16-34, 54-68. The contrary arguments of Kerridge, *Agr. Problems*, pp. 65-93, have failed to win acceptance among legal historians (*The Reports of Sir John Spelman*, ed. J.H. Baker (Selden Soc., vols. xciii, xciv; 1977-8), ii. 184-7).

[2] Raftis, *Tenure*, p. 204 n, interestingly puts the opposite question: given that the copyhold cases brought before the royal courts in the 16th century were often the same sorts of grievance as came before manorial courts, why had the manorial court become inadequate to uphold customary law?

[3] Above, p. 246 (cf. the cases mentioned on pp. 189-90, 192, where the late dates may well be significant); Hilton, *Eng. Peasantry*, p. 69.

Leighton Buzzard.[4] Nearly all copyholders were local tenants whose estates were not only small but transitory; again, it was of a later generation of more firmly established families that it could be said, in 1584, that the copyholders of Hertfordshire were 'the most substantialest men of the country'.[5] Hardly any fifteenth-century copyholders could muster the influence and resources that would be needed to persuade the royal courts to break with precedent and hear their grievances—and therewith diminish the rights and powers of their manorial lords. In any case the growing formality and regular routine of late-medieval manorial courts will have given customary tenants increasing security without recourse to the further protection of the royal courts.

But if Gray's work raised this new problem it at once solved a much larger one posed by the chronology of Savine and Leadam: how was it, if the royal courts and lawyers were already taking an interest in them, that customary tenures had escaped their tendency to define and to standardize and were still at the end of the fifteenth century in such an extraordinary muddle? For muddle is what it was. We have seen this clearly in Berkshire and Bedfordshire and it appears too in studies of other areas in the late Middle Ages.[6] It is not just that there were great differences between one place and another in the forms and conditions of tenures; there was not even any particular consistency in terminology. But this is no more than we would expect, given that the tenures developed locally and that the significance of the words describing them would be called in question only in the manorial court. It follows that we should be careful not to suppose that particular words—tenure by copy, at will, in bondage, and so on—had any sort of technical meaning or connotation outside the manor where they were used; there was no need for them to, for as long as they clearly established that these tenures were customary the local interpretation was all that mattered. Nor, of course, should we fall into the related error of carrying back into late-fourteenth- and fifteenth-century customary tenures the rules that the common law later came to devise for copyhold. They simply did not

[4] Above, pp. 145, 234.

[5] J.E. Neale, *The Elizabeth House of Commons* (London, 1949), p. 96.

[6] e.g. Harvey, *Westminster*, pp. 268-93; Hilton, *Leics. Estates*, pp. 94-105, 121-30.

apply. To the seventeenth-century lawyer copyhold could be extinguished but could never be created or reactivated; there could be no such thing as changing a property into copyhold from some other form of tenure. Yet we find for instance that at Harmondsworth (Middlesex) in 1390 customary land had been converted to freehold, then brought back into customary tenure again.[7]

With all this local diversity it is hardly surprising that it is difficult to see any pattern in the way local tenures developed over this period. Certainly there was regional variation—we have seen this already. By 1300 the rapidly changing structure of local holdings in Norfolk was producing a diversity of tenures that was far from developing generally in the more conservative fourteenth-century Berkshire and Bedfordshire, while in Durham we are in a different world again, an area where there were scarcely more than minimal changes in tenures between 1350 and 1500. We may reasonably suppose that what happened in Norfolk happened elsewhere in East Anglia too; that what we find in Berkshire and Bedfordshire was common to much of midland and central southern England; and that the manors of three estates in south-east Durham epitomize far wider areas of north-east England. But the extent and significance of such regional variations remain to be discovered.

It would be simpler to identify these variations if we could easily classify the various types of local tenure that we find. But this is harder than might appear. One straightforward distinction that could be made is between contractual and customary tenure—a very basic distinction corresponding to the thirteenth century's separation of free tenure from tenure in villeinage. This distinction was familiar to people at the time; it underlies the late-fourteenth century order to the tenants of Englefield that they should show whether they held 'for life, for years, by copy or by charter' as well as the simple division at Kibworth Harcourt (Leicestershire) between tenants who held in bondage and those who held at will.[8] It is not surprising that

[7] G. Jacob, *A New Law-dictionary* (London, 1729), under Copyhold (cf. E. Coke, *The Compleate Copy-holder* (London, 1641), pp. 52-3); Beckerman, 'Customary Law', pp. 134-5, 367-9 (from PRO SC 2/191/15, m. 7).

[8] Above, p. 151; Cicely Howell in Goody, Thirsk, & Thompson, pp. 132-3.

historians have adopted this classification.[9] In theory it was a very clear-cut one. In the last resort the courts—the royal courts of common law—would have been able to decide to which category any particular holding belonged. But in practice, in everyday local life, the distinction may well have been hard to draw. We have seen how on some holdings on the 1292 Martham survey the distinction between socage and villein tenure was forgotten; how at South Moreton in the late fourteenth century the court rolls fail to show whether particular tenures were customary, with copies, or contractual, with charters; how at Shillington in the late fourteenth and fifteenth centuries leases of parcels of demesne land soon took on aspects of customary tenure, apparently without formal or conscious change; how in fifteenth-century Eastrop it could be recorded that the manorial court, perhaps even the tenant, 'does not know how he holds this land'.[10] We might expect this sort of confusion to arise. By the fifteenth century it seems to have made no difference to the peasant's social status whether his lands were held by contractual or customary tenure[11]—indeed, many held properties both ways. Nor were the free tenures necessarily more favourable economically; on the Durham Cathedral Priory estates the renders due from certain free holdings were so like those of similar customary tenements that they had probably been converted from unfree tenures without alteration of the obligations.[12]

The complexities of local tenures did not start with the Black Death of 1348-9. Certainly we may say that a fairly simple structure of free tenures, villein and cottage holdings was typical of the thirteenth- and early-fourteenth-century manor. But we do not have to look far to find more elaborate arrangements, even outside East Anglia or other areas where rapid division of holdings had complicated the picture. Indeed, the twelfth- and thirteenth-century lawyers and common-law courts, in defining villein unfreedom and its attendant disabilities, may certainly have helped to standardize and simplify local forms of

[9] Thus Hilton, *Decline*, pp. 44-51, and Harvey, *Westminster*, pp. 244-93. On the Westminster Abbey estate this logical and straightforward classification clearly works well—but this was a carefully administered estate and if it works less well elsewhere this is not necessarily for want of Westminster's superb archives.

[10] Above, pp. 71, 137-8, 206 (cf. Davenport, *Norf. Manor*, p. 57), 149.

[11] Thus above, pp. 197, 234.

[12] Above, p. 276.

tenure, but they also introduced further complications; their distinction between villeinage by birth and villeinage by tenure pointed the way to a possible fourfold division between free or unfree land held by free or unfree tenants.[13] Thus on the Peterborough Abbey estates in the late thirteenth century any free lands acquired by a villein tenant would be seized by the abbey and granted back to him to hold at the lord's will; this was quite distinct from the customary tenure by which he held his basic holding in the manor.[14] Again, a survey of Sandal (Yorkshire) in 1309 divides tenants into free tenants, villeins (*nativi*), and customary tenants (*custumarii*); only the villeins paid the servile dues of merchet and leyrwite.[15] There was a similar division at Cadland (Hampshire).[16] In Cornwall and parts of Devon there had developed before the fourteenth century a form of leasing known as conventionary tenure, tenure *per conventionem*, which came to be available to both free and unfree tenants.[17]

But if the Black Death did not begin the process of diversifying local tenures it certainly speeded it up. Some changes were a direct and immediate result of the loss of tenants in the plague. In some places manorial lords tried to attract new tenants by offering lighter rents and other improved conditions of tenure; a neat example is on Eynsham Abbey's manor of Woodeaton (Oxfordshire) where, according to a survey of 1366, 'In the time of the mortality of men, the plague which was in the year 1349, there scarcely remained two tenants in the said manor, and they would have left if Brother Nicholas de Upton, then abbot, had not come to a new agreement with them and other surviving tenants'.[18] Here it was simply a question of altering the obligations of villein tenure. Elsewhere completely new tenures were introduced specifically as a stopgap measure. We have seen how villein tenements at Woolstone in the 1350s were divided between lessees who were to hold 'until a tenant

[13] Hyams, *King, Lords, and Peasants*, pp. 107-19.

[14] King, *Peterborough*, pp. 99-102; cf. Davenport, *Norf. Manor*, p. 70.

[15] PRO DL 43/11/17 (I am grateful to Mr D. Bruynseels and Ms M. Stinson for this information).

[16] D.G. Watts, 'Peasant Discontent on the Manors of Titchfield Abbey, 1245-1405', *Proceedings of the Hampshire Field Club and Archaeological Society*, xxxix (1983), pp. 126-7.

[17] Hatcher, *Cornwall*, pp. 52-3, 71-3.

[18] *Eynsham Cart.*, ii. 19.

shall be found who will perform the due and accustomed services'.[19] These would-be temporary tenures can be paralleled elsewhere. They are found at this time on the Westminster Abbey estate and Barbara Harvey has shown that its policy towards its local tenants in the generation after the Black Death assumed that eventually pre-plague conditions would return; it was only after about 1390 that a greater realism prevailed.[20]

In the longer term the usual way of achieving greater flexibility was by leasing, letting out holdings for fixed periods of years or of lives. This could be common-law leasehold, contractual tenure that might be confirmed by a written agreement. Or it could be within the framework of customary tenure, administered wholly through the manorial court. We can detect a gradual growth in leasing in the half-century before the Black Death and a much faster growth in the half-century after.[21] At first it was often small portions of demesne that were leased out to local tenants,[22] but we increasingly find other holdings being leased as well. At Woolstone leasing had begun by 1345; the stopgap tenures introduced in the 1350s took the form of leases; and in the 1390s entire tenements started to be leased for terms of years. At Englefield by the late fourteenth century hereditary customary tenure had been replaced apparently by customary tenure with a fixed term, of years or of lives. In south-east Durham customary tenure for a single life only was prevalent in the mid-fourteenth century, but over the next century came to be replaced by leases for ever shorter terms of years.[23] These developments were typical, but we have seen that they were not universal; on the estates of Westminster Abbey and the bishopric of Worcester leases gave way to hereditary tenures in the early fifteenth century and there must have been many places like Mackney or Arlesey where throughout the fifteenth century

[19] Above, p. 124.

[20] *VCH Oxon.*, vi. 214 (Islip, a Westminster Abbey manor); Harvey, *Westminster*, pp. 244-5, 247, 268-9. Durham Cathedral Priory provides other examples (Halcrow in *EcHR*, 2nd ser., vii (1954-5), 348-9. Woolstone was a manor of Winchester Cathedral Priory, and it may be significant that these estates all belonged to old Benedictine houses.

[21] e.g. Miller, *Ely*, p. 110; Dyer, *Lords and Peasants*, p. 120; Harvey, *Westminster*, pp. 236, 248-56.

[22] Above, pp. 26.

[23] Above, pp. 125, 151, 311.

all tenures were hereditary.[24] The form and development of local tenures seem to have varied a great deal from one place to another and the easiest—and probably the most revealing—way to classify them is simply by their length: hereditary, or limited to one or more lives or to a term of years, or (as we shall see) held at the lord's will.

All the same some general trends can be seen. The spread of leases, of limited tenures, was a move towards greater flexibility. It was also part of a move towards closer definition of tenures, towards stipulating actual conditions and obligations instead of leaving them to be determined, as occasion arose, as declared local custom in the manorial court. This has already been noticed at Brightwalton, and in fact the entries on manorial court rolls of admissions to holdings usually become more and more detailed, more and more specific, as time goes on.[25] We have seen other instances of the same trend in the precise definition of entry fines at Blunham and (a much more usual development) of heriots of South Moreton and Eastrop.[26] Other developments have been more often discussed: the rapid and almost total disappearance of week-work as a form of rent, the more gradual but steady decline of boon-work obligations, the decline of the various incidents and disabilities of servile tenure. All these are commonplace, though we still have much to learn of their detailed chronology and of their variations between different regions and different estates. Closely related is the decline in the number of tenants whom manorial lords claimed to be villeins by blood.[27] Even in the conservative Durham, where the cathedral priory was still compiling lists of its villeins, there were few of them in the mid-fourteenth century, and a hundred years later hardly any at all.[28] At the other end of England, in Cornwall, villeins are last heard of in 1406.[29] In the manors we have looked at in Berkshire and Bedfordshire

[24] Harvey, *Westminster*, pp. 276-8; Dyer, *Lords and Peasants*, pp. 292-4; above pp. 142, 216.

[25] Above, p. 133. The gradual elaboration of formulas of admission can be followed on the court rolls of the Cambridgeshire manors of Crowland Abbey, 1290-1430, printed in Page, *Crowland*, pp. 331-448.

[26] Above, pp. 198, 138, 148; cf. Dyer, *Lords and Peasants*, pp. 285-6.

[27] Thus Hilton, *Eng. Peasantry*, pp. 24-5; Harvey, *Westminster*, pp. 275-6; Dyer, *Lords and Peasants*, pp. 269-75 (cf. p. 242).

[28] Above, p. 282.

[29] Hatcher, *Cornwall*, pp. 61-2.

the question of personal villeinage seems practically never to have arisen. When manorial lords in the sixteenth century claimed that their tenants were villeins by blood we see either isolated pockets of continuing archaic custom or else deliberate attempts to resurrect a legal concept that was in practice long dead.[30] All these trends form the general background to two further developments in tenures: the spread of one old form, tenure at will, and the emergence of one new one, copyhold.

Tenure at will was known to Bracton, and although it was not common before the late fourteenth century there is no difficulty in finding instances of it much earlier. On the bishopric of Ely estates in 1251 some tenements were held 'from year to year at the will of the lord'.[31] The phrase seems to have meant what it said: tenure could be ended or renegotiated whenever the landlord wished. C.M. Gray has pointed out that this was effectively the position of free men who held villein lands,[32] and it is interesting that, as we have already seen, in the late thirteenth century the abbot of Peterborough applied it to the reverse situation, to free lands acquired by his villeins which he had taken into his own hands and granted back to them again. In the second half of the fourteenth century tenure at will seems to have spread. Thus on the estates of Westminster Abbey and the bishopric of Worcester—to take two recent studies—we find it being used along with leasing as a tenure that offered the flexibility that lords were seeking in the generation or so after the Black Death, but by the early fifteenth century it had declined sharply in favour of more stable, more closely defined, arrangements—a change that can be paralleled by its disappearance from Durham Cathedral Priory's lands at about the same time.[33] In some places, however, what was called simply tenure at will continued throughout the fifteenth century; thus in 1472 the tenants of Tilsworth (Bedfordshire) were divided between those who held freely and those who held at will.[34] There may well have been a regional difference; we do

[30] Hilton, *Decline*, pp. 49-50, gives interesting examples; Kerridge, *Agr. Problems*, pp. 90-2.

[31] Miller, *Ely*, pp. 109-10, 136 n.

[32] Gray, *Copyhold*, p. 6.

[33] Harvey, *Westminster*, pp. 246-8, 276-7; Dyer, *Lords and Peasants*, pp. 292-4; above, p. 311.

[34] Godber, *Beds.*, p. 139.

not find simple tenure at will on any of our Berkshire or Bedfordshire manors, whereas in late-fourteenth- and fifteenth-century Leicestershire it seems to have been quite common.[35]

However, this may reflect a verbal rather than a tenurial difference, for it is not always easy to distinguish simple tenure at will, without further qualification, from what we do find on several of our Berkshire manors and commonly elsewhere too, that is the addition of the phrase 'at the will of the lord' to the description of tenures that are very much more closely and elaborately defined.[36] This too can be found well before the mid-fourteenth century. On the Gloucester Abbey estates in the 1260s some tenants who owed heriot, merchet, and other characteristically villein obligations were said to hold 'at the will of the lord for life by fixed service'.[37] On the published court rolls of Crowland Abbey's three Cambridgeshire manors, starting in 1290, we find the phrase occurring in successive formulas for admission to nearly all customary holdings.[38] Its occasional omission there may be a slip or a change in style by clerk or steward;[39] it is unlikely to reflect any tenurial distinction. By the late fourteenth century the usage was not uncommon—it occurs for instance on St. Albans Abbey's manorial court rolls[40]—and in the fifteenth century we find it applied to tenures variously defined and of varying lengths. We have seen it occurring in tenants' admissions for terms of lives at Coleshill and Brightwalton and for indefinite, hereditary tenure at Brightwalton and Mackney.[41] In Leicestershire it was certainly used for grants for specified terms or by customary tenure as well as for simple tenure at will.[42] The general purport of the phrase was of course to underline the precariousness of the tenure, to show that the tenant had no rights at common law,[43]

[35] Hilton, *Leics. Estates*, pp. 95-6, 122-3, 127; Cicely Howell in Goody, Thirsk, & Thompson, pp. 132-3.

[36] Harvey, *Westminster*, p. 246, gives a very clear account of this distinction.

[37] Hilton, *Eng. Peasantry*, p. 145.

[38] Thus, as instances of successive formulas, Page, *Crowland*, pp. 332 (1290), 355 (1326), 393 (1355: applied to tenure for life, not indefinite), 421 (1394).

[39] Ibid., pp. 333, 346-7.

[40] Levett, *Studies*, p. 153.

[41] Above, Ch. III, section 2.

[42] Hilton, *Leics. Estates*, p. 127; Cicely Howell in Goody, Thirsk, & Thompson, p. 134.

[43] Gray, *Copyhold*, p. 9.

but it would be wrong to see its use—or its absence—as having particular legal meaning. It seems to have been no more than a common form that might or might not be included. In some places a distinction was made between simple customary tenure, which was specifically 'at the will of the lord', and tenure by copy, which was seen as giving greater security; this was the case on the estates of Ramsey Abbey, and we have seen that it may also have been so at Brightwalton.[44] But in some places too, tenure by copy was itself described as being at will; at Cuxham (Oxfordshire) we find in 1488 a tenant who held specifically by copy of court roll at the will of the lord, but the same combination is implied by earlier references there to the customary tenants in general as holding by copy (in 1398) and at will (as in 1458).[45]

We can trace the beginning of copyhold, tenure by copy of court roll, back to the early fourteenth century. But until the Court of Chancery started to take a regular interest in it at the very end of the fifteenth century it would again be wrong to see a tenant's possession of a copy of the court-roll entry of his admission as having any technical significance, or indeed to suppose that there was such a thing as copyhold in the sense of a distinctive tenure except occasionally in a purely local context. In a few places a distinction seems to have been drawn between the more privileged of the customary tenants who were given copies and the less privileged who were not; thus we have seen that at Willington the half-virgaters held by copy.[46] But more often, as at Leighton Buzzard,[47] it was simply a matter of form for most or all of the customary tenants to be given copies, a matter of convenience for tenant and lord alike. Copies might be given for tenures of any length: at Willington they were for a term of years or for life, at Englefield for life, at Leighton Buzzard for indefinite hereditary tenure.[48] The vagueness or inconsistency of the terminology used at Cuxham can be found elsewhere. On the Westminster Abbey estates copies were demonstrably given to customary tenants, possibly regularly and commonly, from the second half of the fourteenth century on, but their tenure is practically never said to be

[44] Raftis, *Tenure*, p. 202; above, p. 134.
[45] MM 5923, m. 7; 5932; 5940.
[46] Above, p. 203.
[47] Above, p. 233.
[48] Above, pp. 203, 151, 233; cf. pp. 192-3.

'by copy' before the sixteenth century.[49] It follows that early references to tenure by copy may be an inadequate guide to the spread of the practice of giving tenants copies of court roll. For what it is worth they suggest that the process was a slow one; we have seen the earliest references to copies at Woolstone in 1372 (of an admission in 1349), at Willington in 1383, at Leighton Buzzard in 1413, at Coleshill in 1528-9.[50] But they can be found much earlier. In a careful and perceptive study of the procedures recorded in manorial court rolls J.S. Beckerman has shown how in the second half of the thirteenth century the practice arose of calling the court's own rolls in evidence in questions arising over particular holdings, this being one aspect of an increasing tendency to refer to written records as proof rather than to manorial juries. Then, in the early fourteenth century, we find not the court's original rolls but tenants' own extracts or copies being used for this purpose—by 1311 on the St. Albans Abbey estates, by 1315 on Earl Warenne's manor of Wakefield (Yorkshire).[51] By 1500 the copy of court roll had already a long history behind it, but copyhold, as the legally recognized tenure to which it gave rise, was only starting to come into being. It is in no way surprising, and may not even be significant, that the fifteenth-century records of south-east Durham do not refer to tenure by copy.

LAND MARKET, LORD, AND TENANT

We have seen that townsmen would often acquire lands either in the fields of the town and its hamlets, as at Leighton Buzzard, or in neighbouring manors, as at Speenhamland and Eastrop, where the land market was all but dominated by tenants from Newbury and Highworth.[1] Earlier in Bedfordshire, in 1297, men of Dunstable were prominent as subtenants in Caddington and Kensworth, a couple of miles away.[2]

[49] Harvey, *Westminster*, pp. 280 n, 284-5; cf. above, pp. 140-1.

[50] Above, pp. 130, 203, 233, 156. On the adoption of written copies cf. pp. 130, 138-9, 148.

[51] *Court Rolls of the Manor of Wakefield*, ed. W.P. Baildon and others (Yorks. Archaeological Soc., 5 vols., 1901-45), iii. 149).

[1] Above, pp. 232, 142-6.

[2] A. Jones, 'Caddington, Kensworth, and Dunstable in 1297', *EcHR*, 2nd ser., xxxii (1979), 324-6; cf. Hatcher, *Cornwall*, p. 248, and Searle, *Battle*, pp. 128-32.

Townsmen might go a little farther afield. A Hartlepool merchant had land at Wolviston, while men of Hitchin had lands at Arlesey and also at Knebworth and Offley—all of them places between 3 and 6 miles from the town.[3] Some of this land may have been acquired by town butchers for grazing; at Adisham (Kent) butchers from Canterbury and Sandwich, both about 6 miles away, would buy land which they would sell again after a short period.[4] But mostly it must have been either to supplement a living gained from a trade or craft or simply as an investment to sublet to local husbandmen. We can see the fifteenth-century Londoners who acquired land at Leighton Buzzard and other places in Bedfordshire in the same light;[5] the city's size and, presumably, the strong competition for lands closer at hand led them to look for investment to more distant centres that were linked to London by trade or direct routes.

With these exceptions it was the local inhabitants who took part in the local land market. In south-east Durham, the one area where we have looked at a complete block of land, tenants do not even seem to have acquired lands in neighbouring vills to add to their holdings.[6] The scattered places investigated in Berkshire and Bedfordshire cannot give us such clear evidence, but deeds of 1350-1500 from both counties give the impression that tenants' lands were mostly in the vills where they lived; typically both parties are described as belonging to the place where the lands lay. This of course could well have differed a good deal from one region to another, but at least it seems likely that the rural land market of this period was mostly intensely local. But this was a time of new or increased mobility; those who were acquiring land may already have been living in the vill or they may have been newcomers, moving in from outside.[7] Conversely, of course, the land may have come on the market because it had been given up by those who had moved

[3] Above, pp. 280, 221; Roden, 'Chiltern Field Systems', p. 276.

[4] Campbell, 'Adisham', pp. 291-2.

[5] Above, pp. 189-90, 234-5.

[6] Above, pp. 314-15; work on the Huntingdonshire manors of Ramsey Abbey mostly points the same way (Dewindt, *Holywell*, p. 203 n; Raftis, *Warboys*, pp. 132-3, where no land transactions appear).

[7] Raftis, *Tenure*, pp. 129-82; Dewindt, *Holywell*, pp. 109-10, 152-5; see above, pp. 184-5, 288-90, 314-15.

away. Again we have been given a fairly consistent picture by our last three chapters. Even in south-east Durham, where there was relatively little movement, we find that particularly the larger holdings in a vill tended to be taken over by newcomers.[8] In Berkshire and Bedfordshire there was more mobility: Brightwalton, South Moreton, Englefield, Coleshill, Blunham, and Arlesey all show many changes of family names between one list of tenants and another, pointing to influxes of newcomers, mostly throughout the whole period.[9] This fully accords with other local studies of the fourteenth and fifteenth centuries, though in detail there may be significant variations; thus the apparent rush of newcomers to Englefield in 1348-9 contrasts with Halesowen (Worcestershire) where, although there was much immigration throughout the fourteenth century, vacancies caused by the Black Death seem to have been filled from among the surviving inhabitants.[10] But in looking at the immigration into these villages we should bear in mind the evidence from Arlesey that the newcomers mostly came from places quite close at hand,[11] and the evidence too from Zvi Razi's family reconstructions at Halesowen that many of those coming from outside to take up holdings were in fact, despite differing surnames, related by blood or by marriage to people already living there.[12]

In 1946 R.H. Hilton showed how local landholding on the estates of Leicester Abbey underwent 'a sort of polarization process' in the late fourteenth and fifteenth centuries: some tenants built up large holdings, while others were reduced to very small holdings or none at all, so that the difference between the richest and poorest increased, and few or none were left with middle-sized holdings—a virgate or half-virgate.[13] The same developments have since been reported elsewhere with occasional variations; thus Christopher Dyer

[8] Above, p. 314.

[9] Above, Ch. III, section 2, Ch. IV, section 2.

[10] Above, p. 150; Razi, *Life, Marriage and Death*, pp. 30-2, 110, 112-13, 117-20. In the Chilterns too the vacancies of 1348-9 seem to have been filled mostly from within each manor (D. Roden, 'Changing Settlement in the Chiltern Hills before 1850', *Folk Life*, viii (1971), p. 66). This may reflect differences in the pattern of deaths in each area.

[11] Above, p. 221; cf. p. 289 and Dyer, *Lords and Peasants*, pp. 366-7.

[12] Razi, *Life, Marriage and Death*, pp. 120-2.

[13] Hilton, *Leics. Estates*, pp. 100-5.

found that on the bishopric of Worcester estates, while large holdings grew, the number of small as well as middle-sized holdings shrank, corresponding rather to the picture of general 'economic promotion' drawn by M.M. Postan.[14] But the accumulation of large holdings and with it an increasing gap between rich and poor tenants has seemed a normal consequence of the workings of a local land market. Edward Miller found this even in the mid-thirteenth century on the East Anglian and other estates of the bishopric of Ely.[15]

In the four studies here we have seen rather more varied developments. In particular two of the three places in thirteenth-century Norfolk present a picture quite different from the Ely estates: the division of holdings and the operation of the land market, combined clearly with population pressure and heavy demand for land, increased the number of tenants and reduced not only the average size of holding but the median as well—the largest holdings became smaller, not bigger. This is what probably happened at Gressenhall and certainly at Martham, but at Sedgeford the land market led some tenants to accumulate large holdings.[16] It is very interesting as it shows that we need still more work on individual vills before we can assess the overall effects of the land market on the tenantry of thirteenth-century East Anglia; it confirms too what John Hatcher found in early-fourteenth-century Cornwall,[17] that a local land market did not in fact always lead to engrossment, the building up of extra-large holdings by a few individuals. This, however, is the consistent pattern given by all three of our later-medieval studies: everywhere we see some tenants using the land market to accumulate holdings on a scale that must have placed them in a dominating position within the local community. On the other hand we see a good deal of variation in what happened to the rest of the tenantry. At Cowpen Bewley we have a very clear case of economic promotion: the largest holdings got larger but so did the smallest ones too.[18] Brightwalton presents an equally clear case of

[14] Dyer, *Lords and Peasants*, pp. 299-301, 312; at Holywell (Hunts.) and Battle (Sussex) the pattern was similar (Dewindt, *Holywell*, pp. 113-15; Searle, *Battle*, p. 377).

[15] Miller, *Ely*, pp. 133-5, 148-51.

[16] Above, pp. 59-60, 70, 95, 101.

[17] Hatcher, *Cornwall*, pp. 56, 100, 229.

[18] Above, pp. 309-11.

simple polarization by the early fifteenth century, and at Blunham an increase in the number of small holdings in the late fifteenth century was a prelude to polarization in the first half of the sixteenth, when large holdings were built up and some of the smaller tenants probably lost their lands altogether.[19] At Sotwell Stonor the growth of five large holdings was accompanied by the virtual disappearance of the standard half-virgate tenants whereas at Shillington, although there too some large holdings had been built up, the middle range of tenants, holding basically a virgate each, still predominated in the first half of the fifteenth century.[20] It is easier to point to these and to other variations than to assess their significance. The trouble is that we know very little about the poorest people in each community; the disappearance or continuance or enlargement of the smallest landholdings is only part of the story. The less successful families may have left the manor altogether, as in fourteenth-century Woolstone.[21] Alternatively they may have been employed on the new large holdings: Hilton points to the significant number of labourers and servants in the Leicestershire poll-tax returns of 1380.[22] Then again, the tenants of the largest holdings may have sublet them instead of tilling them themselves; we have seen this happening at Speenhamland and Eastrop.[23] There is some reason too to suppose that the late fourteenth and fifteenth centuries saw a small but significant influx of craftsmen into many village communities; they would be tenants of no more than a dwelling with little land or none at all, but this would be no guide to their economic status or possible local standing.[24] All this suggests that at the lower end of the scale the realities of village life may have differed a good deal from the formal tenurial picture. What we can be certain of is the accumulation of large holdings in so many places; there are very few communities with records full enough to enable us to go beyond

[19] Above, pp. 130-1, 199-200.

[20] Above, pp. 139, 211.

[21] Above, p. 129.

[22] Hilton, *Leics. Estates*, pp. 100, 163-4.

[23] Above, pp. 143-4, 148-9.

[24] P.D.A. Harvey, 'Non-agrarian Activities in the Rural Communities of Late-medieval England', forthcoming in the proceedings of the 14th Settimana di Studio (1982), Istituto Internazionale di Storia Economica Francesco Datini, Prato [never published].

this. It could, for instance, be misleading to draw a sharp distinction between places of economic promotion, where all holdings got larger, and those of polarization, where the smaller ones declined; similar local conditions could produce either phenomenon, but equally either might be produced by quite varied actual circumstances.[25] It is interesting that at Halesowen, where polarization was increasing the gap between rich and poor tenants after 1350, life expectancy of all tenants increased and the difference between rich and poor in this respect diminished—perhaps a result, as Zvi Razi suggests, of generally raised standards of living.[26]

After the mid-fourteenth century, then, the formation of some large holdings seems to have been a consistent feature of the local land market. But the units from which these holdings were built up were mostly quite small. Where the land market was active it was based not on the transfer of standard holdings or of substantial areas of land, 10 or 20 acres at a time; such transactions there will certainly have been, and they were increasing in number and significance,[27] but far more will have been concerned with much smaller parcels, 2 or 3 acres or less. We saw at Arlesey that 'the scale of the market as a whole was probably linked to the scale of the market in small parcels'; it is likely that this comment can correctly be applied to the medieval local land market in general.[28] Some of these small pieces of land will of course have been free land in the first place; some will often have come from piecemeal reductions (permanent or temporary) of manorial demesnes. But some too will have come from the breakup, the fragmentation, of standard holdings. So much does this seem to have been the norm that we may even take the survival or disintegration of the standard holdings on a particular manor as a rough index to the activity of its local land market. Where there was a really active land market the standard holdings disappeared altogether, as in fifteenth-century Leighton Buzzard, or else they survived

[25] Dyer, *Lords and Peasants*, pp. 300-1, similarly urges caution in interpreting limited evidence on these points.

[26] Z. Razi, 'Family, Land and the Village Community in Later Medieval England', *P&P*, xciii (1981), 31; Razi, *Life, Marriage and Death*, pp. 128-31.

[27] Above, pp. 119-20, 184-5, 238.

[28] Above, p. 220. Harvey, *Westminster*, p. 328, reaches a similar conclusion over the market in free land on parts of the Westminster Abbey estates.

only as spectral relics in administrative documents, as at Martham in 1292.[29] At Arlesey, where there was a moderate land market (four or five transfers a year, on average, between 1377 and 1536), some standard holdings had disintegrated but many remained.[30] Where nearly all the standard holdings survived as intact units, as at Eastrop or Willington or in south-east Durham, there were relatively few transfers of land.[31] We must of course always allow for particular local circumstances. But it looks very much as if in places where the standard holding disintegrated at an early date—as at Halesowen, or on Westminster Abbey's manors at Aldenham (Hertfordshire), Hampstead (Middlesex), and South Benfleet (Essex), or in East Anglia and Kent—we would expect to find a land market already active in the thirteenth century.[32] Where the standard holding survived to the sixteenth century—as at Banbury (Oxfordshire), or on Westminster Abbey's manors at Islip (Oxfordshire), Turweston (Buckinghamshire), and around Pershore (Worcestershire), or in Durham—we would expect that there had been no effective medieval land market.[33] At the same time, even where the standard holding survived we often still find that individuals will have formed large holdings simply by taking over two, three, or more of them. The conditions of the late Middle Ages favoured the formation of large holdings whether this was done through the operations of a local land market or by simple accumulation of tenancies.

We have seen how manorial lords such as Winchester Cathedral Priory at Woolstone or the Grey family at Blunham could, with prudent management, make substantial profits from the local land market.[34] R.M. Smith has even estimated that the fines paid on land transfers provided some three-quarters of the income from one manor in Suffolk between 1259 and 1300.[35] But if the development of a land market involved the fragmentation of standard holdings the manorial

[29] Above, pp. 234, 67-9, 75.

[30] Above, Ch. IV, section 3; Ch. II, section 3.

[31] Above, pp. 146-9, 203-4, 304.

[32] Razi, *Life, Marriage and Death*, p. 8; Harvey, *Westminster*, pp. 206-7.

[33] *VCH Oxon.*, x. 53-4; Harvey, *Westminster*, pp. 286-7, 323-7.

[34] Above, pp. 111, 183.

[35] Macfarlane, *Eng. Individualism*, p. 109 n; cf. Hatcher, *Cornwall*, pp. 87-9, and Bean, *Percy Estates*, pp. 52-4, 60.

lord might lose more than he gained. Where the rents from these holdings were paid entirely in money he would run no risk, for they could be as minutely subdivided as the holding itself and the lord would suffer no loss; thus at Shillington in 1306 a virgate had been divided, through sale, among twelve tenants in unequal portions, but as the service due from it was simply a rent of 5*s*. 8*d*. a year there was no difficulty in sharing the obligation in portions as small as ½*d*. and as precisely defined as 1*s*. 1¾*d*.[36] But harder problems arose where rents took the form of labour services and renders of produce, and we can see why Worcester Cathedral Priory, as manorial lord of Stoke Prior (Worcestershire), took back into its own hands a half-virgate that had been so much split up that the services due from it could not be fully and properly performed.[37] Various solutions were possible. One was to maintain, on paper so to speak, a continuing structure of standard holdings which might bear no relation to the existing pattern of tenures but which served as a basis for allotting obligations to rents and services. This was very common in Kent (as at Gillingham[38]) and in East Anglia (as at Gressenhall[39]), but occurred elsewhere too; on two Hertfordshire manors of Westminster Abbey rents and services were still based on the full virgate at the end of the thirteenth century, but on one, Aldenham, these virgates had entirely disintegrated and bore little relation to the actual holdings in the vill, while on the other, Stevenage, an intermediate stage had been reached, for the tenants all held half-virgates while their obligations were expressed in terms of full ones.[40] A simpler way to meet the problem was for the manorial lord to abandon all attempt to keep track of the fragmentation of each standard holding and simply to make one tenant responsible for all its customary obligations—a method adopted on parts of the bishopric of Ely estates by the

[36] Raftis, *Tenure*, p. 88.

[37] Hilton, *Medieval Society*, p. 163. At Weston (Hunts.) Ramsey Abbey in 1308 required all 14 tenants between whom a 30-acre holding was unequally divided to render the whole of the very light services due from the entire holding (Raftis, *Tenure*, pp. 88-9); on the archbishopric of Canterbury's Kent manors special juries might be empanelled to find how a yoke had been divided so as to allot the rents and services proportionately (Du Boulay, *Canterbury*, pp. 152-3).

[38] A.R.H. Baker in Baker & Butlin, pp. 394-8.

[39] Above, Ch. II, section 2.

[40] Harvey, *Westminster*, pp. 206-7, 210.

early fourteenth century.[41] But the simplest solution of all was to stop the holdings from being split up. We can deduce manorial lords' opposition to dividing customary holdings from the fact that on certain estates free holdings were the first to disintegrate, even if (as on the Peterborough Abbey estates) they tried to restrain even their free tenants from dividing their holdings.[42] The 1297 extents of the St. Paul's Cathedral manors of Caddington and Kensworth (Bedfordshire) are particularly interesting: they show that standard holdings held freely for money rents were divided between many sub-tenants, but those held by customary tenants for labour services have no sub-tenants named. It may be that the customary holdings remained intact, or alternatively that the extents did not record their subdivision.[43] In the long term manorial lords were remarkably unsuccessful in preventing any holdings, free or customary, from disintegrating, but from the thirteenth century onwards we find them making more or less vain attempts to stem the tide. We have seen how at Gressenhall the Foliots may have succeeded in this until the early fourteenth century; the bishops of Ely abandoned the struggle on their manors at about the same time.[44] A fairly effective method used on some Westminster Abbey manors was to impose a new rent on any land subtracted from a holding while still demanding almost the whole of its customary obligations from the original tenant.[45] At Blunham we saw the earl of Kent in his Charter of 1471 regulating his tenants' alienation of customary lands and trying to prevent the complete dispersal of the standard half-virgates and quarterlands.[46]

One reason why the standard holdings at Gressenhall survived as long as they did may have been the persistence of labour services there.[47] It is easy to see how difficult it must have been either to divide labour services, especially week-work, in such a way that they would be efficiently performed, or to exact them in full from the tenant of only part of a divided

[41] Miller, *Ely*, pp. 133, 138-9.
[42] Harvey, *Westminster*, pp. 212-13; King, *Peterborough*, p. 119.
[43] Jones in *EcHR*, 2nd ser., xxxii (1979), 322.
[44] Above, p. 56; Miller, *Ely*, pp. 138, 146-7.
[45] Harvey, *Westminster*, pp. 301-3.
[46] Above, p. 198.
[47] Above, pp. 36, 56.

holding. We might thus suppose that the decline of labour services was a prerequisite for a local land market, and that its spread followed the abolition of week-work. This is possible; and it should be said at once that the detailed chronology of the disappearance of week-work in the thirteenth and fourteenth centuries is less clear than might appear at first sight.[48] But labour services by the thirteenth century were, for lords as well as tenants, an inconvenient and inefficient anachronism, kept in being more for legal than for economic reasons; they were not a strong, buoyant institution.[49] Probably, therefore, we should look at it the other way round, and say that labour services (like the standard holding) could survive only where no local land market developed. What happened at Gressenhall and Blunham and on the Ely and Westminster Abbey estates suggests that manorial lords were less powerful than one might suppose in the face of local pressures for an active land market. This is borne out by the differences we find between one manor and another on a single estate—between, for instance, manors of Bec Abbey in the thirteenth century: Ogbourne St. George (Wiltshire) where standard virgates, half-virgates, and 4-acre holdings all rendered week-work and other labour services, and Blakenham (Suffolk) where the wholly disintegrated holdings rendered money rents and a very few boon-works.[50] If the manorial lord chose to keep labour services on the one manor, why not keep them on the other? The regional pattern is too strong to allow us to suppose that the difference arose from the simple choice or initiative of the manorial lord. It is argued that it was the labour services at Martham that led Norwich Cathedral Priory to retain there throughout the thirteenth century the standard tenements that it had wholly abandoned at Sedgeford,[51] but even at Martham it was only the notional structure that remained—the holdings themselves had been split up between many tenants in the course of the thirteenth century.

[48] Because although a holding's obligation might include labour services, these might be 'sold' (i.e. replaced by payment in cash) either occasionally or regularly (this is usefully illustrated by Kosminsky, *Studies*, pp. 164-5, and Raftis, *Warboys*, pp. 197-8). Thus to discover the actual incidence of labour services on any manor we need to have not only a custumal but also a substantial series of annual accounts.

[49] P.D.A. Harvey, 'The English Inflation of 1180-1220', *P&P*, lxi (1973), 22-3.

[50] *Bec Documents*, pp. 29-36, 92-6.

[51] Above, Ch. II, section 5.

Christopher Dyer has suggested that one way the bishops of Worcester tried to prevent the disintegration of standard holdings was by insisting on impartible inheritance—on primogeniture in fact—for all customary holdings; he shows that partible inheritance (or rather, joint tenure among brothers, a point we shall return to) seems to have operated on parts of the estate in about 1170 but had vanished by 1299.[52] This raises the question of the part local inheritance customs may have played in breaking up standard holdings, a question that has already arisen, particularly in looking at our Norfolk manors. That partible inheritance could rapidly split a holding into many fragments is of course obvious; we have seen this happening at Martham.[53] That partible inheritance was widespread in East Anglia and Kent makes it easy to explain the early decay of standard holdings there, and Rosamond Faith's demonstration that it occurred also, though much more patchily, in many other parts of England as well[54] might make us wonder whether it may not have been a widespread cause of disintegration of standard holdings and therewith of the early development of a local land market. But here our three Norfolk manors are instructive. At Gressenhall standard holdings were still intact units at the end of the thirteenth century, whereas by then at Martham no more than a formal structure remained; yet both were manors of partible inheritance. At Sedgeford on the other hand the local inheritance custom was primogeniture, yet here holdings were scarcely less fragmented than on the other two, any slight difference being best explained simply by its being in an area of poorer soils.[55] Nor is it difficult to find places outside East Anglia where, despite impartible inheritance, standard holdings had collapsed and a vigorous land market had developed before the end of the thirteenth century. The manors in the Chilterns investigated by David Roden, especially Codicote (Hertfordshire), are examples of this; Halesowen (Worcestershire) is another.[56]

[52] Dyer, *Lords and Peasants*, pp. 86, 106; cf. King, *Peterborough*, pp. 117-19.

[53] Above, pp. 71-5.

[54] 'Peasant Families and Inheritance Customs in Medieval England', *AgHR*, xiv (1966), 93-5.

[55] Above, Ch. II, section 5.

[56] D. Roden, 'Inheritance Customs and Succession to Land in the Chiltern Hills in the Thirteenth and Early Fourteenth Centuries', *JBS*, vii (1967), 1-11 (but the point is made more clearly in Roden, 'Chiltern Field Systems', pp. 161-2, 165-7); Razi, *Life, Marriage and Death*, p.8.

It seems, then, as if the standard holdings did not simply disintegrate in the wake of the decline of week-work, nor yet in response to local inheritance customs. What in fact generally underlay the phenomenon was the land market itself—the demand for transfers of land not just on death but *inter vivos*, from one living person to another, whether by sale or by lease or by any other arrangement. Historians sometimes assume that the standard holding in the thirteenth and fourteenth centuries was a strong institution, firmly rooted in past history and basic to current agrarian and tenurial practice, so that its disappearance is viewed with mild surprise. In fact it was a very fragile thing. Once the individual tenant's lands and rights had been precisely defined, as they came to be defined in the twelfth century, the standard holding had lost most of its point and became little more than a historical relic; where it survived to the fourteenth or fifteenth century or later this could only be because no pressures had been brought to bear on it in that particular locality. Conversely the peasant land market is sometimes viewed as a frail, almost alien, growth, springing up as it were in the interstices of the medieval agrarian system in places where 'manorialization' was weakest; whereas in fact all the evidence suggests that by the thirteenth century it was already a very strong force indeed and that wherever it arose it created pressures that manorial lords were powerless to resist.

Why did this land market arise? Although in two of our vills in thirteenth-century Norfolk, probably an area of extreme land shortage, the land market seems to have reduced the size of all holdings, in all other cases we have seen it consistently led to the enlargement—the substantial enlargement—of some tenants' holdings. Engrossment of this sort seems generally to have come about only through the workings of a local land market, though in some places after the Black Death population losses led to tenants acquiring extra holdings simply by taking them over as they fell vacant. We should not dismiss the exceptions as insignificant. All the same it looks as if the land market normally led to the enlargement of some individuals' holdings and as if the land market was the way such enlargement was normally achieved. Essentially it was those who acquired lands who brought the market into being and almost certainly the main prerequisite for this was local prosperity: the means of building up enough resources in money, enough capital, to pay whatever was needed for the purchase or lease of the extra land and to find the stock, the labour, the equipment, to work the enlarged holding. This explains why the land market dealt in small pieces rather than complete holdings: the resources of those acquiring lands were inadequate to cope with substantial additions unless they were acquired piecemeal. It seems unlikely that a land market could develop in places where the level of prosperity was low. Significantly, there were unusually uniform holdings on those villages on the bishopric of Worcester estates that came to be deserted in the late Middle Ages—and we have learned from M.W. Beresford and others that desertion and impoverishment went hand in hand.[57] Of the four Chiltern vills investigated by David Roden Ibstone (Buckinghamshire) was the poorest and was the only one to have no local land market—even though free tenants predominated there and labour services were light, consisting of boon-works rather than week-work.[58] Different levels of prosperity are surely one reason for local and regional variations in the land market. But they are far from the whole story; we have been rightly warned, in looking at Coleshill, that an active land market cannot be taken as a simple index to local prosperity.[59] In some prosperous vills or regions seignorial demands in heavy rents, whether in cash or in labour services, high entry fines, and the like might effectively stop a local land market from developing by the indirect means of preventing tenants from collecting sufficient funds; it is significant here that we find an early land market in Kent and East Anglia, where seignorial demands were mostly fairly light, and that in many places a land market quickly developed after the Black Death had led to reduced tenant obligations.[60] And

[57] Dyer, *Lords and Peasants*, p. 263; *Deserted Medieval Villages*, ed. M. Beresford and J.G. Hurst (London, 1971), pp. 21-6.

[58] Roden, 'Chiltern Field Systems', pp. 236-42; D. Roden, 'Field Systems in Ibstone, a Township of the South-west Chilterns, during the Later Middle Ages', *Records of Buckinghamshire*, xviii, No. 1 (1966), pp. 43-4.

[59] Above, p. 156.

[60] Thus Roden, 'Chiltern Field Systems', pp. 115, 117, notes how at Kings Walden (Herts.) the character of the local land market changed in the mid-14th century: values declined and larger units came to be transferred. Hilton, *Eng. Peasantry*, pp. 196-205, shows how difficult it was for the peasant to accumulate capital in the 13th century and

although neither partible inheritance of customary lands nor a predominance of free tenures was a prerequisite for a land market which could perfectly well develop without either, yet each could serve to prime the pump, giving the market an early impetus in particular places.

Why should the peasant wish to enlarge his holding? The answer is not necessarily self-evident, though simple ambition must often have been the explanation. From available evidence we might well suppose that this was why, say, John Pope of Sotwell Stonor or William Taillour of Leighton Buzzard or William White of Cowpen Bewley was buying land in the fifteenth century.[61] It is interesting that on several of our Berkshire manors families that built up extensive holdings in the fifteenth century failed to maintain their position—a peasant aristocracy that came to nothing.[62] The same thing happened in south-east Durham but at an earlier date: families prominent in the late fourteenth century had disappeared by the mid-fifteenth, though some of their late-fifteenth-century successors prospered throughout the sixteenth.[63] Similar patterns of transient peasant wealth in the late Middle Ages have been reported elsewhere.[64] This may simply reflect the mobility of the late-medieval population, or it may be that the conditions of the time made it difficult to sustain the consistent degree of investment needed for farming on this particular scale—as we have been reminded by the evidence of harvests in Durham,[65] small-scale farming was still a chancy business in the late Middle Ages. In some places, indeed, there seems to have been opposition to any accumulation of lands; it was stated as local custom at Ombersley (Worcestershire) in 1415 that a composite holding should be divided on the tenant's death between his children of both sexes.[66]

how the difficulties eased in the 14th and 15th centuries. Royal demands will have contributed to the difficulties in general, and some areas suffered particularly from purveyance; interestingly, however, it was eastern England that was most affected in the early 14th century (Maddicott, *Eng. Peasantry*, pp. 17-18, 67-70).

[61] Above, pp. 141, 231, 245, 310.
[62] Above, pp. 145, 157-8, 167-8, 173-6.
[63] Above, p. 316.
[64] e.g. *VCH Oxon.*, vi. 226; Harvey, *Westminster*, pp. 289-90.
[65] Above, Ch. V, section 4.
[66] Hilton, *Eng. Peasantry*, p. 41.

But sheer economic ambition need not have been the only reason for acquiring extra lands. Particularly since 1966, when the work of A.V. Chayanov was published in English,[67] historians have been looking in medieval England for the pattern he describes from Russia: as a family itself expanded, producing more mouths to feed or more hands to till the soil, so it would enlarge its holding accordingly, but when the children left home or took up lands of their own the parents would no longer need so much land and would reduce the holding. There are not many places in late-medieval England with records of peasant families full enough to tell whether this was happening or not. Halesowen (Worcestershire) was one, however, and there Zvi Razi has found nothing to suggest that lands were being acquired or disposed of on this basis.[68] On the other hand there may well be some cases of this happening on the estates of Ramsey Abbey and the bishopric of Worcester, and although we have no clear evidence it would at least fit very well with what we have seen at Arlesey, with its pool of marketable pieces of land that moved fairly rapidly from one holder to another.[69] Yet another distinct but related reason for acquiring extra land was to provide for children, especially for those who under local inheritance custom would get no part of the basic holding on the parent's death. We have seen this happening at Sedgeford and it has been found elsewhere too in the thirteenth and fourteenth centuries—at Halesowen and in the Chilterns for instance.[70] Where inheritance customs came to be less strictly applied in the late Middle Ages—as at Sotwell Stonor, or as at Adisham (Kent)[71]—there will not have been the same need to buy land for children who would otherwise be landless, but land might still be acquired to provide adequately for all the children if there were several in the family.

These three reasons why the peasant might wish to enlarge his holding—and thus why the peasant land market grew

[67] *The Theory of Peasant Economy*, ed. D. Thorner, B. Kerblay, and R.E.F. Smith (Homewood, 1966); cf. Hilton, *Eng. Peasantry*, pp. 6-7.

[68] Razi, *Life, Marriage and Death*, pp. 144-6.

[69] Dewindt, *Holywell*, pp. 116-22, 129-33; Dyer, *Lords and Peasants*, pp. 306-8; above, pp. 217-19, 222-3.

[70] Above, pp. 92-4, 100-1; Razi in *P&P*, xciii (1981), 6-8; Roden in *JBS*, vii (1967), pp. 6-7.

[71] Above, pp. 139-40; Campbell, 'Adisham', pp. 251-2, 266.

up—are not mutually exclusive: all three may well be correct. It is not just that one may have operated in one village or region, another in another; all three could perfectly well be found in a single place or in a single individual. The late-medieval peasant may well have been no more capable than his historian 500 years later of analysing his motives when he set about acquiring more lands; economic ambition, immediate provision for an expanding family, provision for non-inheriting children after his death might all have played their part in any one purchase.

The part played by the land market in providing for the landless members of the tenant's family, either immediately or on his death, raises the question of the rôles of the family and the individual in the rural society of medieval England—a question that has been especially to the fore since Alan Macfarlane argued in 1978 that the basic unit of this society was essentially the individual not the family, that the members of the peasant family were, as he put it, ego-centred rather than ancestor-centred, and that the peasant tenant should be seen as a landholder in his own right, not as his generation's trustee for property which belonged to the family as a whole, past and future.[72] It is indisputable that the inhabitants of the medieval countryside were viewed as individuals by royal administrators, as we see in tax assessments, by estate administrators, as we see in manorial court records, and very likely by each other too, though of this we have practically no evidence. Certainly if either the king or the estate owner wished to treat with any part of the rural community corporately it was to the vill, the manor, or the tithing group that he would turn, not to the single family. Throughout this book we have seen peasant tenants buying and selling, leasing and bequeathing land, all as between one individual and another. None of this need imply, however, that a very strong sense of family commitment and of family responsibility was not widespread among these same peasant tenants: there is overwhelming evidence that it was. In the four studies here we have seen this over and again: in the way families provided for all their members in despite or even in defiance of local inheritance customs;[73] in the way a

[72] Macfarlane, *Eng. Individualism*, pp. 80-164 *passim*.

[73] Above, pp. 44-5, 92-4, 100-1, 103, 300-1.

holding, technically divided by partible inheritance, yet remained intact;[74] in the reversions and other devices that were used to keep lands in the family.[75] All this can be paralleled and expanded from other work on the rural society of late-medieval England.[76] Admittedly this sense of family may have been declining by the fifteenth century: one strikingly consistent feature of the late-medieval land market in Berkshire, Bedfordshire, and Durham is the declining proportion of land transfers made between relatives.[77] But Zvi Razi, from his detailed reconstruction of family relationships at Halesowen, has shown that this may be illusory: the records from less well-documented manors are just not full enough to show that people of different surnames might still be related.[78] He is careful to stress that what happened in one place did not necessarily happen in another;[79] but if we can correctly draw analogies from Halesowen it could be that we see a very stable balance in the conflicting claims of the individual and his family throughout the thirteenth, fourteenth, and fifteenth centuries.

Whether we can carry this same stability back into the twelfth century is open to question. Particularly since Rosamond Faith's work on inheritance customs in 1966[80] historians have tended to think that in late Anglo-Saxon England there was widespread partible inheritance among the peasantry, a custom generally changed for impartible inheritance (mostly primogeniture) following the example of the upper ranks of Anglo-Norman society.[81] This may be so; but an ancient

[74] Above, pp. 40-1, 43-7, 74-5.

[75] Above, pp. 140, 142.

[76] e.g. Dewindt, *Holywell*, pp. 133-4, 185 n. For the higher ranks of 12th-century society cf. Milsom, *Legal Framework*, pp. 121-2, 132-7.

[77] Above, pp. 132, 136, 150, 156, 217-19, 237, 248, 296-90.

[78] Razi in *P&P*, xciii (1981), 16-22.

[79] Ibid., p. 36.

[80] In *AgHR*, xiv (1966), 77-95.

[81] Thus, e.g., Harvey, *Westminster*, p. 210. Partible inheritance among the peasantry would accord with the partible inheritance found in the upper ranks of late Anglo-Saxon landowners (T.H. Aston, 'The Origins of the Manor in England' and 'A Postscript', in *Social Relations and Ideas*, ed. T.H. Aston and others (Cambridge, 1983), pp. 20-1, 27-9), as well as with contemporary Welsh customs (G.R.J. Jones, 'Multiple Estates and Early Settlement', in *Medieval Settlement: Continuity and Change*, ed. P.H. Sawyer (London, 1976), pp. 16-17, 35). Cf. the comment of Hyams, *King, Lords, and Peasants*, p. 77 n, that manorial custom was Anglo-Saxon law that 'slid down society with those who used it' after 1066.

custom of joint tenure by all sons who stayed at home is by no means impossible.[82] We have seen how in the twelfth century a change in emphasis in the relations between lord and tenant led to the tenant's rights being viewed more strictly in territorial terms and to a more precise definition of his holding.[83] It would be entirely consistent for this to define not only exactly what each man held, but also exactly which man held it—in other words the introduction of precise inheritance custom to make it perfectly clear who was answerable for every parcel of the rights and obligations that made up a holding: the eldest son, or the youngest son, or all the sons, not jointly but each taking a specific share of the divided property. Which of these customs was adopted may well have made less difference in practice than might appear: Rosamond Faith argued in 1966 that the practical effects of ultimogeniture and partible inheritance were much the same,[84] and we can see that, where a local land market enabled younger brothers to be separately provided for, primogeniture too need not be so very different. The crucial difference was not between one or other inheritance custom, but between defined inheritance custom and undifferentiated joint inheritance. This, of course, is no more than speculation, but it is a possibility that needs to be taken into account in assessing what few facts or hints we have about the succession to peasant holdings before the thirteenth century. It is possible, even likely, that the position varied a good deal from one part of the country to another; it may be that what happened was at most a move towards defined inheritance in places where it was not already established custom. But certainly development on these lines would help to explain why in areas of partible inheritance standard holdings could still be basically intact in the early thirteenth century;[85] it is a problem to which slow

[82] It is interesting that Aston, op. cit., p. 28, suggests the possibility of just such joint tenure in particular circumstances in the late Anglo-Saxon period, and that Rees, *S. Wales and March*, p. 157, refers to it in later medieval south Wales.

[83] Above Ch. I, section 2.

[84] In *AgHR*, xiv (1966), 83-4.

[85] It might also help to explain, on the assumption that joint tenure gave way to partible inheritance, the division of standard holdings in many places, so that virgates became half-virgates, etc. (Rees, *S. Wales and March*, p. 153; Raftis, *Tenure*, p. 17; Harvey, *Westminster*, pp. 299-300). But this is to advance still further into mere speculation.

population growth and the availability of new land for colonization are not very convincing answers.

But if this occurred it cannot have been more than a shift of emphasis. If peasant families in fact held jointly there will normally have been a recognized head of each household: eleventh- and twelfth-century manorial surveys generally name individuals, not family groups, as their local tenants,[86] though, significantly, some obligations lay on the entire peasant household, not on the individual tenant.[87] Inheritance customs may have simply extended, or even just made explicit, the local arrangements governing the family group. Alan Macfarlane hints that the origin of what he sees as very distinctive English individualism lay in the depths of Germanic antiquity; this may be so, but it is at least possible that the twelfth century saw a significant step in local tenures away from the family group and towards the individual as an aspect of the process of definition of tenure, of land, and of law. At the same time it was a shift of emphasis at most. The sharp distinction that Macfarlane draws between family and individual may be helpful in contrasting England's rural structure with that of other parts of Europe;[88] it is less clear that it assists our understanding of that structure in England itself. Throughout the Middle Ages in England we see family and individual playing distinct but interdependent rôles. It is with the family made up of individuals and with the individual within his family that the historian of the medieval English peasantry is concerned.

[86] The joint tenure on the bishopric of Worcester estates *c.*1170 (above, p. 348) is thus exceptional in being so recorded.

[87] e.g. the frequent obligation to perform certain harvest works with the entire household, as on the Burton Abbey estates at Appleby (Leics.), Cauldwell, and Findern (both Derbs.) and on manors of the bishop of Durham (C.G.O. Bridgeman, 'The Burton Abbey Twelfth Century Surveys', *Collections for a History of Staffordshire 1916* (William Salt Soc., 1918), pp. 245, 243, 235; *Boldon Buke, passim*).

[88] Macfarlane, *Eng. Individualism, passim.*

ABBREVIATIONS

AgHR	*Agricultural History Review*
Baker & Butlin	*Studies of Field Systems in the British Isles*, ed. A.H.R. Baker and R.A. Butlin (Cambridge, 1973)
Bean, *Percy Estates*	J.M.W. Bean, *The Estates of the Percy Family 1416–1537* (London, 1958)
Bec Documents	*Select Documents of the English Lands of the Abbey of Bec*, ed. Marjorie Chibnall (Camden 3rd Series, vol.lxxiii; 1951)
Beckerman, 'Customary Law'	J.S. Beckerman, 'Customary Law in English Manorial Courts in the Thirteenth and Fourteenth Centuries' (London University Ph.D. thesis, 1972)
Boldon Buke	*Boldon Buke*, ed. W. Greenwell (Surtees Soc., vol. xxv; 1852)
Campbell, 'Adisham'	Sarah Campbell, 'Some Aspects of the Social and Economic History of the Manor of Adisham, *c.*1200 to the Dissolution' (University of Kent M.Phil. thesis, 1981)
Cart. Mon. Ram.	*Cartularium Monasterii de Rameseia*, ed. W.H. Hart and P.A. Lyons (Rolls Series, 3 vols., 1884–93)
Danelaw Ch.	*Documents illustrative of the Social and Economic History of the Danelaw*, ed. F.M. Stenton (British Academy, Records of the Social and Economic History of England and Wales, vol. v; 1920)
Davenport, *Norf. Manor*	Frances G. Davenport, *The Economic Development of a Norfolk Manor 1086–1565* (Cambridge, 1906)
Dewindt, *Holywell*	E.B. Dewindt, *Land and People in Holywell-cum-Needingworth* (Toronto, 1972)

Dodgshon, *Origin*	R.A. Dodgshon, *The Origin of British Field Systems: an Interpretation* (London, 1980)
Du Boulay, *Canterbury*	F.R.H. Du Boulay, *The Lordship of Canterbury* (London, 1966)
Dyer, *Lords and Peasants*	C. Dyer, *Lords and Peasants in a Changing Society: the Estates of the Bishopric of Worcester, 680–1540* (Cambridge, 1980)
EcHR	*Economic History Review*
Eynsham Cart.	*Eynsham Cartulary*, ed. H.E. Salter (Oxford Historical Soc., old ser., vols. xlix, li; 1907–1908)
Godber, *Beds.*	Joyce Godber, *History of Bedfordshire 1066–1888* (Bedford, 1969)
Goody, Thirsk, & Thompson	*Family and Inheritance: Rural Society in Western Europe, 1200–1800*, ed. J. Goody, Joan Thirsk, and E.P. Thompson (Cambridge, 1976)
Gray, *Copyhold*	C.M. Gray, *Copyhold, Equity, and the Common Law* (Cambridge, Mass., 1963)
Harvey, *Westminster*	Barbara Harvey, *Westminster Abbey and its Estates in the Middle Ages* (Oxford, 1977)
Hatcher, *Cornwall*	J. Hatcher, *Rural Economy and Society in the Duchy of Cornwall 1300–1500* (Cambridge, 1970)
Hilton, *Decline*	R.H. Hilton, *The Decline of Serfdom in England* (London, 1969)
Hilton, *Eng. Peasantry*	R.H. Hilton, *The English Peasantry in the Later Middle Ages* (Oxford, 1975)
Hilton, *Leics. Estates*	R.H. Hilton, *The Economic Development of some Leicestershire Estates in the 14th and 15th Centuries* (London, 1947)
Hilton, *Medieval Society*	R.H. Hilton, *A Medieval Society: the West Midlands at the End of the Thirteenth Century* (London, 1966)

Homans, *Eng. Villagers*	G.C. Homans, *English Villagers of the Thirteenth Century* (Cambridge, Mass., 1941)
Hyams, *King, Lords, and Peasants*	P.R. Hyams, *King, Lords, and Peasants in Medieval England: the Common Law of Villeinage in the Twelfth and Thirteenth Centuries* (Oxford, 1980)
JBS	*Journal of British Studies*
Kerridge, *Agr. Problems*	E. Kerridge, *Agrarian Problems in the Sixteenth Century and After* (London, 1969)
King, *Peterborough*	E. King, *Peterborough Abbey 1086–1310: a Study in the Land Market* (Cambridge, 1973)
Kosminsky, *Studies*	E.A. Kosminsky, *Studies in the Agrarian History of England in the Thirteenth Century*, ed. R.H. Hilton (Oxford, 1956)
Lennard, *Rural Eng.*	R. Lennard, *Rural England 1086–1135* (Oxford, 1959)
Levett, *Studies*	A. Elizabeth Levett, *Studies in Manorial History*, ed. H.M. Cam, M. Coate, and L.S. Sutherland (Oxford, 1938)
Macfarlane, *Eng. Individualism*	A. Macfarlane, *The Origins of English Individualism: the Family, Property and Social Transition* (Oxford, 1978)
Maddicott, *Eng. Peasantry*	J.R. Maddicott, *The English Peasantry and the Demands of the Crown, 1291–1341* (*P&P* Supplement i; 1975)
Maitland, *Domesday*	F.W. Maitland, *Domesday Book and Beyond* (Cambridge, 1897)
Man. Records of Cuxham	*Manorial records of Cuxham, Oxfordshire, circa 1200–1359*, ed. P.D.A. Harvey (Historical MSS Commission, Joint Publications No. 23; Oxon. Record Soc., vol. l; 1976)
Miller, *Ely*	E. Miller, *The Abbey and Bishopric of Ely* (Cambridge, 1951)

Milsom, *Legal Framework*	S.F.C. Milsom, *The Legal Framework of English Feudalism* (Cambridge, 1976)
MM	Merton College, Oxford, muniments
Page, *Crowland*	Frances M. Page, *The Estates of Crowland Abbey* (Cambridge, 1934)
P&P	*Past and Present*
PRO	Public Record Office, London
Raftis, *Ramsey*	J.A. Raftis, *The Estates of Ramsey Abbey* (Toronto, 1957)
Raftis, *Tenure*	J.A. Raftis, *Tenure and Mobility: Studies in the Social History of the Mediaeval English Village* (Toronto, 1964)
Raftis, *Warboys*	J.A. Raftis, *Warboys: Two Hundred Years in the Life of an English Mediaeval Village* (Toronto, 1974)
Razi, *Life, Marriage and Death*	Z. Razi, *Life, Marriage and Death in a Medieval Parish: Economy, Society and Demography in Halesowen 1270–1400* (Cambridge, 1980)
Rees, *S. Wales and March*	W. Rees, *South Wales and the March 1284–1415* (London, 1924)
Roden, 'Chiltern Field Systems'	D. Roden, 'Studies in Chiltern Field Systems' (London University Ph.D. thesis, 1965)
Rowley, *Origins*	*The Origins of Open-field Agriculture*, ed. T. Rowley (London, 1981)
Searle, *Battle*	Eleanor Searle, *Lordship and Community: Battle Abbey and its Banlieu 1066–1538* (Toronto, 1974)
VCH	*Victoria History of the County of . . .*

X

The Peasant Land Market in Medieval England—and Beyond

THE medieval economic historian must sometimes defend himself against the charge of thinking anachronistically. In seeking insights into medieval society by investigating its economic structure and development one can all too easily assume—wrongly—that people in the Middle Ages took the same interest in the economy of their time and that they responded consciously and knowledgeably to such phenomena as fluctuations in the value of money, movements in trade patterns, growth of population. This, at least, is one danger we need not guard against in looking at the medieval land market. The medieval landowner, large or small, did not have the same information at his disposal as his historian today: in some ways he knew more, in some less, about the land market of his own time. But we cannot doubt his interest, his keen interest, in the subject, in England as throughout medieval Europe. When Richard Barking died in 1246 after 24 years as abbot of Westminster the epitaph on his tombstone did not refer to his spiritual attainments or other merits, but listed as his four achievements buying the manor of La Neyte, buying the manor of Castlemorton, impropriating the church of Oakham, and securing from the king 'a charter useful in many ways' (*multis commoda charta*).[1] This preoccupation with land, its acquisition, and its value was shared alike by ecclesiastic and nobleman, magnate and villein. In investigating the land market we are looking at something that lay close to the heart of medieval man.

For some aspects of this land market, there is peculiarly full evidence from England. From the rest of the British Isles the picture is less clear. Though estate-owners from England produced in Ireland and, particularly, Wales the same kinds of record as provide the English evidence their coverage is less complete and the differing social structure and political history of these areas in the Middle Ages mean that their land markets demand separate treatment.[2] And in medieval Scotland we are

[1] Harvey, *Westminster Abbey*, 92–3.

[2] The form of the local land market in Wales is discussed by L. B. Smith, 'The Gage and the Land Market in Late Medieval Wales', *EcHR* 2nd ser. 29 (1976), 537–50.

in an altogether different historiographical world, where the sources for such a study are different in kind, and far sparser, than those from the rest of Britain.

Certainly the English sources for the market in substantial rural holdings—entire manors rather than single acres—and for the market in urban properties can be equalled or bettered in other parts of western Europe. Charters of kings and of magnates, which survive from the late seventh century onwards, give us some glimpses of the traffic in land by which the great estates, lay and ecclesiastical, were built up, but these glimpses give us much less than a complete picture for the Anglo-Saxon period. For the new estate structure of the Norman settlement, however, the greater profusion of charters from the twelfth century onwards shows us much more clearly how land was acquired and transferred by the nobility and gentry. The evidence of these charters can be supplemented, first, from the end of the twelfth century, by the records of the royal courts, in particular the feet of fines which recorded the transfer of property in fictitious legal suits; second, from the 1230s onwards, by the inquisitions post mortem which recorded, on the death of a tenant-in-chief, the lands he held and who was to inherit them; and third, in a growing volume from the early thirteenth century, by the administrative records of the estates themselves. The construction of manorial descents—the successive ownership of landed estates—is a long-established tradition in English antiquarian scholarship, and the many county histories that have been published since the seventeenth century contain a vast mass of evidence for the workings of the medieval land market in its higher reaches. More recently there have been a number of important studies of individual estates in medieval England; some have looked only at the administration of the estate and its revenues, but others have investigated in some detail the way the estate's individual properties were acquired, augmented, and disposed of—notable examples are the work on the estates of Peterborough Abbey by Edmund King, of Westminster Abbey by Barbara Harvey, of Thorney and Crowland Abbeys by Sandra Raban.[3] These studies have concentrated particularly on the monastic estates, for it is from these that fullest records survive. The time is ripe for a general survey of the market in substantial areas of land in medieval England; it has yet to be written.

Nor have we any general work on the traffic in urban properties though

[3] King, *Peterborough Abbey*; Harvey, *Westminster Abbey*; S. Raban, *The Estates of Thorney and Crowland: A Study in Medieval Monastic Land Tenure* (Cambridge, 1977).

there a good deal more research may be needed on individual towns before it can be produced. The evidence is there. For certain English towns surveys, deeds, and court records enable us to compile a tolerably complete list of the owners of each property from the thirteenth—even the twelfth—century to the sixteenth. The possibility of work on these lines was demonstrated long ago from Oxford by H. E. Salter, then from Canterbury by William Urry;[4] more recently it has been achieved for Winchester on a monumental scale by Derek Keene, with a detailed analysis of the urban property market there.[5] At York a similar reconstruction is under way; at London it has begun with Derek Keene's work on the properties of Cheapside; at Southampton the publication of detailed editions of relevant records has brought the same goal well within sight.[6] In all these places archaeology is combining with work on documents to build up a full picture of properties in the towns—and therewith of the traffic in these properties, how they moved from one owner to another. Much less has been done on the smaller towns—the local market centres—and few can provide the abundant evidence that survives from some English medieval cities. But even limited sources can throw some light on the small-town property market: surveys of *c.*1225 and 1441 from Banbury (Oxfordshire), a town from which we have practically no medieval deeds, show clearly the growth of a class of middleman rentiers coming between the occupants of individual houses and their landlord, the bishop of Lincoln.[7] And for some small towns fuller sources make it possible to analyse the local property market in detail. The work of Eleanor Searle on Battle (Sussex) in the thirteenth century is an example; it is based on a substantial collection of charters, rentals, and records of the town court.[8]

However, it is above all from the rural manors of medieval England that we have peculiarly full evidence of the land market, evidence fuller

[4] H. E. Salter, *Survey of Oxford*, ed. W. A. Pantin and W. T. Mitchell (Oxford Historical Society, NS vols. 14, 20; 1960–9); W. Urry, *Canterbury under the Angevin Kings* (London, 1967).

[5] D. Keene, *Survey of Medieval Winchester* (2 vols.; Winchester Studies No. 2; Oxford, 1985).

[6] The work on York and London is as yet unpublished. The Southampton publications are, particularly vols. 15, 19, 20, 24, 25, in the Southampton Records Series: *The Southampton Terrier of 1454*, ed. L. A. Burgess (1976), *The Cartulary of God's House, Southampton*, ed. J. M. Kaye (2 vols.; Southampton Rec. Ser., 1976); *The Cartulary of the Priory of St Denys near Southampton*, ed. E. O. Blake (2 vols.; Southampton Rec. Ser., 1981).

[7] *VCH Oxford*, ed. W. Page *et al.* (in progress, London, 1907–present), x. 6.

[8] E. Searle, *Lordship and Community: Battle Abbey and its Banlieu 1066–1538* (Toronto, 1974), 109–33.

than we have from most other parts of Europe for the market in small pieces of land that were bought and sold by landholders of varying status and wealth but whose estates were to be measured in acres rather than manors, whose annual income was more likely to be expressed in shillings than in tens or hundreds of pounds. The bulk of this evidence is of two kinds, depending on the status of the particular piece of land. If it was in customary tenure, being held—or having been held—by villein tenants bound to the land and to their manorial lord, it could change hands only by passing from the tenant to the lord and thence to a new tenant; this new tenant might be the old tenant's heir, or someone whom the old tenant had nominated (perhaps for a consideration) to succeed him, or someone simply found by the lord. In every case the transfer of the property would be effected by public ceremony in the manorial court and would be recorded on the manorial court rolls. The earliest of such court rolls to survive dates from 1246, and it is likely that this sort of record was then something of a novelty.[9] Indeed, we may wonder whether the manorial court itself had then had for more than two or three generations the form revealed by these records, the form that let it act as a registry of land holding on the manor. Certainly, however, manorial court rolls spread quickly, especially from the 1270s onwards, and from many places in fourteenth- and fifteenth-century England substantial series survive. From certain of these we can reconstruct the succession to every parcel of the manor's customary land, down to holdings of a quarter-acre (0.1 ha.) or less, over a period of two or more centuries, and can reconstruct too the histories of the peasant families that held them—their successive generations, their inter-relationships, the part they played in the village community, their rise or decline in wealth.

In many places most of the lands held by manorial tenants were in customary tenure of this sort. But in many others, especially moving eastward across England, much, most, or all of a manor's tenant lands would be held in free tenure. This could be transferred—inherited, bought, or sold—without reference to the manorial court, indeed without reference to the lord of whom it was held. Conveyance was by public ceremony but from the twelfth century onwards this was ever increasingly recorded, confirmed, and ultimately displaced by written documents, charters which again might be concerned with minute areas of land. Our evidence for the market in land held in free tenure consists of the many thousands of surviving charters of this sort, together with the many more

[9] Harvey, *Manorial Records*, 42.

that were copied into cartularies when the lands in question passed, with their title deeds, into the direct control of a monastery or other substantial estate owner. The pattern becomes more complicated where land in free tenure was acquired by someone who was himself a villein tenant, a bondman. Theoretically such land was automatically forfeit to his manorial lord; in practice its new tenant was normally allowed to keep it, but sometimes only on payment or other conditions, after reporting its acquisition in the manorial court. Thus the court rolls will sometimes give us information about the traffic in free as well as customary land among the manorial tenants. The register in which Peterborough Abbey copied deeds of lands in free tenure acquired by its villein tenants is an invaluable but also, unfortunately, unique record of the mid-fourteenth century.[10]

There are of course other less common records that throw light on the local land market at a particular place and particular time. Two examples are the registers compiled from the manorial court rolls of St Albans Abbey in the mid-fourteenth century and the earl of Kent's charter of 1471 to his tenants at Blunham (Bedfordshire), setting out the rules they were to follow if they wished to alienate any parts of their holdings.[11] In many places the evidence of charters and court rolls can be supplemented by manorial surveys which provide an overall picture of the tenants at a particular date, sometimes at several successive dates; and by manorial accounts which, though purely financial records, may contain entries of payments made on transfers of tenancy.

In all, the historian of the local land market in medieval England has a vast amount of information at his disposal. At the same time it has some serious limitations. One is chronological. Our sources begin, and then rapidly increase in quantity, in the mid-twelfth century (charters) and mid-thirteenth century (court rolls). What happened before that is far from clear; the growth of the local land market may itself have called into being the types of document that record it, or alternatively we may simply be seeing the application of written records to practices that had long been conducted without them. When the written records do appear, they survive particularly from monastic estates, the old-established Benedictine houses being quite disproportionately represented; this is especially true of charters surviving in cartularies, but it applies almost equally to manorial court rolls. This may be less of a problem than it would have appeared a generation ago; we are now more inclined to think that there

[10] *Carte Nativorum*.

[11] Levett, *Studies in Manorial History*, 79–96; Harvey (ed.), *Peasant Land Market*, 198.

was little significant difference between monasteries and other landlords in the way they ran their estates, though monastic administration, more conservative and more efficient than the rest, may have maintained stricter control and thereby left us a fuller record of the local land market than we find elsewhere. At the least, this aspect of our sources is one we should bear in mind. A more serious limitation has appeared in recent years as more work has been done on the local land market and on peasant society in general. We have from many places substantial numbers of manorial court rolls from the late thirteenth century onwards; but there are very few places where the series of rolls is complete, or almost complete, over a period of a century or more. Even quite small gaps in the run of documents drastically reduce their value as evidence of the local land market. J. A. Raftis, E. B. DeWindt and others have done important work in pioneering these techniques from the manorial court rolls of Ramsey Abbey;[12] it is no reflection on them, however, to say that these good, but far from complete, series simply do not permit the comprehensive reconstruction of the village community and its constituent families that has been so successfully undertaken for Halesowen (Worcestershire) by Zvi Razi[13] or for Redgrave (Suffolk) by R. M. Smith.[14] As Zvi Razi has pointed out, it is only when we have the fullest possible series of records that we can identify individual villagers with precision, and our ability to do this is essential to any analysis of family and social relationships; baffling inconsistency in the use of surnames in court rolls can make it difficult to recognize someone who appears under two or more different names or to distinguish between two people who may be given the same name.[15] Another limitation of our sources that has appeared more clearly as work has advanced is that they mostly tell us nothing of sub-tenure. The charters are concerned with tenant and tenant, the court rolls with tenant and manorial lord; neither tell us of any arrangements these tenants may have made for sub-letting lands to other villagers who would actually till them. Yet that such arrangements were commonplace is shown by some references in court rolls, when lords tried to restrict or control the practice, and by a very few surveys that set out the entire structure of sub-tenancies on a particular manor; the amount of sub-letting that these surveys reveal at Caddington and Kensworth (Bedfordshire) in 1297 and

[12] DeWindt, *Land and People*, and Raftis, *Warboys*.

[13] Razi, *Life, Marriage and Death*; Razi, 'Family, Land and the Village Community'; Razi, 'Erosion of the Family-Land Bond', 295–304.

[14] Smith, 'Kin and Neighbours', 219–56; Smith, 'Families and their Land', 135–95.

[15] Razi, 'The Toronto School's Reconstitution', 142–5.

at Havering (Essex) in 1352 suggests that our more usual records of peasant land transactions give us only a partial view of the local land market.[16] A further obvious limitation of our sources is that they tell us less than we would wish about the price of land: what was paid or what exactly was got for the money. This is a point we shall return to. Further research may get round some of the limitations of our sources. These are, however, the difficulties that we must confront.

In 1984 two volumes of studies were published that bear directly on the local land market in medieval England. One was *The Peasant Land Market in Medieval England*, edited by P. D. A. Harvey. This contains studies by four authors, each looking at the way the land market worked in a number of well-documented villages in a particular county: Janet Williamson on thirteenth-century Norfolk, Rosamond Faith on fourteenth- and fifteenth-century Berkshire, Andrew Jones on fifteenth-century Bedfordshire, and Tim Lomas on County Durham in the late fourteenth and fifteenth centuries. An introduction and conclusion by the editor bring together some of the results of this research and look more broadly at certain issues raised. The other book was *Land, Kinship and Life-Cycle*, edited by R. M. Smith. This takes a wider theme—the family, its lands, and its other resources—and covers a longer period, from the thirteenth century to the nineteenth, but the local land market is prominent particularly in its medieval sections, which again consist mostly of detailed studies of individual agricultural communities. These fall into two groups: those by B. M. S. Campbell, R. M. Smith, and J. Ravensdale on places in eastern England (in Norfolk, Suffolk, and Cambridgeshire) in the thirteenth and fourteenth centuries, and by Christopher Dyer and Zvi Razi on places in the west Midlands in the fourteenth, fifteenth, and sixteenth. A further study by Ian Blanchard on industrial employment and rural land market, 1380–1520, looks at the mining communities of Derbyshire and Somerset. In a long and important introduction R. M. Smith, as editor, discusses the book's general themes and pays especial attention to family structure and inheritance customs in the Middle Ages.

The two books are very different—indeed the differences between them are of interest in the context of English medieval historiography. Some of the individual studies would be equally at home in either collection. But in the contributions of the two editors there is a contrast that is more than just a difference of emphasis. *The Peasant Land Market in Medieval England* looks particularly at agrarian and legal forms, using the

[16] A. Jones, 'Caddington, Kensworth, and Dunstable in 1297', *EcHR* 2nd ser. 32 (1979), 322–4; McIntosh, 'Land, Tenure, and Population', 20–5.

approach and the language of traditional historical scholarship. *Land, Kinship and Life-Cycle* is concerned with the family and demography and, while no less firmly based on empirical research, it draws on the methods of the social scientist in its sociological and statistical analysis and in relating its findings to general theoretical work on the family and rural society. But in other ways the two books are alike. Both use the same technique of looking in detail at particular well-documented rural communities which we assume (we can only hope) are typical of many others; in each we meet many individual landholders and their families. Both are much concerned with local inheritance customs and their effects, with the actual (as against the theoretical) differences between partible and impartible inheritance and joint tenure by a whole family. Both are feeling their way towards a pattern of regional variation. But the similarities are unsurprising: both books were produced at the same point in the development of our knowledge and both were bound to reflect current preoccupations of scholars working in this field. Between them they mark a new point of departure for work on the local land market of medieval England. Where, then, do we go from here?

Above all we must advance on a wide front. The subject of *The Peasant Land Market in Medieval England* was defined as 'the lands held by the small-scale landholders, whom we call peasants, and the way these lands moved, by mutual agreement, from one living landholder to another, which we call the peasant land market'.[17] This was a broader definition of the land market than that of some historians, who would use the term only of a substantial volume of transactions or of transfers involving sizeable pieces of land; even so, it excludes from consideration the inheritance of land, manorial lords' leasing of their demesne lands to their tenants, and the enlargement of peasant holdings by simply accumulating holdings directly from the manorial lord as they fell vacant. In practice these matters could not be left out: they are all part of the same picture. *Land, Kinship and Life-Cycle* is, of course, broader still in its terms of reference and sees the local land market only as part of the larger picture of the peasant family and its resources. This certainly is the right approach. We can draw an interesting parallel in the study of deserted medieval villages in England. The work of Maurice Beresford forty years ago sparked off immense interest in the subject, introducing a new dimension to our knowledge of the medieval countryside and rural economy.[18] But the more it was appreciated that depopulation might occur for many different

[17] Harvey (ed.), *The Peasant Land Market*, 1.

[18] M. W. Beresford, *The Lost Villages of England* (London, 1954).

reasons, the more it was realized that the deserted village had to be studied—as Beresford indeed had seen it—in the wider context of settlement patterns and population movement in general. Concentration on the single phenomenon—in that case the deserted village, in this the peasant land market—could then become more of a hindrance than a help to the advance of knowledge.

Maintaining, then, this broad outlook we certainly need more of the detailed local studies that are found in both books. It is abundantly clear that the local land market had very different courses of development in different parts of medieval England. In some areas—as in fifteenth-century Berkshire and Bedfordshire—its size and nature varied a good deal from one place to another; but overall it developed much later in the Midlands than in East Anglia, and in County Durham it was slower still. That there was a regional pattern is certain; what this was, how to identify its regions, is much less clear. The studies comprising *The Peasant Land Market in Medieval England* were adapted from existing doctoral theses, and one reviewer pointed out that while the counties they covered are reasonably distributed across England they do not provide that judiciously selected range of areas and periods that would be the ideal: the book was made up of what already lay to hand.[19] The same comment might be made of *Land, Kinship and Life-Cycle*. Although the contributors here were writing for the book in the first instance, they were writing on areas familiar to them from earlier research; here again the editor drew on pre-existing expertise. And in most of these cases the choice of the area studied may well have owed as much to the individual scholar's personal history and academic career as to the perceived need for work in a particular region or period. There has, of course, been other relevant research outside these two collections; the work of David Roden on the Chilterns is a notable example,[20] John Hatcher's on the leasing and other arrangements on duchy manors in late-medieval Cornwall is another.[21] But the fact remains that when we look at the local land market in the Middle Ages we know far more about East Anglia than about north-west England, far more about the west Midlands than about the counties of the south-west.

[19] M. Stinson, 'The Peasant Land Market in Medieval England' [review], *Landscape History*, 7 (1985), 83.

[20] Roden, 'Inheritance Customs', 1–11; D. Roden, 'Fragmentation of Farms and Fields in the Chiltern Hills: Thirteenth Century and Later', *Mediaeval Studies*, 31 (1969), 225–38; etc.

[21] J. Hatcher, *Rural Economy and Society in the Duchy of Cornwall 1300–1500* (Cambridge, 1970), 52–173.

But beyond this there are some general problems that cry out for investigation. Some are centred on the prehistory of the local land market: what can we learn about it before the mid-thirteenth century, when our sources start to give us a reasonably full picture of its operations? The answer may turn out to be, not very much; but some lines of research are certainly worth trying. One is an analysis of estate surveys from the late eleventh century to the early thirteenth to see what obligations specifically lay not on the individual local tenant but on his household as a whole—harvest works with his whole family for instance—and to see if they fall into any recognizable pattern, geographical or chronological. This might well give some clue to how far customs of joint tenure (as against impartible or partible inheritance) prevailed and, therewith, how far there was a need, as clearly there often was later, for transfers of land between members of a family.

Another area of research, uninvestigated and potentially very interesting, is the rapidly spreading use of seals in the late twelfth and early thirteenth centuries: who owned a seal and by what date? If, as we suppose, monasteries led the way in written conveyance, themselves drawing up charters to record the grants of even small amounts of land that they received (a point that itself needs investigation through analysis of diplomatic formulas peculiar to particular houses) did the monasteries also have seals made for the grantors to authenticate these acts? Did other large landowners do the same? Two agreements of 1217–32 between the earl of Chester and Lincoln and groups of his tenants over pasture in Westfen (Lincolnshire) bear seals of the tenants that were almost certainly made for the occasion; in each case they are uniform (round on one, pointed ovals on the other) but with minor, probably deliberate, differences in design (on one a wheel with six, seven, or eight spokes).[22] Did the possession of seals, thus provided, stimulate local landholders to use written charters in land transfers among themselves? These are interesting questions, and from the answers we might expect to learn something, at least, of the workings of the local land market at this time. But they are difficult questions, too, for these charters are mostly undated and give no direct evidence of the grantors'—the seal-owners'—tenurial or economic standing. We are likely to learn most, at first, by detailed local studies of early peasant seals in particularly well-documented areas.

It is likely too that if we knew more of the way large landed estates

[22] PRO DL 25/2422, 2423. A preliminary discussion of some of these questions is in P. D. A. Harvey, 'Personal Seals in Thirteenth-Century England', in I. N. Wood and G. A. Loud (eds.), *Church and Chronicle in the Middle Ages* (London, 1991), 117–27.

were run in the twelfth century we should know more of the land market at local level. The manors of a large estate were at this time (unlike the thirteenth century) normally farmed out for fixed rents, usually to individuals but sometimes to groups of local tenants. Reginald Lennard in 1958 gave us a careful and interesting analysis of the way the system worked in the late eleventh and early twelfth centuries,[23] and for this period his work has been continued and expanded by other historians.[24] But for the mid- and late twelfth century less has been done and there may well be much for us to learn; it was a period when important changes in estate management were beginning.

When we move to the period, the mid-thirteenth century onwards, for which we have fuller evidence of the local land market we find again that the answers to some general questions might help our understanding. The rule governing the succession to customary holdings on a particular manor may have been primogeniture, ultimogeniture, or impartible inheritance; it is open to debate whether one or other custom served as a stimulus or constraint to the development of a local land market or merely influenced the form it took, but all are agreed that there is a close relationship. We need to carry further still the seminal work of Rosamond Faith in 1966 (and much has been done already since then) in exploring the detailed distribution of these modes of inheritance.[25]

We need too to learn more about the chronology of the decline of week-work in the thirteenth and fourteenth centuries. This was the requirement to work for the lord for so many days each week on whatever jobs needed to be done; it is distinct from the obligation to perform boon-works, specified services—ploughing, reaping, and so on—at particular seasons of the year. Boon-works continued in many places to the sixteenth century or later, but week-work was almost everywhere replaced much earlier by money rents. Week-work was clearly an obstacle to the easy partition and recombination of the holdings that owed it and we may reasonably link its disappearance on a manor with the rise of a local land market there, but which is cause, which effect, is less clear. Given the importance that many historians attach to these labour services in other contexts—social, economic, legal—it may seem extraordinary that little

[23] R. V. Lennard, *Rural England 1086–1135: A Study of Social and Agrarian Conditions* (Oxford, 1959).

[24] e.g. S. P. J. Harvey, 'The Extent and Profitability of Demesne Agriculture in the Later Eleventh Century', in Aston *et al.* (eds.), *Social Relations and Ideas* (Cambridge, 1983), 45–72.

[25] Faith, 'Peasant Families', 77–95. Later work on particular areas is listed in Harvey (ed.), *Peasant Land Market*, 4.

detailed work has been done on the process and the timing of their disappearance. This is because, like the local inheritance custom, what happened on a particular manor can seldom be discovered from a single revealing document. Normally week-work was not suddenly abolished but fell gradually into disuse: every year some or all of the works due would be 'sold' to the tenants who owed them, that is to say, money would be paid instead, while the theoretical obligation to do them remained in force and would be set out in full on any custumal or other statement of services. When in 1327 the abbot of Halesowen commuted into money rents all his tenants' labour services we cannot be certain that they only then actually disappeared: some or all may have been in disuse for many years.[26] Even when manorial accounts record the sale of works the position is not necessarily clear. The archbishop of Canterbury's manorial accounts for 1273–4 mention 7*s*. 7*d*. received at Pagham (Sussex) 'from works released', 40*s*. 2$^3/_4$*d*. 'from works spared', but we are not told how many works or tenants were involved. At Hayes (Middlesex) it tells us more: 8*s*. 6*d*. was paid for 104 works spared between Michaelmas and Christmas.[27] But even here we need a custumal to tell us how many tenants there were and how many works they owed before we can say what proportion the lord had foregone. In any case, what happened one year did not necessarily happen the next: to discover when week-work declined on any one manor we need at least one custumal and a substantial series of accounts covering the significant period.

Both these questions, inheritance and labour services, bear closely on the questions of family structure and family resources that *Land, Kinship and Life-Cycle* explores in such depth. The relevance of inheritance customs is obvious and they are carefully discussed there, and so too is work for wages—what there was and who was to do it—a topic clearly connected to labour services; indeed, the continuance of week-work may itself be more closely linked than has been supposed to the presence in tenant households of surplus members free to perform it. Here, in the family, is of course, the heart of the whole question of the local land market in the Middle Ages; all recent work suggests that immigrant strangers (as distinct from returning natives) took little part in it, and it was not until the late fifteenth or sixteenth century that it began to attract significant attention from outside speculators. Further work on the rural

[26] Razi, *Life, Marriage and Death*, 9. The general question is discussed by Marjory Hollings in her edition of *The Red Book of Worcester* (Worcs. Hist. Soc., 1934–50), pt. iv, iii–vi.

[27] BL Add. MS 29794, mm. 2, 9.

household, its composition and its economic basis, and how these varied in the course of time and from one place—or region—to another, is bound to throw new light on the rural land market. In our present state of knowledge it is probably work on the resources rather than the structure of the family (insofar as they can be separated) that is likely to tell us most. We particularly need to know not only why the local tenant bought land but how: what amounts of money would pass through the hands of a virgater in the course of a year, how (and how fast) he was able to build up the capital needed for land purchase, how far purchase might be facilitated by giving a low purchase price with a permanently fixed annual rent paid to the vendor (a practice illegal after 1290) or by mortgages in the form of a combination of purchase and leasing. Some of these questions have already been discussed, by R. H. Hilton, Christopher Dyer, and others.[28] They are difficult but not unanswerable: various sources potentially throw light on the economy of the peasant household. One example that has been little explored is the lists of corn, livestock, and other goods in each household that were compiled by tax assessors in the late thirteenth and early fourteenth centuries. Most of these lists have failed to survive and only four substantial series have been published (from parts of Suffolk in 1283, of Huntingdonshire in 1290, and of Bedfordshire and Yorkshire in 1297).[29] They are regarded with suspicion, for some are demonstrably formalized in a way that may point to a fictitious element. But, though they are uncommon, we have a significant number of these lists and properly understood and analysed they may well tell us much about the differences in resources between neighbours in a single village, between one community and the next, between one region and another—and therewith about the need and capacity of each to participate in the local land market.

Basic to the economics of the local land market is, obviously enough, the price of land, and systematic research on this is badly needed. English sources for prices paid for small areas of land in the Middle Ages are patchy. We cannot expect to be told how much it cost to acquire customary land as a manorial tenant beyond what was paid to the lord as an entry fine (often a substantial sum). Our record of the transfer is the court roll,

[28] Hilton, *English Peasantry*, 174–214. Dyer, *Standards of Living*, 109–50.

[29] *A Suffolk Hundred in the Year 1283*, ed. E. Powell (Cambridge, 1910); *Early Huntingdonshire Lay Subsidy Rolls*, ed. J. A. Raftis and M. P. Hogan (Toronto, 1976); *The Taxation of 1297*, ed. A. T. Gaydon (Beds. Rec. Soc., vol. 39; 1959); *Yorkshire Lay Subsidy Being a Ninth Collected in 25 Edward I (1297)*, ed. W. Brown (Yorks. Arch. Soc. Rec. Ser., vol. 16; 1894). The most recent discussion of the lay-subsidy records is by J. F. Hadwin, 'The Medieval Lay Subsidies and Economic History', *EcHR* 2nd ser. 36 (1983), 200–17.

which is concerned only with the arrangements first between the old tenant and the manorial lord, then between the manorial lord and the new tenant; what the new tenant paid the old to agree to the transfer and set it in motion was no business of the manorial court, and it could come to its notice, and thus to ours, only through occasional chance reference in, for instance, a subsequent plea of debt. Charters recording transfers of land in free tenure are more informative, however. In the fourteenth and fifteenth centuries they seldom say how much was paid for the property, referring simply to 'a certain sum of money' (*quedam summa pecunie*) or other vague phrase. But earlier charters often record the amount that changed hands. This information may be less useful than appears at first sight. The property may be defined in terms that defy quantification: a bovate, a toft, a named (but unmeasured) piece of pasture. When two thirteenth-century Lincolnshire charters record the sale of seven bovates in Kirkby and Miningsby for £5 and of two bovates in Skegness and Burgh le Marsh for £100 we cannot be comparing like with like; and, still in thirteenth-century Lincolnshire, we are little the wiser about land values for knowing that £4 was paid for a toft, a road 20 feet wide but of unspecified length and the services of five tenants at Wrangle.[30] Again, while telling us the purchase price, the charter may reveal nothing of the rent due to the lord of whom the property was to be held (who might or might not be the vendor) and this could crucially affect the price paid. But many charters define the property, the price, and the rent in precise terms:

4½ acres of arable in Roding, granted permanently, for £20 paid, with 12*d.* annual rent. Early 13th century.

½ acre of meadow in Langham, granted permanently, for 16*s.* paid, with 2*d.* annual rent. Early or mid-13th century.

3 acres of arable and a messuage in Walden, granted for a term of three lives by the manorial lord to its former tenant in villeinage, for 40*s.* paid, with 6*d.* annual rent. Late thirteenth century.

Charters that give this information survive in thousands rather than hundreds. These examples are from a group of Essex charters in the records of the duchy of Lancaster.[31] There is some ambiguity in the word *acre* but this need not be an insuperable problem. More of a drawback to exact analysis is that they bear no dates and are dated here simply from their handwriting; for charters copied into a cartulary we lack even this aid

[30] PRO DL 25/2373, 2401, 2477.

[31] PRO DL 25/1938, 2013, 1945.

to chronology. Tiresomely, the charters recording such transfers start regularly bearing dates at just the point—the end of the thirteenth century—when they stop giving the purchase price.

Despite this it would be of great value if cartularies and collections of charters were systematically searched to list (preferably in machine-readable form) all those that state both purchase price and rent. The resulting corpus of information would tell us much about the price of land and how it varied by region and over time; it would also provide a yardstick to assess what we are told of land values in another important source, the extents that we have from many estates from the mid-thirteenth century onwards. The extent was a form of survey that valued every item in terms of the rent it would produce if it were leased out. The only systematic study so far made of these land values is by J. A. Raftis. In examining the extents attached to inquisitions post mortem from eight east Midland counties from the mid-thirteenth to the mid-fourteenth century he discovered an extraordinarily wide range of values.[32] If manors can be found (and they probably can) where charters record the prices paid for particular pieces of land and subsequent extents assess their annual value we might discover some reasonably consistent relationship between annual values and purchase price. In this case we could begin to use, with more confidence than we can now feel, the values recorded in extents as a general guide to land prices and their variations from one place or region to another and over the course of time. We have only casual sources of evidence for the prices paid even for land in free tenure in the local land market of the fourteenth and fifteenth centuries. We can only regret that other owners of modest estates did not follow the examples of Richard and Thomas Hotot and Henry de Bray in thirteenth- and fourteenth-century Northamptonshire; their memorandum books include records of their successive purchases of land and of the price paid in each case.[33]

The collection of land prices from charters would be valuable too for comparing rural property prices with urban and for comparing the prices paid for small pieces of land by local tenants with those paid by large-scale free tenants and estate-owners for substantial areas, or for entire manors. This too is work that is needed: bringing the peasant land market

[32] J. A. Raftis, *Assart Data and Land Values: Two Studies in the East Midlands 1200–1350* (Toronto, 1974), 11–96.

[33] *A Northamptonshire Miscellany*, ed. E. King (Northants Rec. Soc., vol. 32; 1983), 16–20, 23, 31–3, 36; *The Estate Book of Henry de Bray of Harleston, co. Northants* (*c.1289–1340*), ed. D. Willis (Royal Historical Society, Camden 3rd ser. vol. 27; 1916), 11, 27–8, 36–40, 48–51, etc.

into relation with the property market in general and bringing both into relation with the national (and international) economy as a whole. It has now been firmly shown that the local land market cannot be properly viewed apart from the families and communities that engaged in it. But it has new light to throw on economic as well as on social history and this too needs to be further explored and exploited.

ABBREVIATIONS

BL	British Library
Carte Nativorum	*Carte Nativorum: A Peterborough Abbey Cartulary of the Fourteenth Century*, ed. C.N.L. Brooke and M.M. Postan (Northants Rec. Soc. 20, 1960)
DeWindt, *Land and People*	E.B. DeWindt, *Land and People in Holywell-cum-Needingworth: Structures of Tenure and Patterns of Social Organization in an East Midlands Village 1252–1457* (Toronto, 1972)
Dyer, *Standards of Living*	C. Dyer, *Standards of Living in the Later Middle Ages* (Cambridge, 1989)
EcHR	*Economic History Review*
Faith, 'Peasant Families'	R.J. Faith, 'Peasant Families and Inheritance Customs in Medieval England', *Agricultural History Review* 14 (1966), 77–95
Harvey, *Manorial Records*	P.D.A. Harvey, *Manorial Records* (British Records Association, Archives and the User no. 5, London, 1984)
Harvey, *Westminster Abbey*	B.F. Harvey, *Westminster Abbey and its Estates in the Middle Ages* (Oxford, 1977)
Harvey (ed.), *Peasant Land Market*	P.D.A. Harvey (ed.), *The Peasant Land Market in Medieval England* (Oxford, 1984)
Hilton, *English Peasantry*	R.H. Hilton, *The English Peasantry in the Later Middle Ages* (Oxford, 1975)
King, *Peterborough Abbey*	E. King, *Peterborough Abbey 1086–1310: A Study in the Land Market* (Cambridge, 1973)
Levett, *Studies in Manorial History*	A.E. Levett, *Studies in Manorial History*, ed. H.M. Cam, M. Coate, and L.S. Sutherland (Oxford, 1938)

McIntosh, 'Land, Tenure and population'	M.K. McIntosh, 'Land, Tenure and Population in , the Royal Manor of Havering, Essex, 1251–1352/3', *EcHR* 2nd ser. 33 (1980), 17–31
PRO	Public Record Office
Raftis, *Warboys*	J.A. Raftis, *Warboys: Two Hundred Years in the Life of an English Mediaeval Village* (Toronto, 1974)
Razi, 'Erosion of the Family-Land Bond'	Z. Razi, 'The Erosion of the Family-Land Bond in the Late Fourteenth and Fifteenth Centuries: A Methodological Note', in Smith (ed.), *Land, Kinship and Life-Cycle*, 295–305
Razi, 'Family, Land and the Village Community'	Z. Razi, 'Family, Land and the Village Community in Later Medieval England, *Past and Present* 93 (1981), 3–36
Razi, *Life, Marriage and Death*	Z. Razi, *Life, Marriage and Death in a Medieval Parish: Economy, Society and Demography in Halesowen, 1270–1400* (Cambridge, 1980)
Razi, 'The Toronto School's Reconstitution'	Z. Razi, 'The Toronto School's Reconstitution of Medieval Peasant Society: A Critical View', *Past and Present* 85 (1978), 141–57
Roden, 'Inheritance Customs'	Roden, D., 'Inheritance Customs and Succession to Land in the Chiltern Hills in the Thirteenth and Early Fourteenth Centuries', *Journal of British Studies* 7 (1967), 1–11
Smith, 'Families and their Land'	R.M. Smith, 'Families and their Land in an Area of Partible Inheritance: Redgrave, Suffolk 1260–1320', in Smith (ed.), *Land, Kinship and Life-Cycle*, 135–95
Smith, 'Kin and Neighbours'	R.M. Smith, 'Kin and Neighbours in a Thirteenth Century Suffolk Community', *Journal of Family History* 4 (1979), 219–56
Smith (ed.), *Land, Kinship and Life-Cycle*	R.M. Smith (ed.), *Land, Kinship and Life-Cycle* (Cambridge, 1984)
VCH	*Victoria County History*

XI

Personal Seals in Thirteenth-Century England*

[117] English medieval seals are a virtually untapped source of historical information. For centuries seals have been collected, they are often used to illustrate books on medieval history, and particular seals have been discussed in a limited, antiquarian context: successive seals of a magnate, a monastery or a borough. But systematic study of their design, their manufacture, their use, their significance, has largely been confined to the royal Great Seal and its surrogates in various courts and departments. For most other sorts of seal we have not even the most elementary typology, the simplest chronological outline of their development.[1] Given that seals have been the subject of continuous antiquarian interest for at least 350 years it is amazing that we have had to wait until the 1980s for answers to the basic questions on the form and use of the earliest English seals recently discussed by Dr. T.A. Heslop.[2]

But if English medieval seals in general are a little-explored and under-exploited area, personal seals of the non-armigerous classes – or, indeed, any personal seals without coats of arms – are all but virgin territory. Guides to it are lacking. W. de G. Birch's catalogue of seal impressions in the British Museum (now British Library) is by far the largest catalogue of seals to have been published in Britain and is the primary work of reference on the subject.[3] It lists nearly 15,000 English and Welsh seals under headings which include only three sorts of personal seals of laymen other than rulers: 'Equestrian seals', 'Figures [118] of noble and other ladies' and 'Heraldic seals'. As he

---

* The article has been reset with minor changes, but the original page numbers (which have not been used for the index) have been entered in square brackets.

1 Two valuable works do, however, look at these questions and draw attention to their interest and significance: H. Jenkinson, *Guide to Seals in the Public Record Office* (London, 1954), especially pp. 4–9, 56–62, and M.T. Clanchy, *From Memory to Written Record: England 1066–1307* (London, 1979), especially pp. 35–6, 244–8.

2 T.A. Heslop, 'English Seals from the Mid Ninth Century to 1100', *Journal of the British Archaeological Association*, cxxxiii (1980), 1–16, pls. I–III; *idem*, 'Seals as Evidence for Metalworking in England in the Later Twelfth Century', in *Art and Patronage in the English Romanesque*, ed. S. Macready and F.H. Thompson (Society of Antiquaries of London, 1986), pp. 50–60.

3 W. de G. Birch, *Catalogue of Seals in the Department of Manuscripts in the British Museum* (London, 1887–1900).

continues with a further 8,500 entries of Scottish, Irish and foreign seals, Birch can hardly be blamed for excluding the vastly larger number of English and Welsh personal seals that fell outside these categories. However, this limitation is reflected in the only two general books on seals written in English: each has just one chapter, one of the shortest in the book, on the seals of non-armigerous laymen. In J.H. Bloom's work, in 1906, it is called 'Seals of Private Gentlemen and of Merchants' and in Birch's own book, published the following year, 'Miscellaneous Designs and Devices of English Seals'.[4] As Mr. R.H. Ellis pointed out in 1978, in introducing his catalogue of seals in the Public Record Office, 'sigillography ... cannot be fully or systematically studied until our record repositories have compiled, and made accessible, catalogues of all the seals ... which they hold'.[5] Knowing what seals there are is an essential first step; and Mr. Ellis's own work has contributed significantly to it. The first two volumes of his catalogue cover the personal seals on the Ancient Deeds 'S series', those documents selected and set aside from the main classes of Ancient Deeds for the fineness of their seals. In contrast to Birch's catalogue, every kind of personal seal is represented, though as the criteria of selection for the 'S series' included fineness of design and of engraving as well as of condition there is still a disproportionate number of the seals of the armigerous classes.

Among the seals catalogued are those on the 331 'S series' Ancient Deeds in the records of the Duchy of Lancaster (Public Record Office class D 27). The observations that follow on thirteenth-century personal seals derive primarily from work, in continuation of Mr. Ellis's, cataloguing the seals in the other series of Duchy of Lancaster Ancient Deeds (classes D 25, D 26); these comprise 3,757 documents on which there survive impressions of nearly 2,500 different personal seals. The majority date from the thirteenth and fourteenth centuries and they are probably a reasonably typical sample of the many thousands of English medieval seals that survive.[6] Their geographical spread is good: the documents concern transactions in most parts of England and the Welsh Marches, but Essex, north Lancashire and north-east Lincolnshire are particularly well represented. Sixteen per cent of the seals are heraldic, and a further two per cent are equestrian or bear figures of ladies; the remaining [119] eighty-two per cent are types of seal that were excluded from Birch's catalogue,

---

[4] J.H. Bloom, *English Seals* (London, 1906), pp. 177–85; W. de G. Birch, *Seals* (London, 1907), pp. 189–93.

[5] R.H. Ellis, *Catalogue of Seals in the Public Record Office: Personal Seals* (London, 1978–81), i.vii.

[6] The primary listing of the seals on PRO DL 25/1–640 was by Mr. R.H. Ellis and I am grateful for his permission to draw on this work; I accept full responsibility, however, for all descriptive references made here to these documents and their seals.

the seals confined to one small chapter in his and in Bloom's monographs. The potential interest of these seals is considerable. They probably have much to tell us about the craftsmen who made the dies as well as about the men and women who owned and used them. In 1953 Professor R.H. Hilton discussed the significance of a group of peasant seals from thirteenth-century Gloucestershire, but then – and still now – there was no way of seeing them in their sigillographic context.[7] What is offered here is no more than a first scratching of the surface: certain points that leap to the eye in working through all the seals in this group. If it is as good a sample as it seems, these points will be generally valid, but they will need testing against other substantial collections of original thirteenth-century deeds. At best they can be no more than a starting-point for the detailed investigation that is needed of every aspect of the manufacture and use of personal seals in medieval England.

'It was not the custom in the past', said the king's chief justiciar, 'for every little knight to have a seal.' He was speaking perhaps in the 1160s.[8] By the end of the twelfth century the use of seals had spread far beyond the humblest knights, beyond the ranks even of those whom J.H. Bloom called 'Private Gentlemen'. The land market might involve anyone who held land that could be sold and, though the detailed chronology has still to be elucidated, it was commonplace for transfers at least to religious houses and other important landholders to be confirmed with written charters. Being a party to a charter meant owning a seal. It was practically unknown in the early thirteenth century for a grantor, having no seal of his own, to use someone else's or, like one late twelfth-century grantor in Nottinghamshire, to impress the wax with his wife's key instead.[9] Almost certainly the number of people owning seals increased in the first half of the thirteenth century as it became more and more usual for all transfers of free land, as well as many other transactions, to be confirmed by written charters. It is less certain that seal ownership descended any further down the social scale, for it may effectively have had no further to go: those who had seals in 1200 may already have included villeins as well as townsmen and the smallest freeholders. In fact we seldom learn the legal or economic status of those who sealed documents. It is most unusual for a charter to identify as a villein – or as a free man – the man who sealed it, as one Gloucestershire lease

---

7 R.H. Hilton, *The English Peasantry in the Later Middle Ages* (Oxford, 1975), pp. 153–5 (reprinting 'Gloucester Abbey Leases of the Late Thirteenth Century', *University of Birmingham Historical Journal*, iv [1953], 1–17).

8 *The Chronicle of Battle Abbey*, ed. E. Searle (Oxford, 1980), pp. 214–15.

9 *Documents Illustrative of the Social and Economic History of the Danelaw*, ed. F.M. Stenton (British Academy, 1920), p. 275.

[120] did in 1274.[10] And in our present state of knowledge we cannot begin to correlate particular sorts of seal with the standing of their owners beyond what is self-evident: that only the armigerous had heraldic seals and only the well-to-do had elaborate, finely-engraved ones.

One change that occurred in the thirteenth century is very clear: in 1200 a personal seal almost always named its owner, in 1300 it did not necessarily do so. At some date between 1217 and 1232 the grant to Ranulf, earl of Chester and Lincoln, of 500 acres of marshland at Freiston and Butterwick, Lincolnshire, was confirmed in a document bearing seals of fifty of the local tenants; most survive intact and every one of these bears the same name as is written on the tag to which it is attached.[11] In 1274, on the other hand, an indenture leasing two mills at Langham, Essex, was sealed not only by the two lessees but also by their twelve sureties; of the eight seals that survive only two bear the owner's name.[12] There may have been some regional variation in this: of the fifty-seven seals of men from Roos and Rewaynok cantreds in Denbighshire that were once attached to an agreement of 1295 only seventeen survive and are in poor condition, but most, perhaps all, of these bear the name of the signatory.[13] However, the broad trend appears clearly in comparing the thirteenth-century personal seals in classes DL 25 and DL 26 with those of the fourteenth century:[14]

	1201–1300	*1301–1400*
	Per cent	*Per cent*
Equestrian	2	0
Figures of ladies	1	0
Heraldic	9	32
Others with names	80	25
Anonymous	8	42

In the early thirteenth century we very occasionally find armorial seals that instead of the owner's full name have the surname alone or the family

10 Hilton, *English Peasantry*, p. 154.

11 PRO DL 27/270, illustrated in Jenkinson, *Guide to Seals*, pl. II.

12 PRO DL 25/1512.

13 PRO DL 25/2064.

14 Seals on undated documents with estimated date encroaching on two centuries (e.g. 'late thirteenth or early fourteenth') have been counted in both. '0' means less than 0.5 per cent. Seals bearing no name but two initials have been counted with those with names, those with a single initial have been counted as anonymous, but it was not until the fifteenth century that seals with a single letter as the device became common.

motto.[15] They may be seen as precursors of the anonymous [121] seals of the late thirteenth century, but these have a clearer direct line of descent from the seals that important people had as privy seals. This privy seal might be used as a counterseal, impressed on the back of the wax that bore the formal seal on the front, or by itself to authenticate less important documents. Usually it was inscribed 'Secretum' or 'Sigillum secreti' (or similar wording) with the owner's name. Sometimes the privy seal bore a coat of arms and might be not much smaller than a formal seal. An example is that of Robert de Breteuil, earl of Leicester 1191–1204 ('Zecretum Roberti Bretuil'), 30 mm. across and used as a counterseal to his 70 mm. equestrian seal (*plate 1a*).[16] More often the privy seal was an engraved gem set with the inscription around it and probably often worn in a signet ring. The privy seal of John de Lacy, constable of Chester in the early thirteenth century ('Secretum Iohannis de Lacy'), was a gem engraved with a man's head.[17] Occasionally these privy seals made of gems omitted the owner's name: those of William Percy about 1200 (*plate 1b*) and William de Lancaster between 1220 and 1246 read simply 'Sigill' secreti', 'Sigillum secreti'.[18] More unusual is the counterseal of Henry de Beaufou about 1200 (neither a gem nor armorial: a fleur-de-lis) which has no inscription at all.[19] At the same period William son of William de Ashford used as a counterseal a gem with the inscription 'Cela segreta Will'i' (*plate 1c*), while the seal of Osbert, chaplain of William Brewer, again a gem, was inscribed 'Lecta cela nec revela';[20] unlike the others, privy seals of the well-to-do, this last may well have been its owner's sole seal, though, as we shall see, it was not only the great who owned more than one seal at this time. These two last inscriptions are of particular interest. First they remind us that the primary purpose of a privy seal at this period was probably not only to authenticate but also to close securely a private communication.[21] Second they are forerunners of the mottoes inscribed on anonymous seals from the 1270s onwards.

This is the sigillographic background to the appearance of anonymous seals that are neither counterseals nor gems. The earliest that can be dated in

---

15 E.g. PRO DL 25/2388 (surname alone: Richard Bacon, early or mid thirteenth century), DL 25/522 (motto alone: Thomas Hawtrey, temp. King John).

16 Ellis, *Catalogue of Seals*, i, nos P295, P296. Here and subsequently inscriptions from seals are given with lower-case letters where appropriate and with the use of i and j, u and v normalised.

17 *Ibid.*, ii, no. P1639.

18 *Ibid.*, ii, no. P1867; PRO DL 25/578.

19 Ellis, *Catalogue of Seals*, i, no. P61.

20 *Ibid.*, ii, no. P956; i, no. P910.

21 Cf. Heslop, 'English Seals', pp. 14–16.

1a. Privy seal of Robert de Breteuil, earl of Leicester 1191–1204
A large (30 mm. diameter) armorial privy seal, with owner's name in the legend, used here as a counterseal (*PRO E 42/266*)

1b. Privy seal of William Percy, *c.* 1200
A gem with legend, 'Sigill' secreti', not naming the owner, used here as a counterseal (*PRO DL 27/83*)

1c. Privy seal of William de Ashford, early thirteenth century
A gem with legend, 'Cela segreti Will'i', used here as a counterseal (*PRO E 212/68*)

1d. Seal used by William de Micheldever, 1252
An early example of an anonymous seal, inscribed like a privy seal 'Sigillvm secreti' but used by itself to authenticate a formal document (*PRO DL 25/2066*)

the Duchy of Lancaster Ancient Deeds is that of William de Micheldever in 1252 (*plate 1d*),[22] and another two also date from the 1250s;[23] all are inscribed 'Sigillum secreti' (or a variant). There are none from the [122] 1260s, but from 1272 onwards there are many more: over twenty from 1272 to 1280, mostly with a general motto,[24] but at least one with no legend at all.[25] There are of course others on undated documents, and as it was only in the 1270s that it began to be a common practice to give the date on private charters it may well be that the anonymous seal arrived much more rapidly on the scene in the 1250s or 1260s. But at latest by the reign of Edward I the anonymous seal must have been fully accepted as authenticating a document, and in the anonymous seals of the late thirteenth century we find already designs and mottoes (in English and French as well as Latin) that recur over and over again in the next century: the sleeping lion with 'Wake me no man', the stag's head with 'Timete deum', the squirrel, the lamb and flag, the hawk and its prey, the mottoes 'Frange lege tege', 'Mater dei memento mei', 'Sohou' (a hunting-cry) and more besides.[26] From the same period we have a few armorial seals without the owner's name or family motto; the earliest that can be dated in the Duchy of Lancaster deeds is of 1269.[27] Some are inscribed 'Sigillum secreti', some have the same sorts of motto as the non-armorial seals, and some have no inscription at all. If a seal did not bear its owner's name there was no reason why it need be used by only one person. But to judge from the Duchy of Lancaster deeds an anonymous seal by 1300 was normally still as personal to an individual as a seal that bore his name. Where a duplicate impression has been found the signatory has always been the same, even where the charters are years apart; thus Adam le Charman of Rickling, Essex, used in 1296 and again in 1301 an unusual anonymous seal showing a dog and a hare, probably playing dice, and inscribed 'Hasard þin hod is min' (*plate 2a*).[28]

However, it was not necessarily by the seals of the parties to the transaction that a charter was authenticated in the thirteenth century. The seal of a local official might be attached as well, like the seal of the reeveship of Southampton, added to local charters (no doubt against a fee) 'for greater security'.[29] The charter

22 PRO DL 25/2066. It is undated but must have been drawn up not much more than seven weeks after DL 25/2065.

23 PRO DL 25/173, 2102.

24 E.g., PRO DL 25/3423, dated 1272, 'Crede ferenti'.

25 PRO DL 25/1512.

26 E.g., PRO DL 25/1694, 1375, 384, 3190, 1512, 2315, 185, 24.

27 PRO DL 25/2231.

28 PRO DL 25/1684, 1685.

29 E.g., Southampton City Record Office SC 4/2/1, 4/2/9.

for one small grant of land to Beaulieu Abbey before 1230 apparently bore the seal not of the grantor at all, but of his overlord, again 'for greater security'.[30] When the clergy of the deanery of Holderness wrote to the Archbishop of York in 1281 their letter bore the deanery seal instead of (perhaps as well as) their [123] own, 'since our own seals are not known to you';[31] and it was this or some similar phrase that was normally used in the fourteenth and fifteenth centuries if a document was given added weight by bearing the seal of some great person or official besides or in place of the seal of the actual party.[32]

It was not only for corroboration that other people's seals were used in the thirteenth century, for sealing was not always carried out with the careful formality that these examples might suggest. Where husband and wife were a party to a charter it would normally bear both their seals; but one charter of Robert de Veilly and his wife Joan about 1300 has a single seal and says in the sealing clause that it was sealed only by Robert, while another of Robert Acca and his wife Alice in the mid or late thirteenth century, again said to have been sealed by the husband alone, was in fact sealed only by Alice.[33] Rose, widow of John Smith of Pleshey, Essex, again in the mid or late thirteenth century, attached what the sealing clause calls 'my seal' to her charter; but the seal is actually that of John Smith himself.[34] At Poyntington, Dorset, in the late thirteenth century we find a man using his father's seal, apparently during his father's lifetime, and on the face of it the seal used by Gilbert, son of Gilbert, son of Richard of Mablethorpe, Lincolnshire, was also his father's, for it bears the name of 'Gilb' fil' Ric''.[35] It may be, though, that a clerical error gives his name incorrectly in the charter, and a similar mistake may possibly explain why Margaret, daughter of William son of Christian of Skidbrooke, Lincolnshire, used a seal with the name Margaret daughter of John.[36]

Possibly but not certainly, for already in the thirteenth century we occasionally find someone using a seal with another person's name on it, a name that does not appear to belong to a close relative. In the 1230s Roger son of William de Moorcroft used the seal of Hugh de Frodsham.[37] A charter of Alice de

---

30 *The Beaulieu Cartulary*, ed. S.F. Hockey (Southampton Records Series, 1974), pp. 94–5.

31 *The Register of William Wickwane, Lord Archbishop of York, 1279–1285*, ed. W. Brown, Surtees Society (1907), p. 249.

32 E.g., *Calendar of Close Rolls 1364–8*, p. 463 (in 1368); *The Boarstall Cartulary*, ed. H.E. Salter, Oxford Historical Society (1930), p. 199 (in 1452).

33 PRO DL 25/3628, 3111.

34 PRO DL 25/1443.

35 BL Egerton Ch. 1577; PRO DL 25/2768.

36 PRO DL 25/2950.

37 PRO DL 25/594.

la Hill and her son Walter de la Hill bore two seals when it was drawn up in 1274; only one survives and this bears the partly illegible name of Richard –y, clerk.[38] That this last was a clerk's seal suggests that these may be early examples of what demonstrably [124] occurred in later centuries: the clerk who wrote the document owned the seal and used it on the acts of any of his clients.[39] Alternatively Roger and either Walter or Alice may simply have borrowed the seals they used or (perhaps more likely) acquired them secondhand: there are various possibilities. These however were only precursors of what was to come. Before long any remaining inhibitions over using another person's seal were cast aside, whether or not it bore a name. Thus five seals on a charter of 1315 include two impressions of the seal of one William whose surname is obscure on both but who was certainly not one of the five named signatories.[40] Twelve seals on a charter of 1340 include two impressions of each of three seals, all anonymous; of the other six, four are anonymous, one heraldic seal bears a name that does not appear in the charter and only one seal bears the name (and also the arms) of a signatory.[41] From very early in the fourteenth century it seems to have been no longer the norm for a signatory to a document to have a seal of his own. Important or well-to-do persons who often put seals to documents would of course have their own seals, but many lesser people would sign with a seal that the clerk, or perhaps a friend, supplied. The seal-owning proportion of the population indeed increased in the early part of the thirteenth century, but by, say, 1320 it was probably much smaller than a hundred years earlier. This is one reason why the proportion of heraldic seals rose so sharply from the thirteenth to the fourteenth century, in DL 25 and DL 26 from a tenth to a third of the total, as the table above shows: fewer of the non-armigerous now owned seals.[42]

But in the early and mid thirteenth century the humbler person who sealed a document need not own just one seal: he might own two or more, each with his name on it. These were not like the formal and privy seals of the great, each

---

38 PRO DL 25/1380.

39 Thus a seal bearing the (partly illegible) name of Geoffrey Le–ur occurs on two charters written in the same hand, one on 10 January 1308, where it is the seal of Mabel, wife of William ate Noke, and the other on 14 May 1308, where it is the seal of Hugh de Neville (PRO DL 25/1722, 1698). An anonymous seal (but with the letter W as its device) appears on charters written in a single hand on consecutive days in 1465, on 27 November as the seal of William Bernard (two copies) and on 28 November as the seal of Felicia Wood (PRO DL 25/1104–1106).

40 PRO DL 25/1311. The seal may possibly read 'Sigillum Willelmi Syrugoni', i.e. William Surgeon.

41 PRO DL 25/1016.

42 Others include the growing fashion for displaying coats of arms and an increase in the number of people bearing them.

serving a distinct function; they were seals of similar form and identical use, sometimes of similar design, sometimes not. Two seals of John, son of Peter Gall of Saltfleetby, though from different dies, have the same design of a flower with eight [125] petals;[43] of three with the name of William, son of Osbert le French of Saltfleetby, one has a five-petalled flower, one a fleur-de-lis and one a five-rayed star (*plate 2b–d*).[44] These examples, and most others in the Duchy of Lancaster archives, are from the many Lincolnshire charters, especially those from the Saltfleetby area, a large group of conveyances of which most seem to be by very small landholders.[45] But duplicate seals seem to be found among the greater local landholders as well as the lesser and in all parts of the country: examples elsewhere include those of Adam de Marton in north Lancashire and Thomas de Hurtmore in Surrey.[46] It is always possible of course that two seals with the same name belonged not to the same man but to namesakes – the more likely in that in the thirteenth century two brothers might have the same Christian name. Sometimes this was certainly the case. One grant at Saltfleetby was by two brothers; both were named Thomas, son of Hugh son of Godric, on the charter, Thomas, son of Hugh, on their seals, and if the seals had appeared on separate documents one might have thought they were different seals of the same man.[47] But the contents of the documents occasionally show that two seals must belong to the same person – two of the three seals bearing the name of Henry Beck of Lusby are a case in point[48] – and the phenomenon is anyway too common to be explained simply by the existence of namesakes: in the Duchy of Lancaster charters from the Saltfleetby area more than twenty names occur on seals from two, three or even four different dies.[49] And if some seeming duplicate seals were really the seals of namesakes, others owned by a single person may pass unrecognised because an individual might be known by more than one bye-name. William, son of Geva, whose seal is on a charter of William, son of Walter of Saltfleetby, is presumably the same as William, son of Walter, whose seal is on a charter of William, son of Geva, again of Saltfleetby; but if each seal had been attached to the other charter, so that in each case the same name was on charter and seal alike, this would have escaped

43 PRO DL 25/2155, 2441.

44 PRO DL 25/2670, 2807, 2864.

45 The group would repay detailed investigation, along with the Saltfleetby charters in the archives of Magdalen College, Oxford (I am grateful to Dr C.M. Woolgar for details of these).

46 PRO DL 25/443, 445 (Marton); E 210/3559, E 326/4013, 4018 (Hurtmore).

47 PRO DL 25/2886.

48 PRO DL 25/2929, 2930.

49 One on four is William de Redbourne: PRO DL 25/2996, 3020, 3031, 3046.

a

b

c

d

2a. Seal used by Alan le Charman, 1296–1301
An anonymous seal showing a dog and a hare gaming, with the motto 'Hasard þin hod is min' (*PRO DL 25/1684*)

2b–d. Three seals of William son of Osbert le French, mid thirteenth century
Each seal has a different design: a fleur-de-lis, a five-petalled flower and a five-rayed star (*PRO DL 25/2670*, *2807*, *2864*)

us.[50] Seal-dies do not wear out as coin-dies do; nor can they have been lost or thrown away with careless abandon. Why then do we find people owning more than one seal?

A method has yet to be established, the criteria worked out, for identifying the work of a single engraver of these small personal seals. [126] When this can be done we shall be in a position to learn a great deal about a little-known medieval craft. Sometimes, however, two or more seals are so alike in design, style and execution that we cannot doubt that the same engraver made the dies together. Usually there is a clear family connection between the owners of these seals. Among the charters from the Saltfleetby area, for instance, we find several pairs of seals, clearly engraved together, belonging to husband and wife,[51] one belonging to two sisters, one belonging to two brothers.[52] Occasionally, however, a charter bears a group of seals so alike that they must have been engraved together yet with no obvious connection between their owners beyond what was involved in the particular transaction. An example is a document drawn up at Monmouth in 1280 by which the five heirs of Robert le Frere quitclaimed a piece of meadow to Edmund, earl of Leicester and Lancaster. The seals of four survive: Juliana and Margery, the daughters of John Baldekat, Felicia le Frere and Margery widow of Walter (named Margery Frere on her seal). It seems to have been only their kinship to Robert that connected all four, yet their seals have every appearance of having been engraved together as a set: all are pointed ovals, uniform in style, with a different but similar device on each (*plate 3*).[53] Another example is a group of affidavits by which five prisoners held for King John by H. (presumably Hugh) de Neville undertook on the security of their lands not to escape; each is written in the same hand, each has a seal in the same uncoloured wax on the same yellow and red silk cord, and each seal is round, about 35 mm. across and in a uniform style but with a different device.[54] Other groups are on two agreements made between 1217 and 1232 with Ranulf, earl of Chester and Lincoln, over rights of marsh and pasture in Westfen in Lincolnshire. Each has uniform seals. One has four, all pointed ovals with variously elaborated forms of a fleur-de-lis or stylised spray as the device (*plate 4*).[55] On the other, eight seals survive and seven of them are uniform, being round and about 31 mm. across; four of these again have variations of a fleur-de-lis or a spray, while the other three have a cartwheel,

50 PRO DL 25/2786, 2787.

51 E.g., PRO DL 25/1257, 3163.

52 PRO DL 25/2733, 2554.

53 PRO DL 25/2050.

54 PRO DL 25/724, 725, 1343, 1344, 3452.

55 PRO DL 25/2423.

3. Seals of four heirs of Robert le Frere, 1280
All four seals were probably made by a single engraver for this transaction; note, for instance, the reversed S and the E with straight upright and converging bars (*PRO DL 25/2050*)

4. Seals of four tenants at Westfen, Lincolnshire, between 1217 and 1232
A further example of seals probably engraved as a set for a particular transaction; here note the angular S (almost a reversed Z) and the distinctive form of G (*PRO DL 25/2423*)

one with six spokes, one with seven and one with eight. These last three seals may admittedly have belonged to one family: their owners were William, son of Reginald, and Reginald and William, nephews of Toli.[56] But in all there can be little doubt that it was a known practice for seal-dies to be cut and supplied expressly for a particular transaction, a particular act of sealing.

[127] This may explain how people came to possess duplicate seals. In using written charters to record the oral conveyance of small pieces of land it seems to have been the monasteries and perhaps lay magnates that led the way: at the beginning of the thirteenth century the villager selling his land to one of his fellows would do so simply by public ceremony, but if the purchaser was a religious house it would seek to have this confirmed by a charter. It was the monastery that drew up the charter (it was, after all, the monastery that wished to have it) and if the smallholding vendor had no seal of his own one would be supplied, which he would then keep as a perquisite or souvenir. Sometimes a seal may have been supplied inadvertently to a grantor who already had one of his own; indeed, the practice may for a while have hardened into custom, so that the grantor expected a new seal whether or not he already had one. This would explain how a person's two or three seals might all have the same design: the recipient would select his own device and might stick to the same one each time, making it a kind of personal mark which may have been used in other contexts as well – to mark livestock or other property for instance. This is merely guessing; but it is a possible explanation of a rather odd phenomenon.

And we can speculate a little further. Three seals with the name of Henry Beck of Lusby in Lincolnshire all have the same design, a cross with the ends splayed and curled round; the documents they attest are dated 1227, 1231 or are undated.[57] Subsequently the arms of the Beck family (among them Anthony, bishop of Durham 1283–1311) were this same symbol, a cross moline. But need it follow that we can carry this coat of arms back to the early thirteenth century? The personal mark of one generation might, given an appropriate turn in family fortunes, be the coat of arms of the next. Many coats of arms of the later middle ages may have had a rather humbler origin than as a mark of recognition borne by a knight in armour. Social aspiration, good fortune or simply the spread of heraldry will have moved them up in the world. Certainly, however, it was not social pretensions that account for the wide ownership of seals in thirteenth-century England; it was the demand for written charters of confirmation by those to whom land was given or sold.

---

56 PRO DL 25/2422.

57 PRO DL 25/2929, 2930, 2931 (this last seal occurs also on 3213 and 3214).

XII

Agricultural Treatises and Manorial Accounting in Medieval England

Review Article of Dorothea Oschinsky, *Walter of Henley and other Treatises on Estate Management and Accounting*. Clarendon Press, Oxford, 1971. xxiv + 504 pp. £7.

IN 1890 the Royal Historical Society published *Walter of Henley's Husbandry*, edited by Elizabeth Lamond. This contained text and translation not only of Walter of Henley's work, but also of three other thirteenth-century treatises: the anonymous 'Husbandry', the 'Seneschaucie', and the 'Rules' of Robert Grosseteste. In 1934 Eileen Power read a paper to the same society opening a discussion 'On the need for a new edition of Walter of Henley'.[1] The need was by then generally admitted: the discovery of new manuscripts of the text and increasing knowledge of medieval agriculture were making it clear that Miss Lamond's edition was an imperfect guide to the precepts of Walter of Henley and his fellows. As time has gone on the need has become more and more evident. Now, at last, it has been met by the newly published book by Dr Dorothea Oschinsky, Senior Lecturer in Palaeography of Liverpool University; in it she gives us new texts and translations of the four treatises edited by Miss Lamond and much else besides.[2]

It is not difficult to see why we have had to wait so long. Miss Lamond knew of 21 manuscripts of *Walter*, 7 of the *Husbandry*, 7 of the *Seneschaucy* and 6 of the *Rules*; Dr Oschinsky has taken account of 38, 12, 15, and 14 respectively. This larger number of manuscripts would have made little difference to the editor's work had they all been straightforward and accurate copies of their exemplars. This, however, is far from the case. Three of the texts—the *Seneschaucy*, the *Rules*, and the *Husbandry*—were so drastically rearranged by some of their early copyists that even the basic plan of the original treatises is open to question. All four texts, especially *Walter*, have been glossed, miscopied, and mistakenly corrected, often by scribes who understood little of the authors' arguments. The resulting confusion was only partly resolved by Miss Lamond's edition; it has made it hazardous to draw on any of the texts for detailed technical evidence, and it has made Dr Oschinsky's task a peculiarly hard one. What she has tried to do is to give us the text of each treatise in its original state, as near as possible to what was written by its author or compiler. In this she has been brilliantly successful. On some points, to be sure, there is still room for doubt: the best manuscripts of all the treatises are copies at several removes from the originals. Thus in *Rules*, iv, in the phrase "cumben de quarters en semence, e en seriaunz la terre reprendra," "e en seriaunz" reads very like an insertion misplaced in copying; or, again, the surviving manuscripts do not enable Dr Oschinsky to decide whether the proverb in *Walter*, c. 9 ("On yeer other to wronge wylle on honde go ant evere at en hende wrong wile wende") has been omitted from the text by one tradition of copyists or inserted as a gloss by the other.[3] Points such as these can be resolved only in the unlikely event of the discovery of new and significant manuscripts that have eluded Dr Oschinsky; meanwhile she has provided us with texts of the treatises that are probably as good as we

[1] Eileen Power, 'On the Need for a new Edition of Walter of Henley', *Trans. Roy. Hist. Soc.*, 4th ser., XVII, 1934, pp. 101–16.

[2] Henceforth references to the treatises are given in the forms used by Dr Oschinsky: *Husbandry* (i.e. the anonymous 'Husbandry') *Rules*, (i.e. of Robert Grosseteste), *Seneschaucy*, and *Walter* (i.e. Walter of Henley's 'Husbandry').

[3] Pp. 347–8.

shall ever have. Their publication is an important event for medieval economic historians. It is not too much to say that we can now for the first time make proper use of these fundamentally important texts.

Dr Oschinsky backs up her texts with discussion of their transmission and with diagrams showing the relationships of the surviving manuscripts and lost exemplars. The variant readings given in the footnotes and critical apparatus to each treatise are confined to those that are relevant in establishing the text and its descent; minor variations in wording or spelling are ignored and it should perhaps be mentioned that although Dr Oschinsky's transcriptions of the texts are basically trustworthy their spelling very occasionally departs from that of the manuscripts.[1] As in Miss Lamond's edition, each text is accompanied by a parallel translation. For *Walter* Dr Oschinsky has used the translation made by William Lambarde in 1577 and hitherto unpublished. Lambarde used a corrupt copy of the text, and Dr Oschinsky has had to correct his translation in many places, but the idea of using it here is a happy one: it is well written, and it is valuable to have the English phraseology of a writer necessarily more familiar with the techniques of medieval agriculture than we are today. For the other three treatises Dr Oschinsky has provided her own fairly free translations; these are more accurate and more comprehensible than Miss Lamond's, reflecting our increased knowledge over the past eighty years, and they even attain elegance of style, a considerable achievement from such unpromising raw material. It is perhaps a pity that Dr Oschinsky does not take the reader more into her confidence over the various difficulties and perplexities that must have arisen in the work of translation, for there are a few significant points where her interpretation of the texts can reasonably be questioned. These, and some significant manuscript readings that are open to doubt, are listed and discussed at the end of this article. But besides translating the texts Dr Oschinsky provides further elucidation, partly in the critical apparatus, where she explains some points of detail, but mostly in more general discussion of the contents of the treatises in the separate introduction to each text.

It is here that Dr Oschinsky gives us the full benefit of her knowledge not only of the treatises but of medieval agriculture in general. Many aspects of agrarian practice are discussed, learnedly and illuminatingly. Among them is the difficult question of the heaping of corn measures by the threshers and the *incrementa* included in the issue of corn on many manorial accounts. This was important to Walter of Henley and his fellows because it touched on a loophole in the accounting system which might permit the reeve to enrich himself from the lord's corn, and it is important to the economic historian because of its bearing on medieval crop yields. Dr Oschinsky is the first to discuss the problem in a general context and to offer a well-reasoned and convincing explanation of the varied and puzzling evidence of manorial accounts. Her answer may not explain every case—we must always reckon with the strength of local variation in medieval practice—and it will have to be tested against the many references in surviving account rolls.[2] But here as in many other instances she provides a sound starting-point for further investigation, a starting-point very properly based on the rules and advice offered to those in charge of husbandry and accounting at that time.

In the introduction to each treatise Dr

[1] E.g. the manuscript used for the text of the *Seneschaucy* (B.M. Add. MS. 32085, fols. 132–8v) reads, in cc. 1–16: 'meyns' for 'meynz', c. 9, line 3; 'seit' for 'seyt', c. 9, line 6; 'surcarke' for 'surkarke', c. 13, line 2; 'poet' for 'poez', c. 13, line 2; 'lez' for 'les' (twice), c. 15, line 3.

[2] Among local peculiarities may be mentioned the entry of amounts 'de incremento mensure' for corn delivered to the lord, though not for the issue of corn whether threshed by hired or customary labour (Newton, Cambs., 1311–12: muniments of the Dean and Chapter of Ely), and the entry of *incrementum* for issues of all types of corn threshed by hired labour except oats (Froyle, Hants., 1363–4, 1365–6: St Mary's Abbey, Winchester, estates: B.M. Add. Rolls 17479, 17480). Cases such as these make it the more likely that local custom in agreements made with hired threshers lay at the root of the use of heaped measures as recorded on the account rolls rather than, as Dr Oschinsky suggests, the greater ease or speed of measuring by heaped measures. Cf. P. D. A. Harvey, *Medieval Oxfordshire village*, 1965, p. 55.

Oschinsky also discusses its origin, date, and authorship. In the case of *Walter* she is able to demonstrate clearly first that its form is that of contemporary sermons and second that its arrangement so clearly follows that of the *Seneschaucy* that it was probably written as a commentary on what must have been the earlier of the two treatises.[1] Beyond this Dr Oschinsky gives us no indisputable new evidence of the provenance of any of the treatises. She points to references in *Walter* that might suggest associations with Gloucestershire and Herefordshire, and connects this with the presence of a *dominus miles* Walter of Henley in 1266–7 in the following of the Clare family who were, of course, important Marcher lords; the suggestion is a very interesting one, but the local associations are not conclusively proved.[2] Similarly she suggests that the *Husbandry* originated at Ramsey Abbey, primarily because the best (and only un-rearranged) text is in a manuscript with Ramsey associations, but also because the treatise refers to (i) leaving gaps in manorial accounts for the insertion of totals at the audit, (ii) the organization of sheep-farming and wool-marketing for the estate as a whole, not manor by manor, (iii) a method of estimating corn yields that involved recording the issue of individual stacks of unthreshed corn, and (iv) particularly detailed accounting for corn and for harvest expenses; all these features occur in surviving accounts from the Ramsey estates.[3] But in fact (i) is found almost universally in unenrolled manorial accounts,[4] (ii) is very common on large estates,[5] while (iii) is one of only three methods mentioned by the treatise and, like (iv), can easily be paralleled elsewhere.[6] At least five, probably six, of the other manuscripts of the *Husbandry* come from Kent;[7] the emphasis in the treatise on the profits from woodland and pasture accords very well with a Canterbury or other Kentish origin[8] and much less well with a Ramsey one. Here again Dr Oschinsky's suggestions are stimulating and interesting, but are not necessarily conclusive.

The same is true of her discussion of the dates of the treatises. Because *Walter* gives an abstract of the document known as the 'Extenta Manerii' while the *Seneschaucy* does not, Dr Oschinsky suggests that the *Seneschaucy* was written before about 1276, when estate management in general and this document in particular were discussed in Parliament, and *Walter* after that date. The argument would be tentative even if these proceedings in Parliament were well attested, but in fact they are known only from references in the sixteenth century.[9]

[1] Pp. 75–82, 149–52.

[2] Pp. 147–8. The references that Dr Oschinsky associates with later practice in Gloucestershire and Herefordshire are to (i) reaping to leave a long stubble that might be cut for thatch, (ii) the high proportion of butter, as against cheese, produced from the milk, suggesting that the cheese was of low fat-content, and (iii) keeping sheep in houses in winter, and feeding them there on the haulm and pods of peas (pp. 175, 181, 183–4, 376).

[3] Pp. 201–2, 204, 207.

[4] Close examination of most original manorial accounts will show that, as first drawn up, a gap was left for the total of each paragraph of the cash account as well as for the total charge, discharge, and balance; the totals were entered on audit, sometimes by the original clerk, sometimes in another hand. Sometimes gaps were left for totals in the corn account also, but the stock account was normally fully written up from the first, with totals and balances. Occasionally preliminary totals for corn or cash were entered as tiny marginal figures when the roll was first written, e.g. Gisleham, Suff., 1349 (Biskele family estates: B.M. Add. Roll 26085) or Little Hinton and Stockton, Wilts., 1266–7 (muniments of the Dean and Chapter of Winchester).

[5] D. Knowles, *The Religious Orders in England*, 3v., 1948–59, I, p. 42.

[6] E.g. on the estates of Merton College, Oxford, as in the surviving grain-estimate rolls of 1272, 1276, 1282, and 1322 (Merton College muniments, 5529 dorse, 4120, 4122b, 4137).

[7] Pp. 37–8, 201.

[8] It appears very clearly, for instance, on the enrolled accounts for the estates of the Archbishop of Canterbury in 1273–4.—B.M. Add. MS. 29794.

[9] Pp. 70–2. It is difficult to follow Dr Oschinsky's argument on the significance for dating of Walter's inclusion of the 'Extenta Manerii'; as she herself mentions (p. 68n.), the particular form of the document that he uses had been included in Bracton's 'De legibus' (ed. G. E. Woodbine, 1922, II, p. 219) probably at least twenty years before 1276.

Building on this foundation, Dr Oschinsky argues that the *Seneschaucy* was written after about 1260, because it was omitted from a manuscript compiled then, and noting that *Walter* speaks of 24 weeks between Easter and Michaelmas she suggests that this refers to 1285 when Easter fell on 14 April, exactly 24 weeks before 29 September.[1] Combining this with a note in one manuscript that Walter of Henley 'primes fu chivalier e puis se rendist frere precheur', she suggests that the author, whether or not the former retainer of the Clares, became a Dominican about 1280, wrote *Walter* in the sermon form that he now learned, and died before about 1290–1300, the date of a copy of the treatise that refers to the author in the past tense.[2] Clearly all this is very hypothetical indeed, and Dr Oschinsky claims no more for these arguments than that they point to dates that on general grounds are not unlikely.

But there is room for more work that would enable these two treatises—and the *Husbandry*—to be dated more exactly. Dr Oschinsky gives precise dates for very few of the manuscripts she has used; so many are involved and so much detailed work would be needed to date any one of them that this is hardly surprising. But now that she has pin-pointed the key manuscripts in the transmission of these treatises it could well be that careful study of their hands and of the associated texts could produce useful results. The treatises themselves also provide some internal evidence for their dates of composition. For instance, the *Seneschaucy* envisages the rendering of manorial accounts by bailiff and reeve jointly. This was a practice only of the earliest period of written accounts, and although on some estates it lingered on into the fourteenth century by the end of Henry III's reign it must have seemed a rather old-fashioned form;[3] accounting by the reeve alone is a normal feature of the slicker methods typical of estates where the earliest written accounts date from after 1250, as at Westminster Abbey or Merton College. Even if we accept Dr Oschinsky's *terminus a quo* for the composition of the *Seneschaucy* this points at any rate to a date nearer 1260 than 1276. Other features of the treatise also point to an early date: the direct responsibility of the *famuli* rather than of the reeve on their behalf, and, indeed, the role of the steward himself as an estate supervisor.[4] In turn, an early date accords better with one

[1] Pp. 89, 144–5.

[2] Pp. 145–6. Dr Oschinsky does not mention the possibility that casting the treatise in the form of a sermon may have been slightly humorous in intention; the use of ecclesiastical or solemn forms for very mundane purposes was a frequent form of humour in the Middle Ages. It certainly seems bizarre, to say the least, that a new entrant to a landless Order should use the homilectic style that he now learned to write what is by and large a very worldly treatise on estate management. There is no evidence that Walter of Henley became a Dominican apart from the note in this single manuscript, and it is not impossible that this too is no more than a humorous reference to the style of the treatise, or even to the fact that it takes the form of a sermon. Humour —and humour of this sort—does occur in texts of this type: the accounting formulary in B.M. Add. MS. 45896 concludes with the mock-serious colophon 'Explicit Rotulus Nobilissimus secundum vsum Ihonis de Barton', and contains the obviously comic entry of the death of a sow by illness, attested 'per visum coronatoris'.

[3] In *Accounts and Surveys of the Wiltshire lands of Adam de Stratton*, ed. M. W. Farr, Wilts Arch. & Nat. Hist. Soc., Records Branch, vol. XIV, 1959, we actually see the change taking place: the accounts for 1269–70 and 1272–3 are presented by bailiff and reeve together, those from 1273–4 on by reeve alone. At Peterborough Abbey, however, where old styles of accounting long continued in use, bailiff and reeve were still normally accounting together in 1310 and at Winchester Cathedral Priory the change was not complete until the 1320's (Northants. Record Office, Fitzwilliam (Milton) MS. 2389; J. S. Drew, 'Manorial Accounts of St Swithun's Priory, Winchester', *Eng. Hist. Rev.*, LXII, 1947, pp. 22–3). The reference to joint accountability might help to provide evidence of the place, as well as the date, of the composition of the *Seneschaucy*; the system was widespread (it is found, for instance, at Canterbury, Peterborough, Winchester, and Worcester), but even in the mid-thirteenth century was far from universal (thus it does not appear on the estates of Crowland or Ramsey Abbey, nor on the royal estates). Dr Oschinsky (pp. 96, 234) may be right in saying that the bailiff was held especially responsible for a joint account's accuracy, but the financial responsibility seems always to have been the reeve's alone (Drew, *loc. cit.*, pp. 26–7; N. R. Holt, in *The Pipe Roll of the Bishopric of Winchester, 1210–11*, 1964, p. xx, finds the same use of singular, not plural, verb to express the final debt).

[4] Dr Oschinsky (pp. 98–9) notes that interest in the *Seneschaucy* soon waned, as it was hardly ever copied after the early fourteenth century. This could well be because the type of organization it describes was now out of date.

piece of internal evidence from *Walter*, which, as Dr Oschinsky shows, must have been written later than the *Seneschaucy*. It is an interesting fact, and may well be a significant one, that *Walter* nowhere suggests that the reeve or bailiff of each manor should present written annual accounts. If *Walter* was written in, say, the last decade of Henry III's reign this would not be surprising, especially if (as will be suggested) the treatise was written for lay owners of small estates: written accounts were probably known then only on a limited number of ecclesiastical and large lay estates. But if *Walter* was written in the mid-1280's the omission would be surprising. Written accounts of that date survive from a wide variety of estates; their use was spreading fast,[1] and they represent just the sort of up-to-date, provident technique that would appeal to Walter of Henley. This, of course, is arguing from negative evidence, and it would be wrong to attach too much weight to it. It may serve to show, though, that we have more to learn about the dates as well as the provenance of these treatises. The question is an important one, for the differences between the four treatises are interesting and may be seen as significant if they can be placed in their proper chronological context. The same applies to the other texts on accounting and estate management that Dr Oschinsky discusses. Probably almost all were written over a relatively short period,[2] a period when the use of written accounts was spreading rapidly and when other new techniques and new methods of organization were being adopted.

Here we touch on the remaining contents and the overall plan of the book. Besides giving us a full edition of the four treatises, Dr Oschinsky deals with other contemporary texts on estate management and accounting: the detailed list of manuscripts that forms the bulk of the book's first chapter comprises all manuscripts containing this type of material, a long chapter is devoted to analysing and discussing the treatises on accounting, and in an appendix nine unpublished texts of this sort are printed in whole or in part. The contents and arrangement of the book are dictated by the conclusions that Dr Oschinsky has reached, first on the purpose of the four treatises that she has edited in full and secondly on their relationship to the development of manorial accounting. These conclusions are basic to her work, and as much of what she says stems from them they deserve very careful examination.

"This book," Dr Oschinsky writes, "by the nature of our principal treatises has become a study of didactic literature for the training of officers employed on non-monastic estates."[3] Because many copies of the treatises are found in compilations of legal texts, she argues that they were intended for the instruction of future estate officers—stewards and bailiffs—who would be trained in estate management while acquiring a working knowledge of common law. In the mid-thirteenth century these officers had no need of legal education and they probably learned their job simply by practice and experience—grew into it as it were—but "the legal demands under Edward I made it essential for stewards and bailiffs to have a good legal education," and, from the evidence of the manuscripts, "at least some lawyers received, in addition to their legal training, instruction on estate management, conveyancing, and accounting."[4] It is in this light that Dr Oschinsky views each of the treatises. Thus of the *Seneschaucy* she writes, "The stress laid on legal training and the legal demands to be made on steward and bailiff suggest that the compiler had himself been trained in law."[5] The sermon form of *Walter* suggests that this was composed as a lecture: "The delivery of the lecture will very likely have taken place in Oxford. Pre-

[1] By 1325 at Langenhoe, Essex, we see written accounts being prepared for the owner of a single manor (R. H. Britnell, 'Production for the market on a small fourteenth-century estate', *Econ. Hist. Rev.*, 2nd ser., XIX, 1966, p. 380).

[2] Probably very few of these texts were composed later than 1300. For one (Gloucester, Library of the Dean and Chapter, MS. 33, fols. 48–55v), for which Dr Oschinsky suggests a mid-fourteenth-century origin, a mid-thirteenth-century date seems more likely on internal evidence: (i) the growing of winter barley, (ii) the description of all cash discharge as *expense necessarie* except forinsec expenses and liveries, and (iii) the use of *estricum* and *hop* as measures of capacity for corn and flour (pp. 249, 469–75).

[3] P. 257. [4] Pp. 61–5. [5] Pp. 88–9.

scribed courses in business training are known to have existed there, which provided instruction in the art of letter writing, the formulation of writs, deeds, and accounts. Within such a course instruction in estate management would merely have been an extension of the syllabus." Soon afterwards a copyist of *Walter* revised it, by introducing chapter headings and altering some of the more rhetorical turns of phrase, so that it could be used more easily as a textbook; it is this 'didactic version' that is represented by most surviving manuscripts.[1] The *Husbandry* was compiled as a work of reference for an auditor, probably from an existing file of notes and memoranda, arranged in the order in which the matters would arise on audit;[2] in this form, Dr Oschinsky argues, the work would be of limited use, as auditing procedures varied from one estate to another, but all but one of the surviving copies are of a very much revised version, a 'didactic adaptation' that arranged the material in the order of a manorial account so that it could be used more generally by clerks or bailiffs as a guide to their composition[3]. The *Rules* too were, of course, not drawn up as a work of general instruction; they were compiled, perhaps by Robert Grosseteste himself, between 1240 and 1242 for the specific guidance of the newly widowed Countess of Lincoln. But from the surviving manuscripts Dr Oschinsky argues that some fifty years later they were "adopted by the legal profession as a textbook for the training of estate personnel" and they thus acquire a sort of honorary status as a didactic treatise for general instruction.[4]

The suggestion that there existed a regular programme of instruction for future estate officers and that these treatises played an important part in it, were, indeed, mostly written for this purpose, is a very interesting one that deserves careful consideration. We should be wary, though, of regarding it as proved fact. We know little enough of how students learned common law in the thirteenth century, but it is far from certain that any form of instruction was offered them beyond attendance in the courts and gathering up what crumbs of wisdom, experience, and anecdote their elders let fall. To add instruction in estate management to a formal training in the common law is to add hypothesis to hypothesis. Certainly, though, the presence of the treatises in legal manuscripts points to their being used for guidance (which is not the same thing as formal training) by men with knowledge of the law whose duties included estate administration. These men would be the estate stewards or seneschals; what is normally meant by a bailiff, an officer having the oversight of not more than a very few manors, would hardly have needed any form of specialized legal knowledge.[5] The mid-thirteenth-century steward, however, was expected to supervise the running of an entire estate, which would include holding the court for each manor; this would clearly call for a knowledge of more than local customary law, and the *Seneschaucy* (c.1) specifically says that he "deit saver la assise du regne pur foreyne bosoynes defendre, e pur lez baillifs ke desuz

[1] Pp. 117–20, 148.

[2] This suggestion can be supported by the curiously contradictory passages in the treatise (e.g. c. 28 and c. 32, or c. 35 and c. 36).

[3] Pp. 201–6. It is worth noting, though, that very often—perhaps normally—the same order was followed in writing manorial accounts as in auditing them, i.e. the dorse with the corn and stock accounts would be written first, and the cash account on the front last. This would be the most efficient order of procedure for the clerk as well as for the auditor; it can be demonstrated from accounts where there is a change of hand partway through the writing of the account or where alterations made in the corn or stock account have already been taken into account in the cash account, e.g. Whitchurch, Hants., 1260–1 (muniments of the Dean and Chapter of Winchester), Melbourne, Cambs., 1317–18 (B.M. Add. Roll 25866). Cases where it can be demonstrated that the cash account was written first, the corn and stock accounts last, seem much rarer, but Cuxham, Oxon., 1290–1 (muniments of Merton College, Oxford, 5809), and possibly also Broughton, Hunts., 1378–9 (Ramsey Abbey estates: B.M. Add. Roll 39535), are examples.

[4] Pp. 5–6, 196–9.

[5] Dr Oschinsky mentions (p. 89) the legal demands made on the bailiff by c. 21 of the *Seneschaucy*; but this, effectively, requires him to hand over to the steward's judgement any decision requiring a knowledge of more than the local custom of the manor.

luy sunt en lur dutes certefier e aprendre." Significantly, perhaps, towards the end of the thirteenth century, as more emphasis was placed on their duty of holding courts and on their legal qualifications, the stewards played less and less part in the actual running of estates.[1] However, there is no reason to doubt that, as its name implies, the *Seneschaucy* was actually compiled to assist men of legal background who found themselves taking up posts that involved a knowledge of estate management.[2]

But to find *Walter* bracketed with the *Seneschaucy* as composed for the instruction of estate officers is surprising. Even Dr Oschinsky's interesting demonstration that it is a commentary on the *Seneschaucy* cannot alter the fact that, *prima facie*, it gives advice not on how to run someone else's landed property but on how to extract—honorably and honestly—the greatest profit from one's own. There is no mention of a steward; throughout it is implied that the owner of the estate, presumably a fairly modest one, will oversee it in person. When the author mentions bailiffs (c.33) it is not to instruct them in their duties but to give advice on how they are to be selected. Of course, this could well be a literary device: by being told how to manage landed property from the point of view of its owner *Walter's* audience or readers would learn how to run it on his behalf. But for this suggestion to carry weight much more evidence is needed. Even if it is accepted, it is still difficult to accept the further suggestion that *Walter* was compose as a lecture for students of estate managemen at Oxford. Mr H. G. Richardson has shown from surviving collections of specimen documents, that training in the *ars dictamini* known to have been taught at Oxford by th mid-fourteenth century, was probably available there at least a century earlier.[3] This *ar dictaminis* was the basic training of a clerk. I included letter-writing and the drafting of con veyances and other legal documents and it i very likely that it came to include training i compiling accounts; there are reasons to sus pect that certain specimen manorial account originated at Oxford.[4] Instruction in draftin accounts would naturally include the tech niques of auditing, for, as Dr Oschinsky show the clerk writing a manorial account playe almost the part of an intermediary between th officer rendering the account and the auditors but as far as we know the actual running of landed estate would normally lie quite outsic his duties.[6] To suggest that instruction wa offered in estate management is to assume th presence at Oxford of a quite different group students, the future stewards, and this is some thing for which we have no evidence. On th whole it seems simplest and soundest to regar *Walter* as written for the edification and prof of small estate owners, who were too lowly t employ an experienced or professional stewar and supervised their own property; it wa written to be read as a book, not heard as lecture,[7] for if an audience of would-be esta

[1] See, for example, Jennifer C. Ward, 'The Estates of the Clare Family, 1066–1317', *Bull. Inst. Hist. Re.* XXXVII, 1964, p. 115.

[2] It is interesting to note that the proportion of surviving copies in legal manuscripts is higher for th *Seneschaucy* than for the other three treatises (Oschinsky, p. 61, n.1).

[3] H. G. Richardson, 'An Oxford Teacher of the 15th Century', *Bull. John Rylands Library*, XXIII, 193 pp. 436–57; 'Business Training in medieval Oxford', *Amer. Hist. Rev.*, XLIV, 1940–1, pp. 259–80; 'Th Oxford law school under John', *Law Quarterly Rev.*, LVII, 1941, pp. 319–38; *Formularies which bear on t history of Oxford, c. 1204–1402*, ed. H. E. Salter, W. A. Pantin, and H. G. Richardson, Oxf. Hist. Soc., N. IV, V, 1942, II, pp. 331–54.

[4] Dr Oschinsky (p. 251) mentions the possible Oxford associations of the text in B.M. Harley MS. 27 fols. 29–49. Several texts of the 'Forma Compoti' group (notably B.M. Add. MS. 45896, B.M. Egerton M 2360, and P.R.O., E.163/24/34) contain references to places in Oxfordshire, or are linked with texts th suggest an Oxford origin; this was probably the most popular of the specimen manorial accounts that we in circulation from the late thirteenth century onwards.

[5] Pp. 231–2.

[6] Dr Oschinsky, however, argues otherwise (p. 234). We badly need some case-studies of the careers individual clerks; these would be difficult to compile, but not impossible in areas where surviving recor happen to be abundant.

[7] The changes from rhetorical to less rhetorical style that Dr Oschinsky notes (pp. 118–19) among t

stewards is unlikely, an audience of trainee gentlemen farmers is unthinkable.

If we do not accept that *Walter* was written as a didactic guide for estate officers, it follows that of the four treatises only the *Seneschaucy* was originally compiled for this purpose. This makes it harder to accept the distinction drawn between these texts and two other contemporary treatises on estate management. One of these is the rules in French compiled for the Mohun family's estates centred on Dunster, written in their fourteenth-century cartulary.[1] Dr Oschinsky prints the text as an appendix, but without translation or discussion of its contents; it has never been printed before and although it was written for a particular estate (so too were the *Rules* and the *Husbandry*) and contains some purely local detail this seems all the more reason for fuller elucidation. The other treatise is the Latin text headed 'Scriptum quoddam' in the Gloucester Abbey cartulary published in the Rolls Series.[2] As Dr Oschinsky points out, this was probably written as a general treatise, not necessarily at Gloucester, as there is another copy in a manuscript from Luffield Priory,[3] and yet a third manuscript, of uncertain origin, includes passages from the text translated into French.[4] The text is a very interesting one; its bureaucratic, almost military, rules for the governance of a manor are very different in tone as well as in content from the other treatises. In 1934 the late Mr R. V. Lennard called attention to the importance of the text and the need for a critical edition.[5] It is difficult not to regret that Dr Oschinsky's view of the purpose of the four edited texts should have prevented her from placing these two other treatises alongside them.

Besides determining her choice of the texts to edit, Dr Oschinsky's view of the purpose of the treatises has governed her treatment of those selected. Aspects of the treatises and their history relevant neither to the establishment of the texts nor to their presumed purpose of training estate officers have been omitted. Thus there is no discussion of those points of the *Rules* dealing with the regulation of the household; indeed Dr Oschinsky almost apologizes for including them in the edited text.[6] She alludes once or twice to the 'sermon' adapted from the opening chapters of *Walter* (i.e. omitting all the agricultural contents), but tells us no more about it, not even mentioning the curious fact that in one version it is written in rhymed couplets,[7] and for the late medieval English translation of *Walter* we must still use Miss Lamond's edition. It is a pity that Dr Oschinsky should have limited the scope of her edition for reasons that are so open to question.

But the plan of the book has been determined also by Dr Oschinsky's view of the relationship of the four edited texts to the development of manorial accounting. The *Husbandry*, of course, is clearly and closely linked with the system of presenting and auditing manorial accounts, a system that was in general use by the time it was written at the very end of the thirteenth century. But it is argued that in the *Seneschaucy* too "special emphasis is laid on accounting responsibilities and on the importance and methods of checking the accounts of manorial offices,"[8] and indeed that a similar emphasis on accountability to the lord is to be

alterations made for the 'didactic version' of *Walter* are very interesting, but do not amount to proof that it was originally intended for oral delivery; in one or two cases the revised version seems to use a more high-flown style than the original (as 'vos fenz facet unyr' for 'facez aunir vos fens').

[1] B. M. Egerton MS. 3724, fols. 19–20.

[2] *Historia et cartularium monasterii sancti Petri Gloucestriae*, ed. W. H. Hart, Rolls Series, XXXIII, 3 v., 1863–7, III, pp. 213–21.

[3] Pp. 30, 256.

[4] In Paris, Bibl. Nat., MS. Fr. 400, at the end of the conflation of *Walter* and the *Husbandry* that Dr Oschinsky describes (p. 49; L. Lacour, 'Traité d'économie rurale composé en Angleterre au treizième siècle', *Bibliothèque de l'École de Chartes*, 4e sér., II, pp. 375–7 (sections 34, 35)).

[5] *Trans. Roy. Hist. Soc.*, 4th ser., XVII, p. 115.

[6] P. 198.

[7] Pp. 21–2, 36, 113. The verse version is in B.M. Egerton MS. 3724, fols. 39–40; the text in Cambridge Univ. Lib. MS. Hh. III. 11, fols. 82–3, is in prose and seems to have much less obvious connection with *Walter*.

[8] P. 98.

found in both *Walter* and the *Rules*. Dr Oschinsky sees the development of manorial accounting in thirteenth-century England as lying at the heart of the system of demesne farming that created the need for the treatises on estate management. The existence of a well-developed royal accounting system that private estate owners could adapt to their own needs was, she suggests, an important reason why demesne farming flourished in England, and she draws a contrast with Germany, where it failed to develop despite similar economic conditions.[1] It is on this basis that she devotes a substantial part of the present book to accounting treatises and formularies.

For this we have every reason to be grateful. Dr Oschinsky has already given us some valuable work on the accounting treatises;[2] she now provides the fullest discussion we have yet had of this interesting and important class of text. At the same time, it is not to be compared with her work on the four edited treatises as final or definitive treatment. Almost certainly we have more to learn about the treatises on accounting. For one thing, there are more manuscripts to be discovered; the following additions to those listed in the book are probably no more than a tithe of those that have still to come to light:[3]

British Museum, Sloane MS. 513, fols. 3r–10v. Specimen account of Winchester form, with interspersed rules (without heading), opening with the verses 'Quid de quo quantum . . .' (see Oschinsky, p. 241), and continuing 'Notandum quando aliquis debeat ordinare compotum alicuius manerii neccesse est quod habeat rotulum reddituale min quo sunt nomina terre tenencium . . .'. Hand of late 13th cent.; the manuscript miscellanea mostly of the 15th cent., came from Buckfast Abbey, Devon.[4]

B.M., Add. MS. 45896 (lost; but a photocopy forms MS. Facsimiles 504). Roll giving specimens of a manorial court roll (dated 1348) and an account with interspersed rules (heading missing). The rules and headings of the account, but not the contents of the paragraphs, are those of the 'Forma Compoti' (see Oschinsky, pp. 247–9). Hand of mid-14th cent.; probably compiled for use on the estates of the Harcourt family.[5]

B.M., Add. MS. 48978. Specimen accounts for a single unspecified year (probably about 1260–70), with interspersed rules, etc., for all manorial officers and obedientiaries of Beaulieu Abbey, Hampshire; Bodleian Library MS. Barlow 49, fols. 58–114, (see Oschinsky, p. 254) is largely a copy of the sections of this text relating to Faringdon. Hand of mid- or late 13th cent.[6]

Bodleian Library, MS. Oseney Roll 2. Roll giving specimens of manorial court rolls (of which one is dated 1333) and an account (imperfect: latter part of stock account only) endorsed with a single rule. The account is one of the 'Forma Compoti' group (see Oschinsky, pp. 247–9). Hand of mid-14th cent.; from the archives of Oseney Abbey, Oxfordshire.

Muniments of King's College, Cambridge. Ministers' accounts before the time of the College, Ogbourn 17 Edward II. Specimen account roll, without rules, dated 1322–3. The account is one of the 'Forma Compoti' group (see Oschinsky, pp. 247–9). Hand of early to mid-14th cent.; from the archives of the English estates of Bec Abbey.[7]

[1] Pp. 72–3, 213–14.

[2] D. Oschinsky, 'Medieval Treatises on Estate Accounting', *Econ. Hist. Rev.*, 1st ser., XVII, 1947, pp. 52–61; 'Notes on the editing and interpretation of Estate Accounts', *Archives*, IX, 1970, pp. 84–9, 142–52.

[3] Besides these additions to the list of surviving manuscripts, one item can be added to the note of those now lost (Oschinsky, p. 10, n.2): Royal MS. 11 A. XVIII, which contained a 'Formula Computus Servientis in Manerio pro A.D. 1377'. This manuscript was probably in existence in 1698, when it was mentioned in Bernard's *Catalogi manuscriptorum Angliae*, but by 1734 it had disappeared (G. F. Warner and J. P. Gilson, *Catalogue of western manuscripts in the Old Royal and King's Collections*, 4v., 1921, I, p. 343).

[4] N. R. Ker, *Medieval libraries of Great Britain*, 2nd edn., 1964, p. 14.

[5] Described in the British Museum's *Catalogue of Additions to the Manuscripts, 1936–45*, 2v., 1970, I.

[6] I am grateful to Mr M. A. F. Borrie for helpful discussion of this manuscript.

[7] It is interesting that here and in the previous item we see the specimen account for a lay-owned manor being used on ecclesiastical estates. The Bec version shows no obvious adaptation, retaining even the headin

Muniments of the Dean and Chapter of Canterbury: Rural Economy 65. A single membrane from a roll probably of three membranes, giving a specimen account, without rules, in somewhat disarranged order, with specimen deeds on the dorse. The account seems unrelated to other specimen accounts known. Hand of late 13th or early 14th cent.

The last three items are significant as each occurs in a collection of actual manorial rolls, its status as a specimen unrecognized; many others must still be lurking among the thousands of surviving manorial accounts of the thirteenth, fourteenth, and fifteenth centuries. The discovery of further texts and, even more perhaps, detailed study of the forms of actual accounts in relation to the specimen texts have almost certainly much to tell us about the spread of written accounts and accounting techniques in general, as well as about wider questions of estate organization. Further work in collating the various specimen accounts may throw light on their provenance and, in turn, on the training of clerks; those of the 'Forma Compoti' group are particularly promising here as some have wildly corrupt texts suggesting that they were copied and re-copied many times.[1] Here Dr Oschinsky's work opens the way to further research. The chapter on the accounting treatises is very far from being an unimportant part of her book.

At the same time, we may still question the reasoning that led to its inclusion in a work on agricultural treatises. Apart from the *Husbandry* the edited treatises say nothing of the techniques of manorial accountancy. The *Seneschaucy* (c. 81) refers once, almost in passing, to account rolls, as do the *Rules* (vii), while *Walter* does not mention them at all. The chronology of the widespread adoption of demesne farming is one of the most important outstanding questions in medieval English history, a question with far-reaching implications to which we urgently need an answer. In the present state of our knowledge it cannot be assumed that the spread of demesne farming went hand-in-hand with the adoption of written accounts. Manorial accounts, as Dr Oschinsky points out, are known from few estates, and those mostly very large ones, before the second half of the thirteenth century;[2] the references to them in the *Rules* and the *Seneschaucy* are not surprising, for both these texts were probably written for very large estates indeed. But it is not improbable that the leasing of estates gave way to the general adoption of demesne farming in the twenty or thirty years after 1200.[3] Of course demesne farming implied some form of account-keeping and audit: the lord of the manor and his reeve or bailiff would have to know the state of account between them, how much money was owing from one to another at any particular time. But this need not imply the compilation of written accounts. There is no reason why proceedings at audit should not have been conducted entirely by verbal explanation aided by tallies and counters; indeed, this would seem the obvious and natural method. It was only in the reign of Edward I that the sophisticated techniques of written accounts became general, a process in which the accounting treatises and specimen texts clearly played an important part. But this had no necessary connection with the good husbandry, the close watch on reeve and *famuli*, the eye for profit, that were necessary to efficient demesne farming; there is no essential link between the treatises on husbandry, written for estate owners and their stewards, and the

Expense Domini et Domine'; the cash account is missing from the Oseney version. Dr Oschinsky's insistence on the distinction between monastic and other estates in auditing and accounting techniques (as on pp. 212–3) is interesting, and it is to be hoped that she may develop it further; it is not immediately apparent from the manorial accounts that survive from both types of estate.

[1] Cf. p. 176, n. 4

[2] Pp. 3–4, 64, 233. Dr Oschinsky's contention (p. 4) that on monastic estates accounting developed late, and then mostly on large estates, is hardly borne out by surviving accounts: I know of no small lay estates to be compared with Little Dunmow Priory, Essex, or Southwick Priory, Hants. (to name only two) from which written accounts survive from before 1250. But I would not disagree with her general argument that manorial accounting developed primarily on very large estates.

[3] I would agree with Dr J. Z. Titow (*English Rural Society, 1200–1350*, 1969, pp. 43–4) in associating the growth of demesne farming with the rise in prices of the late twelfth and early thirteenth centuries.

treatises on accounting, written for the clerks who drew up the manorial accounts.

Whether or not we agree with all Dr Oschinsky's conclusions, it cannot be questioned that she has given us a most important and valuable book. In some ways it is not easy to use, as information on a particular topic is apt to be scattered in several places. It will be a great pity if future students of medieval agriculture allow themselves to be daunted by this: the book is learned and thorough, a mine of erudite information and of interesting, stimulating ideas, and it will be the starting-point for researches in this field for many years to come. We may be certain that forty years hence we or our successors will have no reason to meet for a discussion 'On the need for a new edition of Walter of Henley'.

APPENDIX

Questioned Readings and Translations [Godefroy: F. Godefroy, *Dictionnaire de l'ancienne langue française* (10v., 1880–1902)]

Seneschaucy

c. 18. par la perche mesure: by the measured perch. A preferable translation might be 'measured by the perch'.

cc. 19, 48. custumers, custumeres, custumes: customary tenants. As they are differentiated from the boon-workers, who were themselves customary tenants, 'customary workers' (i.e. customary tenants doing their week-work) might be a better translation.

c. 21. vyleyn aleger: remove villeins from the lord's jurisdiction. Perhaps, more precisely, 'manumit villeins'.

cc. 26, 36, 52, 56. The translation does not follow the author's distinction between *affres*, 'plough-horses', and *avers*, 'cattle', especially 'draught-cattle' (the usage differs from that of the glossarial note cited on p. 27, n.1, where *averium* is applied to any farm animal). In c. 31 'affres' is almost certainly a scribal slip for 'avers', which occurs in the same context in c. 57.

cc. 27, 62. mors, morz: killed. Probably 'bitten' (Godefroy, s.v. *mors* (1)).

c. 38. lez crappes: the chaff. Probably 'the sweepings of straw' (cf. Godefroy, s.v. *crape*); the chaff would be 'lez remesilles de ventur' that are mentioned next.

c. 38. ventresse: winnowers. Strictly, 'winnowing-women'; winnowing was, of course, normally women's work.

c. 38. huise: boots. Thus Godefroy, s.v. *heuse* (1); but the word is a cognate of the English *hose*, i.e. stockings or leggings, and this fits the context better. The author occasionally uses English words, as 'croddes' in c. 43.

cc. 39, 74. tascurs: stackers. Almost certainly 'threshers', and probably 'hired threshers' (corresponding to the Latin *tascatores*); the implied contrast with *baturs*, the threshers in general, is significant in view of variation between them in records of heaping measures (cf. Oschinsky, pp. 170–1, and p. 171, n. 2 above). But in some areas *tasker*, in English, seems to have been synonymous with *thresher* from the 14th century to the 19th (B. Sundby, *Studies in the Middle English dialect material of Worcestershire records* (1963), pp. 235–6; *The Countryman*, lxxiii, no.1 (Autumn 1969), p. 63; *English dialect dictionary*, ed. J. Wright (6v., 1898–1905), s.v. *task* (3)).

c. 41. forage: hay. More likely straw used as fodder (presumably mostly oats straw), in contrast to *litere*, the straw that could be used only as litter.

c. 43. de plaider nul play: to begin an action at law. But 'to hear a plea' (cf. the use of *pleyder* in c. 21) fits the context better.

cc. 53, 54. The translation does not bring out the distinction between the *tenurs dez charues*, the men who hold and guide the ploughs (Latin *tentores*, etc.), and the *charuurs* the men who drive on the oxen when ploughing (Latin *fugatores*, etc.); the description of their duties shows that these are the meanings intended.

c. 61. lez hetches e lez thetches reparayller e

fere: repair and do hedging, fences and hurdles. 'thetches' is obscure, but it can equally be read as 'checches' on the manuscript (B.M. Add. MS. 32085, fol. 137), and it is tempting to see it as a scribal error for 'crecches' (cf. 'cresches' in *Walter*, c. 103); in this case the meaning would be 'make and repair the hay-racks and mangers', which fits the context admirably. Alternatively *thetch* might be taken as an English word, in which case the phrase might mean 'make and repair the doors and thatched roofs'.

c. 71. e fere lez amendes: assess the fines. Simply 'make good their loss' would fit the context better.

Walter

c. 47. estreit: narrow (Lambarde's translation). Dr Oschinsky suggests (pp. 165, 321 n.) that this should be 'straight', but this is a scarcely admissible translation of *estreit*, and 'narrow' makes good sense: the furrow between the selions should be as narrow as possible (to gain land and save wasted seed) and to reduce its width the small hillock of soil that runs down its centre (since it was formed by two separate traverses of the plough in opposite directions) should be turned over to one side with spades after the plough has passed.

c. 55. gerner: barn. Correctly, 'garner' (as in the translation of the *Seneschaucy*, cc. 39, 40).

c. 64. e pus estramer sure: and after cast upon it. I.e. spread straw on it (Godefroy, s.v. *estramer*).

c. 65. en wasseaus: in muckhylles. Correctly, 'in wet places' (as on p. 177, and in the translation to the *Seneschaucy*, c. 23; Godefroy, s.v. *gacel*).

c. 66. sans feyntyse: without loytering. Better perhaps, 'without deceit'.

c. 106. ii des meudres: two of the lesser (amended by Dr Oschinsky from Lambarde's 'two of the greatest'). It is difficult to accept this amendment on which much of the discussion on pp. 185, 382, is based: *meudre* is a normal comparative of *bon* (Godefroy, s.v. *meillor*), and though a few manuscripts of *Walter* give instead some form of *menor* (e.g. B.M. Lansdowne MS. 1176, fol. 136v, clearly reads 'meindr's') these are not the best texts, nor does this make such good sense. The translation should almost certainly be 'two of the best'.

Rules

i. vos foreyne terres: those of your lands which are not part of your demesne manors. 'those of your lands which are not part of your manorial demesnes' might be a clearer interpretation, for the contents of rules i and ii suggest that it is meant to include tenants' lands on all manors, even those in demesne; rule ii envisages a further survey of the demesne manors, but this time restricted to the lands of the manorial demesnes.

xviii. aline: soul. Probably 'alme'; the manuscript (B.M. Harley MS. 1005, fol. 53) admits either reading.

Husbandry

c. 9. aioynz: mated. Alternatively, 'added to the older stock' (Godefroy, s.v. *ajoindre*).

c. 12. veux: aged cattle. Probably 'calves'.

c. 19. dedenz la cort: on the manor. Perhaps 'in the farmyard', the reference being to all the poultry and smaller livestock that would normally be kept in or around the farm buildings.

c. 37. des acres de rente: acres put out to rent. Probably 'increases of rent' (Godefroy, s.v. *acrois*).

c. 39. clonnes: nails. Probably 'clouues', meaning either nails or cloves; the manuscript (Bodl. MS. Ashmole 1524, fol. 25) admits either reading, and both interpretations fit the context.

c. 51. chevestres: harnesses. Or, more specifically, 'halters' (Latin *capistrum*).

c. 58. espessement: sparsely. Literally, 'thickly' (Godefroy, s.v. *espessement* (2)), but in the context this does not affect the sense.

Appendix I

p. 459. carta forinseca (*bis*; and quoted thus on p. 236). Correctly 'carca forinseca', i.e. forinsec charge; the manuscript admits either reading, but the latter makes better sense.

p. 459. Finito compotus. Correctly 'Finito

compoto': the manuscript reads 'Finito compot''.

Appendix III

p. 463. titulus. Or, more likely, 'tunc': the manuscript reads 't'c' or 't't'.

p. 463. The table at the end of Appendix III is an editorial illustration, not part of the text.

Appendix IX

p. 476. caefe. Probably 'aese', though the manuscript is not very clear.

NOTE: *Coronator* occurs more often than I thought when I wrote p. 173, note 2, as a form of *cronator* or *caronator*, inspector of carcasses (*Dictionary of Medieval Latin from British Sources*, ed. R.E. Latham *et al*. [London: 1975–], *sub* cronator); all the same, I suspect that any clerk who spelled the word thus had his tongue in his cheek.

XIII

Mid-13th-Century Accounts from Bury St Edmunds Abbey

BL MS Harley 645, a register from Bury, contains copies of sixty-six accounts of obedientiaries, manorial officials and others. Fifty-five date from 1247–50, eleven from 1256–61, and they may have been prepared in order to give new abbots (Edmund 1248, Simon 1257) a detailed view of the abbey's finance. They mostly predate the adoption of formally structured written accounts at Bury and are of especial interest in throwing light on a stage in the development of estate accounts of which little evidence survives elsewhere. They also provide interesting details of the system of food farms that still operated in Bury.

FORM AND ORIGIN

Harley MS 645 in the British Library is a volume from Bury St Edmunds known in the 17th century as the Kempe register. It comprises copies of a great variety of documents — charters, rentals, tax assessments and many more — and was probably made up from several different books or parts of books in the late 14th century, when its folios were numbered throughout. More material was later written on blank pages, and ten leaves were added at the beginning.[1]

Dispersed in the latter part of the book are sixty-six accounts of obedientiaries, manorial officials and others; they are fully listed in the appendix to this article and reference is made to them by their numbers in this list. They were written on parchment bifolia that eventually found their way into this volume with other leaves interspersed and with other material written on blank pages; these bifolia were not formed from rolls but a few had to be slightly trimmed to fit the book.[2] The accounts fall into four groups, the first three dating from 1247–50 and the fourth from 1256–61:

- I The cellarer's department, 1247–50: the cellarer's manorial receipts, subcellarer, cellarer's manors and receivers of their cash (nos 1–36). Written originally on an eight-leaf gathering with the sixth leaf mostly blank and the seventh and eighth leaves either blank or (as they are now) cancelled. Fols 193–95, 197–99 (fol. 196 has been intruded into the gathering).[3]
- II The chamberlain's department, 1248–49: the chamberlain's manorial receipts, subchamberlain, pittancer and chamberlain's manors (nos 37–49). Written originally on a four-leaf gathering with the second leaf mostly blank, the third leaf blank or (as now) cancelled and the fourth leaf blank. Fols 252, 253, 259 (fols 254–58 have probably been intruded into the gathering).
- III The sacrist's department, 1248–50: the sacrist's manors, sacrist and moneyer (nos 50–55). Written on the first leaf of a single bifolium, but there must have been at least one other bifolium to make a gathering of four or more leaves as it begins near the end of an account; the notes following this account read as if it was a manor not of the sacrist but of the cellarer. Fols 260, 261.

IV Miscellaneous accounts, 1256–61: four of the cellarer's manors, the moneyer, subcellarer and kitchener (nos 56–66). Begun on the second leaf of the bifolium used for group III, continuing on two successive single leaves, the second mostly blank.[4] Fols 261, 262, 265.

All the accounts are written in hands of the mid 13th century, groups I to III probably by six hands in all, group IV by three; which wrote which account is shown in the appendix. The arrangement of the accounts within groups I to III tells us something of how and when they were written. In group II the accounts of the chamberlain's department follow a clearly ordered sequence: the chamberlain's manorial receipts, the junior obedientiaries, the individual manors belonging to the office. Group III is differently arranged, but in Group I the accounts of the cellarer's department follow this same pattern, except that after the accounts for the individual manors there are more accounts, of the subcellarer and of the manorial receivers (nos 32–36). However, as the hand that wrote those of the subcellarer (hand D) added a cross-reference to them before the accounts of the manors,[5] they can be seen as additions, like the last account for Warkton, closing on 28 July 1250 (no. 31), which is written, in a hand that wrote no other full account, in space fortunately (or deliberately) left blank at the foot of the page. Of the other accounts the latest closes on 20 July (no. 1). It looks, then, as if they were first set out in late July 1250 or soon after, and that a few further accounts for recent periods were added as they came to hand. This is borne out by the pittancer's account in group II: a first version, closing in March, was struck out and a new one, closing in early July, was squeezed into the available space, again at the foot of a page (nos 40, 41). It is likely, then, that the moneyer's account, which does not close until December 1250 (no. 55), should be seen as an addition to group III, again in a blank space at the foot of a page.

It is most unusual to find manorial or monastic accounts in a cartulary or, indeed, set out on parchment leaves and gatherings in book form. The Bury accounts as we have them played no part in the practical business of accounting and auditing. This appears from the sequence of successive accounts for the same place or office audited on different occasions — at Warkton, for instance, on 28 April, 16 May and 28 July 1250. But it is clear from the accounts themselves that they are not the records from which the auditors passed financial judgement but copies made later. Thus at the end of the subcellarer's account for sales of corn in 1249–50 there is a note of certain debts 'that were mentioned at the time of his audit' (no. 35),[6] and at the end of the Whepstead accounts we are asked to 'note that not all the details of the rolls are set out here but just the biggest and most noteworthy items' (no. 27).[7] There are other references to rolls from which the accounts were drawn, but it is not entirely clear how far the notes and memoranda that follow the final balance on many of the accounts were copied from the audited record or how far they were added here from other sources of information.

Why were these copies made of the accounts? They are so irregular in structure and in the periods covered that they cannot be seen as a normal systematic enrolment of accounts after audit, such as those made for the bishopric of Winchester and some other large estates at this period, nor yet as a model or standard for future accounting, of the kind that was drawn up at Beaulieu Abbey in 1269–70.[8] They are far from comprehensive; as we shall see, only ten of the sixteen manors rendering food farms are represented. Perhaps we should look to earlier events at Bury. When Samson became abbot in 1182 he visited all the manors of both abbot and convent and had a survey made of the abbey's estates.[9] When Abbot Hugh took office in 1215 he ordered a survey of lands, sown and unsown, and an inventory of the stock held on all the manors and in all the monastic offices, and visited the manors of both abbot and convent.[10] The summer of 1250 was less than two years after the consecration of Abbot Edmund of Walpole; it would be in keeping with these precedents if he asked for the audited accounts of all manors and offices for at least the most recent full year to be shown to him — a snapshot of the state of current finance across all the abbey's departments. Perhaps we should also link them with the statute of the Benedictines' provincial chapter in October 1249 which required an abbot to hear the account of the obedientiaries' receipts and expenses every year.[11] However, an abbot of Bury might reasonably expect a rather more formal, more careful and neater presentation of the financial conspectus of his house than we find in Harley MS 645. In fact, as we shall see, some editing would be needed to provide this sort of overview from the working records that were available. The surviving accounts of 1247–50 were probably drafts, culled from these working records, from which a more stately document — roll or book — was produced for the abbot's inspection.

This is no more than conjecture — but conjecture can be carried further to link group IV of the accounts with the consecration of Abbot Edmund's successor, Simon, in 1257 and to suppose that a few years later he, too, demanded a detailed view of finances throughout the abbey. By 1260 or 1261 the original accounts may well have been more orderly than ten years earlier, so that what we have in group IV may be just the few that needed some sort of editing before they were shown to the abbot. But this is the merest guess. What is certain is that all four groups of accounts are of extraordinary interest, not least for the light they throw on mid-13th-century accounting methods and on the way Bury St Edmunds ran its estates at this time.

ACCOUNTING METHODS

Accounting techniques and practice probably varied from one estate to another much more in the mid 13th century than they did fifty years later.[12] On some large estates and a few smaller ones a system of well organized, formal written accounts was long established: after regular, annual audit the year's account for each manor was set out on a roll for future reference. The pipe rolls of the bishopric of Winchester from 1208–9 onwards are the most notable accounts we have of this kind, but stray rolls surviving from other estates show that the same system was in use elsewhere — the archbishopric of Canterbury, Ramsey Abbey, the honour of Gloucester are examples. The formality and method of these enrolled accounts came to be applied to the documents used in the process of auditing — whether from the start or as a later development we do not know — producing the same structured written accounts for individual manors, of which the earliest known, from Froyle in Hampshire, dates from about 1233. Before long most estates that enrolled their accounts after audit began instead to keep these individual accounts, sometimes filing them together in a bundle for each year, thus saving a great deal of trouble in copying them all out. Ramsey Abbey and Winchester Cathedral Priory are examples of estates that made this change in the mid 13th century, and it was in this form of formally structured accounts for individual manors, presented at audit as the basis for the auditors' investigations, that written manorial accounts spread rapidly from the 1250s onwards, until by the end of the 13th century they were produced all but universally, on estates of all sizes and of every kind of structure or lordship. An estate that was in the vanguard of this development was Westminster

Abbey, where we have no enrolled accounts but where the earliest accounts for individual manors date from the 1250s. We should of course be wary of equating the earliest surviving document with the innovation itself, but here the form of the record suggests that it represents new practice, not just the first evidence of what had long been done. Alongside Westminster we may place Bury St Edmunds: the earliest account roll for a Bury manor is from Hinderclay in 1250–51.[13]

However, the introduction — and the preservation — of formal written documents as the basis of accounting did not mean that these estates were now adopting accounting procedures for the first time. Since the early 13th century it had been normal for estate owners to put their properties in the charge of their own officials instead of lessees, a change that occurred early at Bury on at least the abbot's manors, for Jocelin of Brakelond tells how Abbot Samson took them all into his own hands in 1182.[14] This so-called demesne farming absolutely required some form of accounting, a regular and detailed check on what the local official had received and spent, so as to determine how much he owed the estate owner, or vice versa. Our knowledge of how this was achieved before the advent of formal written accounts must be inferential, but it seems that the official would present his accounts orally to the auditors, with the help of the tallies he had collected and other more or less informal documents; the auditors might keep no more elaborate accounts than would be needed to record the situation for the next audit. It is the peculiar merit of the accounts in Harley MS 645 that they provide an insight, possibly unique, into the way a large estate maintained demesne farming before it started to keep formally structured written accounts.

Certainly they dispel any illusion that estate accounting was then in a state of primitive simplicity. We get glimpses of a lack of system that produced convolutions and complications all its own, and we realize that the introduction of formal accounts, far from imposing bureaucratic complexities, must often have brought in a sensible streamlined efficiency, hitherto unknown. At Bury, instead of an account for each office or manor that covered one year, all ending on the same day and audited soon after, we have accounts for various periods with no uniformity of terminal date. Certainly many are for about a year, ending in July or August, and some cover a shorter period simply because the official has changed. Thus at Beccles the account for September 1248 to July 1249 was rendered by Alexander and Ralph, *servientes*; the manor was then run for a few weeks by the pittancer, then throughout August by Stephen of Caister and Alexander, and each of these short periods produced a separate account (nos 44–46). On the other hand there is no apparent reason why the bedel at Whepstead submitted two successive accounts between March and June 1250 (nos 26, 27), and although all other accounts were for a year or less the subcellarer in 1250 accounted for the sale of corn over fourteen months (no. 35) and the moneyer in 1258 presented a single account covering two years (no. 58). Not many of the accounts note the date when they were audited, but those that do are noted in the appendix. They show that two groups of auditors systematically visited some of the manors of the cellarer and the chamberlain in mid July 1249,[15] but they also show that some accounts were audited irregularly and only after long delay, as at Southery, where accounts that ended in August 1248 and August 1249 were both audited in February 1250 (nos 21, 22). Nor was there complete consistency in the units for which accounts were rendered: for the period from September 1249 to March 1250 we have separate accounts of the subcellarer for the general management of his nearby lands, for the cash received and spent on livestock and for the bailiwick and mill (nos 10–12), but for the next four months all three aspects were subsumed in a single account (no. 33).

This was no more than a single quirk, but in their contents — what they report and how — the accounts differ greatly. We have seen that some detail was specifically left out when some of the accounts were copied, but a lot is left in, and when, as at Whepstead (nos 24–27), the account gives no more than the totals of cash charge and discharge with the balance and a few further notes, we can be fairly sure that this was because the original told no more, not because it has been severely abridged. Sometimes the notes after the balance include information we would expect to find in the account itself: thus at Pakenham (no. 18) we are told that the local officials held 20*s*. from the levy (*auxilium*), 2*s*. 1*d*. from *wardsilver* and have taken nothing for peat (*de turbis*). A few of the accounts begin with the total of the cash charge, then list the items that produced it, followed by the total discharge, similarly itemized, and the balance; Mildenhall (no. 28) is arranged this way and so, for the discharge, is the subchamberlain's account (no. 39). The subcellarer's two accounts from August 1259 to June 1261 account first for corn that was held, then for the cash (nos 65, 66) — the normal order of a formally structured manorial account was cash, corn, stock — but his account ending in May 1258 omits the corn altogether (no. 61) and in fact few of these Bury accounts formally set out charge and discharge for each type of corn. At Horringer in 1249 the notes about corn that follow the cash balance are so full that they practically amount to a proper corn account (no. 14), but at Hinderclay and Stanton there is nothing about the corn in store apart from a note of the amount of threshed corn in the granary (nos 42, 43). There are no stock accounts, but the 1248–49 account for Aylsham (no. 53) and all the manorial accounts from 1257 to 1260 conclude with lists of the stock on the manor — for Horringer the list was added in the margin (no. 59). Two of these later accounts in group IV — from Semer and Groton (nos 56, 57) — also include regular corn accounts and we may see them, along with the surviving original 1250–51 account roll from Hinderclay, which includes a regular stock account as well,[16] as evidence of the move towards standard accounting forms and techniques at this period. But that stock was accounted for earlier and just not recorded in writing, is shown by the note after the Nowton account from 1249, that the officials 'have not yet answered for the hens, the eggs, the geese or the works' (no. 23).[17] The reference to works — that is, the tenants' labour services — is interesting, for in fact they seldom appear even on formally structured manorial accounts at this period,[18] and it is a reminder that even these may not record all the matters that were enquired into at audit. They are mentioned again at Aylsham, where the officials 'answer for the works only in bulk' (no. 53).[19]

The copied accounts contain enough references to the original rolls (they are always *rotuli*) to show that it was here, not in the copyists' editing, that these inconsistencies arose. Probably some of the copied accounts were made not from one original roll each but from several, and some of the accounts for short periods and the subcellarer's separate cash accounts for livestock and mill may appear simply because the copyist had failed to combine accounts that were elsewhere amalgamated. Successive accounts of, probably, the cellarer's barnkeeper for short periods in 1249 (nos 5–7) are specifically taken from three different rolls but are followed by a note of the corn used for malting based on all three,[20] while on the unidentified manor that now opens group III of the accounts 13*s*. 8$\frac{3}{4}$*d*. was spent for the abbot by one roll, 10*d*. by another (no. 50). A roll of collection of rents (*rotulus de collectione*) is mentioned (no. 1) and a reference at Nowton to the local official's separate roll for *metecorn*, the mixed corn given in wages to manorial employees (no. 23), may explain why a note about *metecorn* is appended to many of the accounts.

That the original rolls were not only complex but sometimes inadequate or defective is shown by several comments in the accounts. At Hinderclay, they tell us, the amount of beans and peas sold differed by 2 seams in the roll and the tally (no. 42); at Icklingham (no. 51) the official did not know how much he had received for his upkeep (*de manutenemento*). The subcellarer's accounts for 1249–50 (no. 4), written by hand A, have marginal notes 'defect in the account' (*defectus compoti*) by another copyist, hand C, to draw attention to problems in two entries: payments to the kitchener and cellarer as entered on the cellarer's roll were significantly less than the record of the tallies, while the cellarer and subcellarer had both claimed for the same expenditure on three of the manors. Most telling of all, at Beccles it was noted that 'this account was drawn up from the details in a certain old roll and a certain new roll, written badly and across, but they were rehearsed before Alexander the bedel, Philip and Henry, who all gave their agreement, and the details are entered in the rolls with arabic figures' (no. 45).[21]

Certainly it would be misleading to stress the deficiencies or eccentricities of these accounts. Some of their individual features can be found on formal written accounts elsewhere; at Westminster Abbey, for instance, each manor closed its annual account at a different date until 1284, when a uniform financial year starting at Michaelmas was introduced, while accounts for the Bec Abbey estates still did not include formal stock accounts in the 1280s.[22] In some ways the Bury accounts exemplify what contemporaries would consider excellent practice. The revenue from Chepenhall was checked with the extent (no. 15), the rents at Aylsham with the previous year's account (no. 53). The bedel of Warkton was amerced for concealment (no. 31) and when Stephen of Caister could account only for 27*s.* 1¼*d.* of the 27*s.* 10½*d.* that the rolls said was owing to him, this was all he was given, 'and thus he loses 9¼*d.*' (no. 49).[23] The subcellarer's account of sales of corn anticipates what only later became normal in manorial accounts in recording the fluctuations of prices throughout the year, though only in broad bands: for the first nine weeks the sale price of wheat was from 3*s.* to 4*s.*, for the next six weeks from 4*s.* to 5*s.*, and so on (no. 35). The abbot and monks of Bury might reasonably feel that they were carrying out the provincial chapter's injunction in the spirit as well as the letter.

At the same time we get some hints that they had not entirely grasped the concepts behind this form of accounting. They were not fully at ease with a firm cut-off date for the end of the account and were tempted, for one reason or another, to include later transactions or omit earlier ones. At Mildenhall the sale of new corn was excluded from the account, which ended on 11 October, and entered in a separate note (no. 28); at Warkton an account starting on 29 September included 13*s.* 6*d.* harvest expenses which were to be deducted because they were in the previous year's roll (no. 30); at Beccles the local official's final credit balance of 6*s.* 3½*d.* was to go to boost next year's account (*ad auxilium anni subsequentis*) along with receipts from amercements, from sales of hay and from some other items omitted this year (no. 46); the subcellarer's account for sales of corn in 1249–50 closed on 14 July for malt, 29 June for everything else (no. 35). Again, outstanding debts that the officials had failed to collect were noted, but not charged against them, as at Southery, where a list was appended of tenants who had not yet paid their rent and the officials were enjoined to do their best (*diligenter inquirendo*) to collect it (no. 22); under good accounting practice these debts would be formally charged against the official, whose responsibility was thus made clear, though a good manorial lord would not wash his hands of them but would still keep an eye on what happened.[24] Still more significant are the cases where the account, instead of adding up a total cash charge of all sums received and a total discharge of all expenditure, explains what happened to the money received for each particular transaction. The account of the manorial receivers in 1249–50 is entirely in this form: the considerable sum acquired from each manor is immediately followed by a note of what was done with the money (no. 36). This was perhaps a special case, but we get occasional hints elsewhere that any sum of money that came to an official was seen as a discrete unit, to be accounted for as such and not combined with others in an integrated whole. Thus at Caister the 24*s.* 5*d.* owed to the bedel at the end of the account is explained as 18*s.* for a millstone, 2*s.* 9*d.* for boards, 6*d.* for shoeing and so on (no. 47), and on the subcellarer's local properties arrears carried over from the previous account are defined as so much from the stock, so much from the wool (no. 33). Among the notes at the end of the Mildenhall account is a tiny account of the hayward (*messor*) in two sections: what he received from rent and how it was spent, what he received from perquisites and how it was spent (no. 28). The subcellarer's separate accounts for cash from stock and cash from mill may be evidence of the same kind of thinking (nos 11, 12). It is all rather closer to the countryman remembering what he got from a particular source and remembering how he spent it than to the accountant's neatly classified categories of charge and discharge.

RUNNING THE ESTATE

Equally, however, we may wonder whether the monks of Bury had fully grasped the concept — or at least the implications — of demesne farming in general. Two features of the accounts raise reasonable doubts. The first is a small but possibly significant detail. Notes following the accounts from Horringer for 1247–48 and 1248–49 compare what has been received from the sale of stock with the amount spent in buying replacements (nos 13, 14); thus in 1247–48 'Note that this year £3.11*s.* was taken from the stock, and in buying stock only 13*s.* was laid out'.[25] This is the language and the way of thought of the 12th century, when manors were normally leased to farmers; the value of the livestock on the manor was carefully watched because this directly affected the amount the lessee could be asked to pay. Diminishing the value of the stock diminished the value of the manor, and the nature of the stock — whether it was oxen or horses, sheep or pigs — was a secondary consideration. Demesne farming effectively changed this. The value of the stock on the manor now scarcely mattered. What was important, what the estate-owner ought to be asking, was how much and what kind of stock could best be placed there to secure the greatest profit. It may be a change of little more than emphasis, but it represents a different and new approach to estate management.

A less subtle, much more obvious archaism was the food farms that dominate so many of these accounts. The food farm belonged not to the 12th century so much as to the 11th or earlier. The local farmer held a manor in return not for the cash payments that were to become normal but for regular deliveries of specified produce; these were organized so as to keep the estate-owner constantly supplied throughout the year. In Domesday Book we see vestiges of this system in the 'farm of one night' (*firma unius noctis*) due from certain manors on the royal estates. At Bury in the 11th century the food farms were organized on a basis not of single days but of months, and by the late 12th century, instead of a system of leasing for cash, the food farms were still in place but on a weekly basis: each week throughout the year one of sixteen of the convent's manors delivered to Bury fixed amounts of wheat, barley malt, oats malt and meat.[26] There were slight differences in the amounts of these weekly deliveries, and the number demanded in the course of the year differed between the manors, from one to eight. With the introduction of demesne farming the system lost all point as far as drawing profit from the estate was concerned: through its local officials the abbey now took all the profit and could take all the produce from its demesnes, not just these predetermined amounts. But in the internal economy of the monastery the system was too deeply entrenched to be easily abandoned: the work of the cellarer and the kitchen was arranged around these weekly deliveries. The change in estate management produced no corresponding change here and the weekly food farms, now simply fixed deliveries from manors held in demesne, continued unabated.

Elsewhere in Harley MS 645 the scheme of food farms is set out by a mid-13th-century hand,[27] and in the accounts we see it in operation: they include accounts from ten of the sixteen contributory manors, which were all in the cellarer's domain. Just how important the system was to the cellarer's department is shown by the opening lines of the subcellarer's and kitchener's accounts, which after giving the date define it in terms of the week's food farm: 'Barton before Herringswell coming into the kitchen' — 'before Herringswell' to distinguish this from Barton's other weeks in the course of the year (nos 63, 64). Supplying the food farm was clearly troublesome to the manorial officials. The accounts record arrears of farm owed from the manors: Whepstead was 6 seams of wheat short in June 1250 (no. 27), Horringer in May 1258 owed 17 seams 6 bushels of wheat from its four weeks' farm (no. 59). One manor might be called upon to make up the deficiency in the farm of another: Nowton in 1259–60 contributed 12 seams of wheat to Pakenham's farm, 6 seams to Whepstead's (no. 60). The corn was not all produced on the demesnes, and the notes following some of the accounts show where it came from. Thus at Pakenham in 1248–49, of 192 seams of oats malt due in the farm, 64 came from the demesne granary, 109 from the tenants' rents, 55 were bought and 33 were still owing (no. 18). Other accounts, too, show that corn might be bought for the farm, but accounts of the subcellarer show large amounts of corn being sold from the granary that received the food farms — over 77 seams of wheat in four months of 1250, for example (no. 32). From the evidence of the accounts we cannot rule out the possibility that, simply to meet the food farms' traditional requirements, corn was bought on the manors, carried expensively to Bury and then sold; but probably we can acquit the abbey of this absurdity. When we are told that a manor owes corn for the farm we are often also told its value: the 6 seams of wheat at Whepstead were worth 18*s.*, the 17 seams 6 bushels at Horringer £8 5*s.* 5½*d.* It seems likely that the accounts had already achieved the sophistication of fictitious purchases and sales, that buying the farm's corn on the manor and selling it at Bury simply meant that cash, not corn, was transferred. On the other hand it may be that the only commutations of the farm for cash are what the subcellarer obscurely entered as exchanges (*de escambiis*): on one account 3*s.*, on another £4 16*s.* 8*d.* from sixteen farms (nos 35, 32).

Further work on the accounts might resolve this problem. They need to be investigated along with all other evidence of the way the abbey and its estates were administered at this time. They are not an easy source to use: the writing and layout of the manuscript present some difficulties, their purpose and their relationship to the original rolls are uncertain and we have no similar accounts elsewhere with which they can be compared. In the history of estate accounting they are valuable relics, telling us of a little-known and interesting stage of development, but this is certainly not all they have to reveal.

ACKNOWLEDGEMENT

I am grateful to Dr Antonia Gransden for first telling me of these extraordinarily interesting documents, now some ten years ago, for encouraging me to write this account of them and for much patience over my delays in producing it.

APPENDIX
Accounts in Harley MS 645

The dates, given here in modern form, in the text often refer to the abbatial year. That this was reckoned, as we would expect, from the abbot's consecration, not his election, is shown by the account for Barton, from St Peter in Chains (1 August) 1248 to St Margaret (20 July) in the first year of Abbot E. (no. 17): Edmund of Walpole was elected on 7 July 1248, consecrated on 27 September. In a few cases there is possible ambiguity in the text's form of the date; what is given here seems the most likely interpretation.

It is difficult to compare the handwriting in different parts of the text and some of the identifications of hands are open to doubt.

Group I: the cellarer's department, 1247–50 (nos 5–9 are reproduced in plate XXXII)

1 [rents and other payments from manors], 20 July 1249–20 July 1250. Hand A. Fol. 193
2 [rents and other payments from manors], unknown dates. Hand A. Fol. 193^v
3 subcellarer, 29 Apr.–29 Dec. 1249. Hand A. Fol. 193^v
4 subcellarer, 29 Dec. 1249–2 Mar. 1250. Hand A with marginal notes in hand C and followed by a note in hand D. Fol. 193^v
5 barnkeeper(?), 21 Feb–23 May 1249. Hand A. Fol. 194
6 barnkeeper(?), 23 May–20 July 1249. Hand A. Fol. 194
7 barnkeeper(?), 20 July–1 Aug. 1249. Hand A. Fol. 194
8 barnkeeper(?), 1 Aug.–1 Sept. 1249. Hand A. Fol. 194
9 barnkeeper(?), 1 Sept. 1249–8 Mar. 1250. Hand A. Fol. 194
10 subcellarer, local properties (general), 1 Sept. 1249–8 Mar. 1250. Hand A. Fol. 194
11 subcellarer, local properties (stock), 1 Sept. 1249–8 Mar. 1250. Hand A. Fol. 194
12 subcellarer, local properties (bailiwick and mill), 1 Sept. 1249–8 Mar. 1250. Hand A. Fol. 194
13 Horringer, 29 Sept. 1247–25 July 1248, rendered 3 May 1250. Hand A. Fol. 194^v
14 Horringer, 25 July 1248–1 Aug. 1249. Hand A. Fol. 194^v
15 Chepenhall, 1 Aug. 1248–1 Aug. 1249, rendered 17 July 1249. Hand B. Fol. 195
16 Southwold, 1 Sept. 1248–17 July 1249, rendered 17 July 1249. Hand B. Fol. 195
17 Barton, 1 Aug. 1248–20 July 1249, rendered 19 July 1249. Hand B. Fol. 195
18 Pakenham, 1 Aug. 1248–20 July 1249, rendered 19 July 1249. Hand B. Fol. 195
19 Bradfield, 20 July 1248–20 July 1249, rendered 20 July 1249. Hand B. Fol. 195^v
20 Rougham, 20 July 1248–20 July 1249, rendered 20 July 1249. Hand B. Fol. 195^v
21 Southery, 2 Feb.–1 Aug. 1248, rendered 13 Feb. 1250. Hand B. Fol. 195^v
22 Southery, 1 Aug. 1248–1 Aug. 1249, rendered 13 Feb. 1250. Hand B. Fols 195^v, 197
23 Nowton, 9 Apr.–1 Aug. 1249. Hand B. Fol. 197
24 Whepstead, 25 July 1248–25 July 1249. Hand B. Fol. 197
25 Whepstead, 20 Nov. 1249–27 Mar. 1250. Hand B. Fol. 197
26 Whepstead, 27 Mar.–15 May 1250. Hand B. Fol. 197
27 Whepstead, 15 May–29 June 1250. Hand B. Fol. 197
28 Mildenhall, 17 Oct. 1247–11 Oct. 1248. Hand B. Fols 197^v–98
29 Warkton, 29 Sept. 1248–29 Sept. 1249, rendered 28 Apr. 1250. Hand B. Fol. 198
30 Warkton, 29 Sept. 1249–5 May 1250, rendered 16 May 1250. Hand B. Fol. 198
31 Warkton, 5 May–28 July 1250, rendered 28 July 1250. Hand C. Fol. 198

XIII

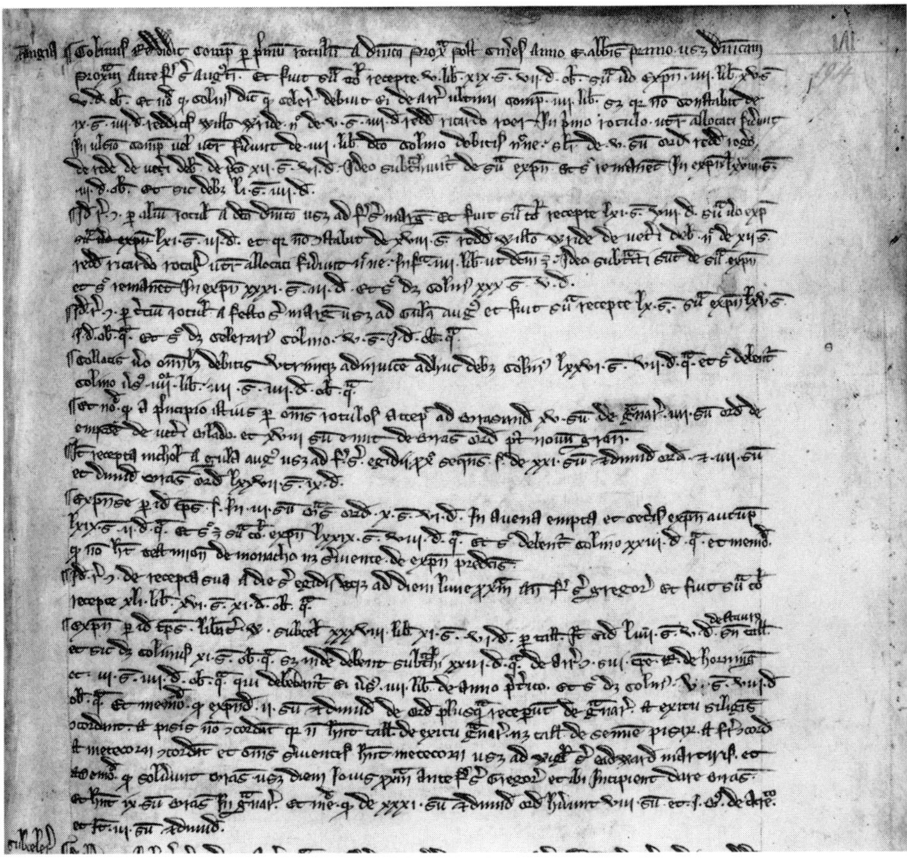

XXXII. Accounts of Colin and Nicholas, successive keepers of, apparently, the cellarer's barn at Bury, for five periods from 21 February 1249 to 8 March 1250 (accounts nos 5–9, written in hand A). British Library, Harley MS 645, top half of fol. 194

32 subcellarer, 2 Mar.–29 June 1250. Hand D. Fol. 198^{v}
33 subcellarer, local properties, 15 Mar.–4 July 1250. Hand D. Fol. 198^{v}
34 subcellarer, fodder bought, 15 Mar.–4 July 1250. Hand D. Fol. 198^{v}
35 subcellarer, sale of corn, 29 Apr. 1249–29 June 1250. Hand D. Fol. 198^{v}
36 [manorial receivers], 24 July 1249–20 July 1250. Hand E. Fol. 199

Group II: the chamberlain's department, 1248–49

37 manorial receipts, unknown date to 23 June 1249.[28] Hand E. Fol. 252
38 manorial receipts, 23 June 1249 to unknown date. Hand E. Fol. 252
39 subchamberlain, 1 Sept. 1249–28 Feb. 1250, rendered 8 Mar 1250. Hand E. Fol. 252
40 pittancer, 29 Sept. 1249–12 Mar. 1250, rendered 8 Mar. 1250 (account struck out). Hand E. Fol. 252
41 pittancer, 29 Sept. 1249–7 July 1250, rendered 7 July 1250. First two lines hand E, the rest hand D. Fol. 252
42 Hinderclay, 29 Sept. 1248–17 July 1249, rendered 17 July 1249. Hand B. Fol. 252^{v}
43 Stanton, 29 Sept. 1248–18 July 1249, rendered 18 July 1249. Hand B. Fol. 252^{v}
44 Beccles, 29 Sept. 1248–7 July 1249, rendered 16 July 1249. Hand B. Fol. 252^{v}
45 Beccles, 7 July–1 Aug. 1249. Hand B. Fol. 252^{v}
46 Beccles, 1–29 Aug. 1249. Hand B. Fols 252^{v}–53
47 Caister St Edmunds, 7 July–1 Aug. 1249. Hand B. Fol. 253
48 Caister St Edmunds, 1–29 Aug. 1249. Hand B. Fol. 253
49 Caister St Edmunds, 29 Aug.–29 Sept. 1249. Hand B. Fol. 253

Group III: the sacrist's department, 1248–50

50 [unidentified manor], unknown dates (concluding sections only). Hand B. Fol. 260
51 Icklingham, 29 Sept. 1248–22 Feb. 1249, rendered 18 Mar. or 9 July 1250. Hand A. Fol. 260
52 Icklingham, 22 Feb.–29 Sept. 1249. Hand A, except a few items in the expenses in hand C. Fol. 260
53 Aylsham, 29 Sept. 1248–29 Sept. 1249, rendered 30 Nov. 1249. Hand B, except a note added to the receipts in hand F. Fol. 260^{v}
54 sacrist, 28 Feb.–29 Sept. 1249, rendered 14 Dec. 1249. Hand B. Fol. 260^{v}
55 moneyer, 29 Jan.–20 Dec. 1250, rendered 19 Dec. 1250. Hand B receipts, hand C total of receipts and all expenses. Fol. 260^{v}

Group IV: miscellaneous accounts, 1256–61

56 Semer, 20 July 1257–15 May 1258. Hand G. Fol. 261
57 Groton, 20 July 1257–15 May 1258. Hand G. Fols 261, 261^{v}
58 moneyer, 11 Apr. 1256–19 Apr. 1258, rendered 17 May 1258. Hand G. Fol. 261^{v}
59 Horringer, 3 July 1257–18 May 1258. Hand G. Fol. 261^{v}
60 Nowton, 1 Aug. 1259–7 July 1260. Hand G. Fols 261^{v}–62
61 subcellarer, 4 July 1257–29 May 1258. Hand G. Fol. 262
62 kitchener, 6 July 1259–4 July 1260.[29] Hand G. Fol. 262
63 kitchener, 25 Aug. 1258–24 Aug. 1259. Hand G. Fol. 262
64 kitchener, 24 Aug. 1259–22 Aug. 1260. Hand G. Fol. 262^{v}
65 subcellarer, 22 Aug. 1259–24 Aug. 1260. Hand H. Fol. 262^{v}
66 subcellarer, 24 Aug. 1260–2 June 1261. Hand I. Fols 262^{v}, 265

REFERENCES

1. It is described briefly in G. R. C. Davis, *Medieval Cartularies of Great Britain: a Short Catalogue* (London 1958), 16–17 (no. 124), and in greater detail in Thomson, *Archives*, 127–29 (no. 1282)

2. The loss of the headings at the top of fol. 193 and fol. 193^{v} seems to be due to wear, not trimming, but fols 195, 252 and 260 have all been slightly trimmed at the foot and fol. 260 also on the outer edge.
3. When Harley MS 645 was rebound in 1969 its gatherings were sewn on to guards and it can thus be collated with reasonable confidence.
4. As there is no break in the text between fol. 261^{v} and fol. 262, the presumed other bifolium of group III must already have been lost, or its second leaf cancelled. Alternatively, fol. 262 may be the second leaf of this presumed bifolium (and even fol. 265 the second leaf of another), the first leaf (or leaves) of group III being cancelled then or later.
5. At the foot of fol. 193^{v}: 'Quere compotos de Granario et de Grangia et de prebenda empta in quarto folio sequenti'.
6. 'de quibus facta est mentio in tempore compoti sui' (fol. 198^{v}).
7. 'Et notandum Quod omnes particule Rotulorum non sunt hic specificate sed grossiores et notabiliores' (fol. 197).
8. *The Account-Book of Beaulieu Abbey*, ed. S. F. Hockey (Camden 4th Series, XVI, 1975), 8–9.
9. *JB*, 28–29, 31.
10. *Election*, ed. Thomson, 165–67. When Hugh became bishop of Ely in 1229 he ordered a full survey of the bishopric's manors (ibid., 195).
11. *Chapters*, ed. Pantin, i.36.
12. A general account of manorial accounts in the 13th century is in P. D. A. Harvey, *Manorial Records* (British Records Association, Archives and the User, V, 1984), 25–31, reprinted with postscript in *Accounting History: Some British Contributions*, ed. R. H. Parker and B. S. Yamey (Oxford 1994), 91–100, 115.
13. University of Chicago Library, Bacon Collection, no. 405. Thomson, *Archives*, 86, follows the Library's own list in dating this simply mid 13th century; in fact it specifically covers the period from St Peter in Chains to the Translation of St Benedict in the 3rd year of Abbot E[dmund], i.e. 1 Aug. 1250–11 July 1251.
14. *JB*, 32.
15. It was not a single audit of the properties of both obedientiaries, for although a single group could just have covered both Southwold and Chepenhall (12 miles apart) on one day, 17 July, they could hardly have also taken in Hinderclay (18 miles further on).
16. University of Chicago Library, Bacon Collection, no. 405.
17. 'Et non dum Responderunt de gallinis neque de ouis neque de aucis neque de operibus' (fol. 197).
18. *Manorial Records of Cuxham, Oxfordshire, circa 1200–1359*, ed. P. D. A. Harvey (Historical Manuscripts Commission, Joint Publication, XXIII; Oxfordshire Record Society, L; 1976), 64–65.
19. 'de operibus non Respondent nisi in grosso' (fol. 260^{v}).
20. 'per primum rotulum', 'per alium rotulum', 'per tertium rotulum', 'per omnes rotulos' (fol. 194).
21. 'Et sciendum quod iste compotus requiritur per particulas cuiusdam Rotuli ueteris Et cuiusdam noui male et etransuerso scriptas sed Recitate fuerunt coram alexandro Bedello et philippo et Henrico qui omnes perhibebant testimonium et signate sunt particule in rotula (*sic*) per syfras' (fol. 252^{v}).
22. Typed catalogue of manor and court rolls at Westminster Abbey; 'Computus Rolls of the English Lands of the Abbey of Bec (1272–1289)', ed. Marjorie Chibnall, in *Camden Miscellany vol. XXIX* (Camden 4th Series, XXXIV, 1987), 17–185.
23. 'Et sic amittit ix d' q'' (fol. 253).
24. P. D. A. Harvey, *A Medieval Oxfordshire Village: Cuxham 1240 to 1400* (London 1965), 66–68.
25. 'Et Memorandum quod hoc anno ceperunt de stauro lxxj s' et non reposuerunt ad staurum emendum nisi xiij s'' (fol. 194^{v}).
26. *Kalendar*, ed. Davis, l.
27. Fol. 225^{v}.
28. That nos 37 and 38 both cover much less than a full year is shown by the two lists, added in the margin, of receipts from the same manors for complete years, one simply 'anno integro', the other 'anno j^{o}', presumably 1248–49.
29. There is clearly a mistake in the dates of nos 62–64: all three cover the same kinds of item in the same way. No. 64 certainly begins at the point where no. 63 ends, and the date on each refers to the year of grace. One possibility is that no. 62, dated from Sunday before the Translation of St Benedict in the 2nd year to the same day a year later ('anno reuoluto') really belongs to the abbacy of Edmund, not Simon; in this case its date would be 10 July 1250–9 July 1251.

ABBREVIATIONS

BL	British Library
Chapters, ed. Pantin	*Chapters of the English Black Monks 1215–1540*, ed. W.A. Pantin, 3 vols (Camden Soc., 3rd ser., XLV, XLVII, LIV, 1931–37)
Election, ed. Thomson	*The Chronicle of the Election of Hugh Abbot of Bury St Edmunds and Later Bishop of Ely*, ed. and trans. R.M. Thomson (Oxford 1974). [Composed between *c.* 1222 and 1229]
JB	*The Chronicle of Jocelin of Brakelond,* ed. and trans. H.E. Butler (Nelson's Medieval Classics, 1949)
Kalendar, ed. Davis	*The Kalendar of Abbot Samson of Bury St Edmunds and Related Documents*, ed. R.H.C. Davis (Camden Soc., 3rd ser., LXXXIV, 1954)
Thomson, *Archives*	*The Archives of the Abbey of Bury St Edmunds*, ed. R.M. Thomson, Suffolk Records Soc., XXI (Woodbridge 1980)

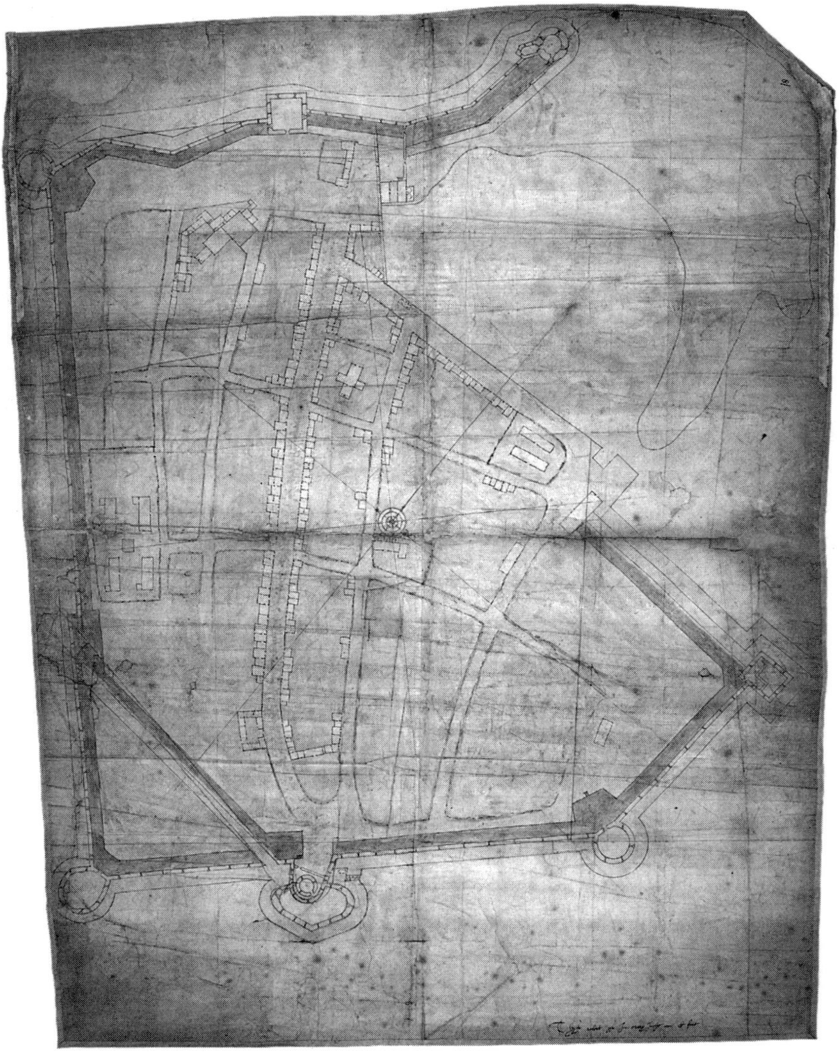

Fig. 4 MAP OF PORTSMOUTH drawn in 1545 with some pencilled alterations in 1546

XIV

THE PORTSMOUTH MAP OF 1545 AND THE INTRODUCTION OF SCALE MAPS INTO ENGLAND

The earliest surviving map to show Portsmouth in any detail is British Library Cotton MS. Augustus I.i.81, reproduced here as Fig.4. The original is very much larger, about 31 by 23 inches (78 by 59 cm.); it is drawn in ink, but a dull olive wash has been added to mark the tops of the town walls and (much paler) the areas without buildings inside the town, while pale grey flecks have been painted on the High Street, the other streets being left uncoloured. It has no inscriptions apart from a note of the scale, one inch to 100 feet (1:1200) in the bottom right corner. In his catalogue of *Maps of Portsmouth before 1801* Mr D. Hodson has recently discussed the date and purpose of the map on the basis of work by Professor Martin Biddle, as yet unpublished, on Portsmouth's defences in the sixteenth century.[1] For it is essentially a map of the town's fortifications, and although it shows the layout of streets and houses as well this is subsidiary to its main purpose, which is to set out a scheme for improving the defensive works around the town. Those we see on the map are partly what was there already, partly proposals for alterations that were never carried out in full. The scheme and the map date from 1545, when a French attack and landing on the Isle of Wight at a time when King Henry VIII was himself in Portsmouth drew attention to deficiencies in the local defences. Some pencilled alterations on the map, showing alternative proposals, were probably made in the following year.

All this has been elucidated and discussed in detail by Professor Biddle and Mr Hodson. As Mr Hodson's work makes clear, the map is important in the local context: a map of the shore-line defences south-east of the town is a few years earlier,[2] but otherwise it is the first of the many plans of the fortifications and dockyard which had by the end of the eighteenth century made Portsmouth one of the best mapped towns of its size anywhere in the world. The present note will examine the map from a different viewpoint, looking at its place in the general development of mapping in sixteenth-century England and Europe; in this context too it occupies a significant position.

The first half of the sixteenth century saw rapid and important changes in topographical maps — that is, maps of small areas, such as an individual might draw from his own knowledge.[3] In the middle ages such maps were practically unknown; although it is difficult for us to appreciate this, the habit of mind that enables us easily to envisage routes and landscape in cartographic form was alien to our medieval ancestors. By the late fifteenth century in a very few areas there had developed something like a local tradition of drawing and using maps: particularly in Lombardy and in towns elsewhere in Italy, but also in south-west Germany, along the coast of the Low Countries and possibly in part of eastern England around the Wash. Even here not many maps can have been produced, and outside these areas it seems as if the few local maps that survive to us from the twelfth to the fifteenth century were all individual productions, each drawn for a particular occasion by someone who had hit by chance on the idea of putting topographical information in the form of a map or diagram. Somehow — and it is not always easy to see how — people in the middle ages managed without even the simplest sketch map for the many purposes for which nowadays we take the use of maps for granted.

But although there was nothing remotely like a conscious tradition or style of topographical maps in medieval Europe, nearly all those made were maps of one particular sort. They were picture-maps: there was no attempt to draw to a consistent scale — the concept of scale was almost unknown — and, if the map showed anything beyond boundaries, roads, watercourses or other ground-level features, all detail appeared in the form of pictures (whether stylized or realistic) as though seen either from the side or obliquely from above. By the end of the fifteenth century the most elaborate of these picture-maps — those of Italian cities — were taking the form of realistic bird's-eye views; and indeed the bird's-eye views and the topographical maps of the sixteenth and seventeenth centuries and later had a common origin in the picture-maps of the late middle ages. Picture-maps of this kind, whether sophisticated works of art or the simplest sketches, are not peculiar to medieval Europe; they represent a common stage in the development of cartographic ideas before maps came to be drawn to scale, and they are found in many other cultures and traditions — ancient Egypt, Aztec Mexico, early China and so on.

The first and most obvious change that came over topographical maps in early-sixteenth-century Europe was that far more were produced than before. An appreciation of the use and value of maps seems to have spread very rapidly soon after the turn of the century. From England we have about ten local maps surviving from each half-century between 1350 and 1500;[4] from 1500 to 1550 we have probably between 150 and 200.[5] There was a similar upsurge of production in France, Germany and the Low Countries, and probably elsewhere in western Europe as well. But at first there was no change in cartographic concepts or techniques. The maps that were now drawn in greatly increased numbers were picture-maps just as before. One of Brighton showing the French attack on the town in 1514 is a roughly drawn bird's-eye view as seen from the sea; the town appears as several rows of houses, many of them in flames, on the downs behind we see windmills and a warning beacon burning, and in the foreground are the French ships from which the troops are being landed.[6] A map of Hull and its fortifications in 1537–9 is a far more realistic and carefully drawn bird's-eye view, by an artist who probably also drew a similar, but rather cruder, view of the town and castle of Scarborough.[7] Watermills and streams at Gussage in Dorset appear on a diagrammatic map which shows the three mills as vast structures while the churches or villages of Gussage St Michael and Gussage All Saints are represented by far smaller buildings in the foreground.[8] When parts of Dumfriesshire were in English occupation in 1547 a map of the area was prepared by a military surveyor, Thomas Petyt, for presentation to the duke of Somerset; this has been identified with a surviving fully pictorial map which shows towns and fortresses in bird's-eye view surrounded by impressionistically painted hills.[9] In this case there was demonstrably no intention to draw the map to a particular scale, for lines joining the various towns and fortresses have notes of the distances between them, which do not correspond to their lengths on the map; these sixteenth-century picture-maps were no more concerned with consistent scale than their medieval precursors had been.

But the map drawn to consistent scale followed in the wake of the picture-map's rapid advance. The earliest certain scale-map from medieval Europe is a plan of Vienna which dates from the mid-fifteenth century but was probably copied from an original drawn about 1422;[10] it is just possible that there was a much older tradition in Italy of city plans drawn to scale. By the end of the fifteenth century a few scale-maps of particular regions had been produced in central Europe: of Germany by Nicholas de Cusa, of

Germany and north Italy by Erhard Etzlaub, and of Switzerland by Konrad Türst. But the idea of drawing local maps to scale spread through western Europe only in the first half of the sixteenth century. It spread far more slowly than the use of the picture-map. It was, after all, not simply a matter of applying new techniques of measuring and drawing; it was rather the introduction of an entirely new concept of cartography, a concept which seems a very natural and straightforward one to us but which was grasped only with some difficulty by people introduced to it for the first time. We see this in the phrases used in Germany and the Low Countries to describe a map drawn to scale; it was said to be made 'by the little measure' or 'by the little foot', meaning the tiny distance on the map corresponding to the length on the ground, expressions that would occur to people to whom the idea of scale drawing was a novelty.[11] We see it too in the clumsy phrases used to define the scale on the earliest English scale-maps. The note on the 1545 map of Portsmouth reads 'Thys plat ys In every Inche on C fote' (a *plat* was any map, plan or bird's-eye view); on others we find, for instance, 'In thys plat 3 inches contayneth one mile', 'This platte is made after ye rate of 32 foote to thynche', 'The ynche stondeth for C fote', 'made by the measur of ten foot to the inche'.[12] Not only was there no generally accepted formula for expressing the scale of a map; it was necessary to use a form of words that would explain the notion of scale to users of the map unfamiliar with it.

Maps 'by the little foot' were being drawn in the Netherlands by 1537.[13] When did the idea of the scale-map reach England? What is the earliest surviving English scale-map? These questions are harder than one might suppose. Even in the few instances where a date appears on a map it is not necessarily trustworthy and, as Mr Hodson's work on the maps of Portsmouth demonstrates, most early maps can be dated only from internal evidence, from features appearing on them to which a date can be assigned from other sources. It follows that to discover exactly when a map was drawn — if it is possible at all — may involve extended work on the local history of the area it covers. Recently many of the relevant maps have been securely dated by Mr H.M. Colvin and his collaborators for *The History of the King's Works;*[14] but there are others for which the accepted dates are open to doubt and which may be significantly redated by further research. Thus no more than a provisional chronology can be given for the introduction of scale-maps into England; it may well have to be revised in the light of future work.

It seems though as if the earliest surviving maps to be drawn to scale by English map-makers were produced about 1540, not in England itself but on the English territory around Calais. Between 1539 and 1541 a good deal of work was being done to strengthen the castle and other defences of Guines, six miles south of Calais and not far from the border with France. Of the various surviving maps and drawings made for these works four are plans drawn to consistent scale, in two cases explicitly so: on one there is a small (now nearly illegible) scale-bar, while the other has a note 'The Inshe conteynyth L fotte' and is almost fully framed by a continuous scale-bar, marked off in twenty-foot divisions.[15] A map of a larger area drawn to scale has been dated 1541 by Mr Colvin, who identifies it as showing a project for settlement on a newly drained area near Calais; no scale is stated on the map, but it is of particular interest because several of the roads it shows have very exact notes of their direction by compass, such as 'est & ij pontes [*i.e. points*] by southe' and 'it lackes a di. pont [*i.e. half a point*] of the southe west'.[16] Drawing maps to scale involved exact measurement on the ground not only of lengths but of angles as well.

It was not long before scale-maps were being drawn in England itself, if, indeed, this had not already occurred. Plans of fortifications near Hull were drawn soon after October 1541. One showing a line of defences along the shore has a scale-bar, and so has another which covers a very much larger area, showing settlements on both sides of the Humber as well as roads and watercourses; a third, showing the part of the town fronting on the River Hull and the fortifications opposite, is also drawn to scale though not explicitly.[17] Other scale-maps that can be dated to the early fifteen-forties show works at Dover, Carlisle and Harwich.[18] More were produced in the territories around Calais between 1545 and 1547, when the defences of Boulogne and Ambleteuse were being strengthened.[19] In all we have some forty scale-maps of English origin dating from before 1550, of which about half are of places in England, half from the lands in Picardy.

One thing all these scale-maps have in common: they were all drawn for official purposes by engineers — military engineers — in royal service. Like the Portsmouth map of 1545 they were drawn to illustrate proposals for new or altered fortifications and other works. Orders, payments and other records often enable us to identify the engineers in charge of any particular operations. The work at Guines was supervised by Richard Lee, assisted by John Rogers. John Rogers, alone among these Tudor engineers, has been the subject of detailed study; this is by Professor L.R. Shelby, who shows from close work on the handwriting and style of draughtsmanship that Rogers was definitely responsible for the Hull maps of about 1541, and that either he or, just possibly, Lee drew three of the scale-maps that come from Guines in 1539–41.[20] It seems quite likely that one of these men was the first Englishman to draw topographical maps to a consistent scale.

It was pointed out by the late Dr R.A Skelton that military engineers played a crucial role in the development of accurate scale-maps in sixteenth-century England and in introducing new techniques in surveying.[21] The contrast with the Low Countries is interesting: there private surveyors seem to have been among the first to draw maps to scale, 'by the little foot'.[22] But it has also been suggested that in England architectural practice and tradition may have played a part; that the engineers were adapting to military use a technique of drawing plans to scale that had long been known to the architects and masons of other sorts of building.[23] This is not impossible, but from the little evidence available to us it seems unlikely. Certainly drawing plans was a regular practice among builders in England from at least the late fourteenth century, for we find references to them in building contracts. These plans were drawn to show to the clients who commissioned the buildings; others may (but need not) have been drawn to guide the masons on the site.[24] We have one surviving building plan from medieval England; it shows the side of one court in Winchester College, with a doorway and staircase, and Dr J.H. Harvey has dated it about 1390, when this part of the College was built.[25] It has a note of measurements on it, but the lines on the plan are not proportionate in length — it is not a scale drawing. The same is true of a group of four plans of building plots in London, Southwark and Deptford in the fourteen-seventies; full measurements are given, and the outlines of the plots are drawn roughly the right shape, but there is clearly no attempt to draw them to an exact scale.[26] We find the same method used on the very similar plans of London building plots that accompany two conveyances of 1534 in the archives of the Drapers' Company; on a more detailed plan of about the same period showing buildings in Cornhill, London, belonging to the Charterhouse; and on an outline plan of Wallingford Castle in 1555.[27] Most impressive of all is an elaborate plan of the Exchequer at Calais, made in 1534, in preparation for a proposed meeting there between the kings of England and France; drawn on a piece of parchment over four feet square, it shows the detailed layout of some sixty rooms and notes a great many measurements, but it is not drawn to scale.[28] These examples do not add up to anything like proof. But they strongly suggest that the plans (unlike some of the elevations) drawn by the architects or master masons of early-sixteenth-century England were not scale drawings but were approximate outlines with notes of measurements. These, after all, would give both the practitioner and his client all the information they would need in a form that was simple and intelligible. Interestingly, what seem to be the earliest scale-plans of a non-military building in England are a set of three showing the ground floor and two alternative schemes for the first floor of the royal manor-house at Hull; they were drawn in 1542 or 1543 to a scale of about sixteen feet to the inch and the draughtsman was the military engineer John Rogers.[29] It seems quite likely that, in England at least, it was the civilian architect who learned the art of drawing scale-plans from the military engineer, not vice versa.

If, then, it was not from the master masons, from whom did the English military engineers learn the idea and the technique of drawing maps and plans to scale? Almost certainly it was from the military engineers of Italy, engineers who were often civilian architects as well, for it was only in the second half of the sixteenth century that the two branches of building diverged to form distinct specialized crafts.[30] It was primarily from Italy that the English engineers of Henry VIII's reign drew their style and methods of fortification. The system of angle bastions that in the new age of gunpowder offered a surer defence than the high thick walls of the medieval castle or city may have originated in fifteenth-century Italy. Certainly after the disastrous French invasion of Italy in 1494 had shown the inadequacy of traditional fortifications Italian engineers rapidly took the lead in developing this new approach to defensive works.[31] Despite the important treatise of Albrecht Dürer, *Etliche Underricht, zu Befestigung der Stett, Schloss, und Flecken (Some Directions for the Fortification of Cities, Castles and Towns),* published at Nuremberg in 1527, the principal exponents and practitioners of the science of fortification in the first half of the sixteenth century were mostly Italians: Francesco di Giorgio Martini, Giovan Battista della Valle di Vanafro, Antonio Sangallo the younger, Michele Sanmicheli and others.[32] When Francis I of France in 1544 set about building a fortress town, Villefranche-sur-Meuse, on his new border with the Empire, it was to an Italian, Girolamo Marini of Bologna, that he turned.[33] It seems likely that these Italian architects and engineers had been accustomed to using maps and plans drawn to scale from the turn of the century. In 1502 Leonardo da Vinci entered the service of Cesare Borgia as an engineer, and when in the following year this involved him in taking part in the war against Pisa he drew scale-maps of the theatre of war in Tuscany.[34] Da Vinci was an inventive genius, but in the following decades we find plans of buildings and fortifications which make it clear that other Italian architect-engineers were familiar with the idea of drawing plans and maps to scale.[35] We should not wholly rule out the possibility that the scale-map was introduced to England from Germany: German influence has been seen in Henry VIII's coastal forts,[36] and Dürer's treatise shows that scale-plans of fortifications were in use there. But it seems very much more likely that the idea of the scale-map reached England from Italy, like the methods of fortification that the earliest English scale-maps illustrated.

But if Italy, directly or indirectly, was the source of the English scale-map it is not at all clear by what means the idea was transmitted. There is no difficulty in finding engineers from Italy and other countries who were brought to England in Henry VIII's reign to superintend or help or advise in works of fortification. Some of them drew maps that survive. We have three plans of Carlisle, its castle and walls, signed by Stefan von Haschenperg who was in charge of the royal works there from 1541 to 1543; sometimes referred to as Stephen the Almain, he came in fact from Moravia.[37] Inscriptions in a mixture of Spanish and French on a map of Guines have been attributed to a Portuguese engineer who angered Henry VIII by reporting in April 1541 that it could never be made a strong fortress because it was so easily commanded from two hills outside the walls.[38] A map of Boulogne and nearby forts has all its inscriptions in Italian.[39] It is anonymous, but the Italian authors of some other maps can be identified: Antonio Bergamo signed alterations made in 1545 to a plan of Wark-on-Tweed Castle, while maps of Newcastle-upon-Tyne and Tynemouth drawn also about 1545 with inscriptions in Italian are attributed to Gian Tomasso Scala.[40] Giovanni di Rossetti produced plans of fortresses at Ardres in Picardy and (in 1547) at Broughty Craig in Angus which have the appearance of being true to scale;[41] apart from these two, none of these maps and plans, the products of imported expertise, is drawn to scale.

If the first appearance of the scale-map among English engineers has been correctly dated here, then on purely chronological grounds it is possible that the idea and the technique were brought to England by Girolamo da Treviso, who entered the service of Henry VIII in 1538. Born in 1497 he had worked as an artist in various cities of north Italy; but like Da Vinci and others he was an architect and engineer as well, and Vasari, in his *Lives of the Painters,* says that Henry VIII employed him 'not so much as a painter, but rather as an engineer'.[42] We have one picture which he is thought to have painted in England between 1542 and 1544, an anti-papal allegory showing the four evangelists stoning the pope and figures representing greed and hypocrisy, suggesting that he was prepared to go to some lengths to please his royal employer.[43] No surviving maps have been attributed to him, though we know that one he drew of Montreuil had its accuracy called in question in 1544, and it is possible that this is to be identified, or at least associated, with a surviving scale-map of the fortifications of Montreuil that must date from about this period.[44] He died in 1544, killed while leading a troop of arquebusiers at the siege of Boulogne. His connexion with the appearance of scale-maps in England rests entirely on the possible coincidence of dates; it must be regarded as the merest guess.

Another route by which the scale-map may have been brought from Italy is suggested by a work by Niccolo Tartaglia, *Quesiti et Invenzioni Diverse (Problems and Various Discoveries),* first published at Venice in 1538 and reissued there in 1546. This takes the form of a series of imaginary conversations on aspects of mathematics, mechanics, fortification and so on. Its fifth book describes how to make scale-maps, and here the dialogue is with 'Mr Richard Wentworth, gentleman of his majesty the king of England';[45] in addition the entire work is dedicated to Henry VIII and the dedicatory preface also mentions Richard Wentworth. The identity of this Richard Wentworth is not clear. A Richard Wentworth who was made a Knight of the Bath at Henry VIII's coronation in 1509 was prominent among the gentry of Suffolk until his death about twenty years later; he served in France in 1512, 1513, 1523 and 1526, and in 1520 he accompanied the king to the Field of Cloth of Gold.[46] However, there is no record of his visiting Italy, he had died some ten years before Tartaglia's work was published, and one would have expected him to be called knight (*cavaliere*) rather than gentleman (*gentil'huomo*). But the reference in Tartaglia's book at least suggests the possibility that there were one or more agents of the English crown in Italy in the fifteen-thirties learning the latest techniques of fortification and allied skills that included the construction of scale-maps.

This then, as far as it can be reconstructed at present, is the sequence of cartographic development that lies behind the

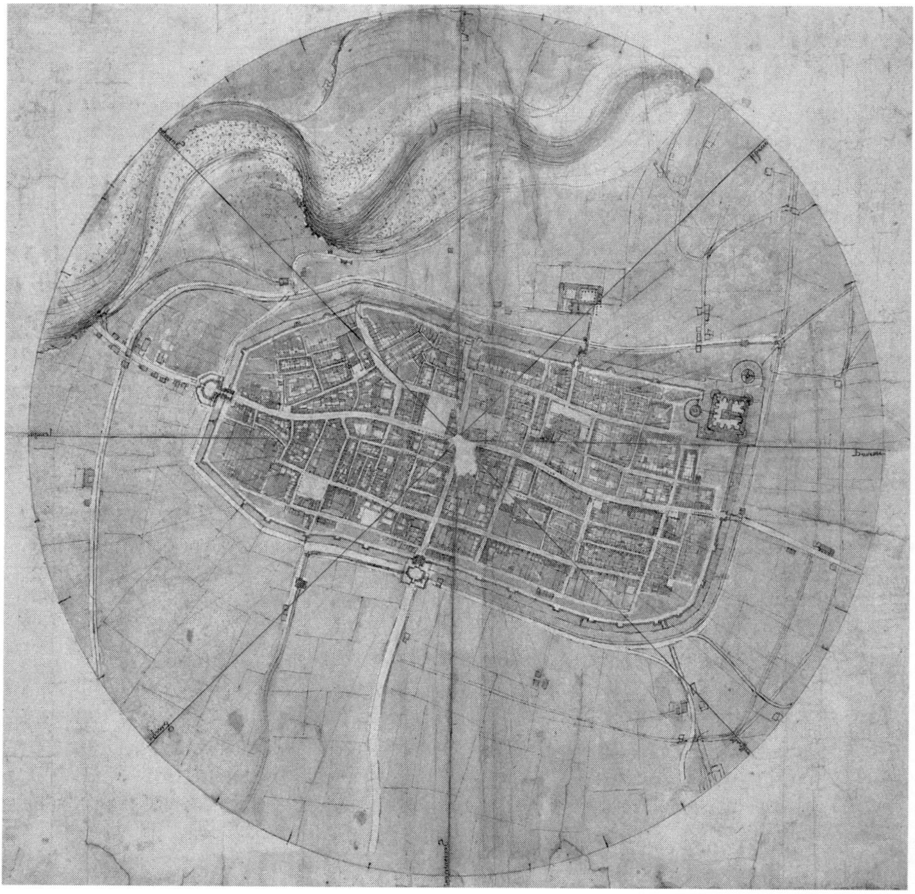

Fig. 5 MAP OF IMOLA, NORTH ITALY. drawn by Leonardo da Vinci in 1502 or 1503

Portsmouth map of 1545. Though not the first, it is still a very early example of an English scale-map; the idea of drawing maps and plans to scale may have been introduced to England no more than four or five years before. It is significant that it is drawn to scale whereas the earlier map (possibly about 1539) of the shore south-east of the town is not, just as John Rogers made scale-maps of the fortifications at Hull in 1541 whereas the map of the town made between 1537 and 1539 is a bird's-eye view. But it is not only as an early English scale-map that the Portsmouth map of 1545 is of interest in the history of cartography. It is in fact the most advanced in concept of all the English scale-maps before 1550, for it shows the entire town in ground-plan; not a single feature is shown pictorially in elevation or bird's-eye view. This distinguishes it from the other scale-maps of the fifteen-forties. Certainly where these show a single structure, such as the manor-house at Hull, or a single set of fortifications, as the walls of Guines in 1539–41, simple plan is used throughout. But where more than this appears on the map the draughtsman of this period would normally use the conventions of the picture-map and superimpose pictures of features (stylized or realistic) on his scale outline. Thus when John Rogers drew his map showing the fortifications along one side of the River Hull and the town of Hull on the opposite bank, he drew the fortifications in strict plan but showed the town as a row of houses in bird's-eye view (probably drawn from life) with the street pattern in plan behind them.[47] The draughtsman of the Portsmouth map, on the other hand, has shown the houses and other buildings of the town simply by their outline ground-plans. There is no trace here of the picture-map. What we have is a true plan of the town, what has been called an ichnographic plan.[48] It is the earliest ichnographic plan of any English town.

It is interesting to compare it with the earliest known ichnographic map from renaissance Italy. This is of Imola, near Bologna, drawn by Leonardo da Vinci in 1502 or 1503 in his capacity as a military engineer; it is reproduced here as Figure 5. It is much more elaborate than the Portsmouth map: Imola was larger than Portsmouth and the map shows in full detail not only the individual houses but also the plots of land belonging to them. The same basic technique was of course used to compile both maps: the map-maker made a perambulation of the town walls, carefully measuring each length and, with a magnetic compass, the angle of each change of direction (we have Da Vinci's notes from a similar perambulation of the city walls of Cesena). Within this framework further measurements would determine the lines of the streets and buildings (for these we have Da Vinci's sketches and measurements from Imola itself). At Portsmouth, at least, individual houses were probably simply sketched in, but the greater elaboration of the Imola map suggests that there still more detailed measurements may have been made. But, more important than the similarities of technique, the same cartographic concepts lie behind both maps; and just how advanced those concepts were in the early and mid-sixteenth century is underlined by the creative eminence of the author of the Imola map and by the fact that even in Italy no other ichnographic city plan survives from before the middle of the century.[49] In fact the Portsmouth and Imola maps are extraordinarily alike. The similarity may not strike a reader unfamiliar with the general run of maps of the time, for they are so like the maps we are used to today that they may both seem very normal and unremarkable works; but in the context of the topographical mapping of the first half of the sixteenth century their individuality and the similarity of the two stand out clearly. It cannot be supposed, of course, that there is any immediate connexion between them, that the draughtsman of the Portsmouth map had seen the map of Imola; but that there is some indirect connexion (even at many removes) can hardly be doubted. It makes it the more likely that it was from Italy that the scale-map came to England.

Who made the Portsmouth map is not known. Sir Richard Lee (he had been knighted in 1544) was inspecting and reporting on the fortifications there in August 1545, the month after the French attack, and it is possible that it was drawn by or for him.[50] There is no internal evidence of authorship. The map's only inscription, the note about its scale which may or may not have been written by the draughtsman, is in a hand that has not been found on other maps of the period, nor does its particular turn of phrase, 'ys In every Inche on C fote', occur elsewhere. The way it shows the bridges outside the town gate and at the end of the wall across the east corner of the town can be paralleled only approximately, on a map of Ambleteuse by John Rogers in 1546;[51] a map of the Orwell estuary in 1543–4, which is explicitly by Lee, shows such bridges in a different way.[52] Nor have clear parallels appeared for the symbols that mark three cardinal points on the compass rose (that for the east is lost as the map is slightly damaged); but the direction symbols found on early maps have never had the detailed general study that they deserve.[53] Two idiosyncrasies of the map's draughtsmanship should be mentioned. One is that the gun-ports in the walls are mostly shown in a simplified form consisting of an elongated diagonal cross; the other is that the gaps that mark the entrances to houses have a distinctive tiny circle at the end of the line on each side. Neither of these characteristics has been found on a contemporary map.

In our present state of knowledge the conclusions reached here can only be provisional. Briefly summarized they are as follows. All known scale-maps of English origin before 1550 were drawn by men who were working primarily as military engineers. They first appear about 1540 for sites in the English territory around Calais and the earliest known from any place in England seem to be those drawn at Hull in 1541–2. It seems unlikely that they represent simply an extension of ideas and techniques already familiar to non-military builders in England, for English architectural plans before this date seem to have consisted only of outlines drawn approximately to shape, with measurements written in. Instead the scale-map was probably brought to England from Italy; Girolamo da Treviso and an unidentified Richard Wentworth may (either or both of them) have played a part in its introduction. The Portsmouth map of 1545 is in concept the most advanced of any English map surviving from before 1550; it is the earliest ichnographic map we have of any town in England.

NOTES

1 D. Hodson, *Maps of Portsmouth before 1801* (Portsmouth Record Series, IV, 1978), no. 30. It is appropriate to refer here to the debt that the Portsmouth Record Series owes to Dr Dymond: as Chairman of its Editorial Board from 1970 to 1975 she guided the Series through the publication of its first two volumes and the planning of many more. I myself, as General Editor of the Series, warmly appreciate the wise counsel and careful scholarship which she has brought to bear on its problems and which have left their mark on the Series as a whole.

2 Hodson, *Maps of Portsmouth*, no. 183.

3 The topics touched on in this and the next three paragraphs — the picture-maps of medieval Europe and the spread of picture-maps and scale-maps in the sixteenth century — are discussed in P.D.A. Harvey, *The History of Topographical Maps* (London, 1980), pp. 55–103, 153–68.

4 *Local Maps and Plans from Medieval England,* ed. R.A. Skelton and P.D.A. Harvey (Oxford, 1981), Introduction, Ch. 4.

5 I.e. including maps drawn in the English territories around Calais and a very few of places in Scotland under short-lived English occupation.

6 B(ritish) L(ibrary) Cotton MS. Aug(ustus) I.i.18; reproduced to accompany J. Gairdner, 'On a Contemporary Drawing of the Burning of Brighton in the Time of Henry VIII', *Transactions of the Royal Historical Society,* 3rd ser. I (1907), frontispiece.

7 B.L. Cotton MS. Aug. I.i.83; reproduced and its date discussed by G. de Boer, 'The Two Earliest Maps of Hull', *Post-Medieval Archaeology,* VII (1973), pp. 79–83, pl.X.

8 Nottingham University Library M.i.P.14. I am grateful to Professor M.W. Barley and Mr A. Cameron for telling me of this map and supplying me with a photograph.

9 Hatfield House (Herts.) C(ecil) P(apers) M(aps) II.27; reproduced and discussed by M. Merriman, 'The Platte of Castlemilk, 1547', *Transactions of the Dumfriesshire and Galloway Natural History and Antiquarian Society,* 3rd ser. XLIV (1967), pp. 175–81, pl.XV.

10 Harvey, *History of Topographical Maps,* pp. 80–1, ill. 41; the most recent detailed discussion of the map is by M. Kratochwill, 'Zur Frage der Echtheit des "Albertinischen Planes" von Wien', *Jahrbuch des Vereins für Geschichte der Stadt Wien,* XXIX (1973), pp. 7–36.

11 The German usage appears in Albrecht Dürer's work on fortifications, *Etliche Underricht, zu Befestigung der Stett, Schloss, und Flecken* (Nuremberg, Oct. 1527), sigg. Ciii recto and verso, Ei verso, Fi recto, where the titles of both plans and elevations include such phrases as 'mit dem kleinen schuch gemessen', 'nach dem verjungten schuch gemessen', 'auffgerissen doch den schuch grosser genomen' (i.e. drawn to a larger scale). In the Low Countries, where the usage probably derived from Germany, the phrases used were 'op te cleyne maten', 'up ten cleynen voet', etc. (P.S. Teeling, 'Oud-Nederlandse Landmeters', *Orgaan der Vereniging van Technische Ambtenaren van het Kadaster,* VII (1949), pp. 160–1).

12 Hatfield House C.P.M. I.61, II.28; B.L. Cotton MS. Aug. I.i.59, Cotton Roll XIII.43.

13 Teeling, *Orgaan der Vereniging van Technische Ambtenaren van het Kadaster,* VII, p. 161.

14 *The History of the King's Works,* ed. H.M. Colvin (London, 6 vols, 1963–). I am most grateful to Mr Colvin, as well as to Professor M. Biddle and Dr

M.H. Merriman, for information about the dates of maps discussed in vol.4, as yet unpublished.

15 B.L. Cotton MSS. Aug. I.ii.51, I Suppl.14; the two without explicit statement of scale are Aug. I.ii.52, I suppl.2. The origin and dates of the maps are discussed by L.R. Shelby, *John Rogers, Tudor Military Engineer* (Oxford, 1967), pp. 15–22 (where all four are reproduced, pl.1-3, 5), and by Colvin, *History of the King's Works,* 3 (1975), pp. 366–9, 401–2 (where there is a full account of the works at Guines).

16 B.L. Cotton MS. Aug. I.ii.69; Colvin, *History of the King's Works,* 3, pp. 374, 402, pl.29.

17 B.L. Cotton MSS. Aug. I.ii.49, I Suppl.20, I Suppl.4; Shelby, *John Rogers,* pp.24–34, pl.8, 9.

18 Dover: B.L. Cotton MS. Aug. I.i.26 (reproduced to accompany W. Minet, 'Some Unpublished Plans of Dover Harbour', *Archaeologia,* LXXII (1921–2), pl.XXXVII). Carlisle: B.L. Cotton MS. Aug. I.i.11, Hatfield House C.P.M. II.28. Harwich: B.L. Cotton MS. Aug. I.i.61, I.i.56 (where the date 25 Henry VIII is a mistake, probably for 35 Henry VIII: I am grateful to Mr H.M. Colvin for this information).

19 E.g. the plans that are reproduced in Shelby, *John Rogers,* pl.14–21.

20 Shelby, *John Rogers*, pp.7–34, 158–63.

21 R.A. Skelton, 'The Military Surveyor's Contribution to British Cartography in the 16th Century', *Imago Mundi,* XXIV (1970), pp.77–83; R.A. Skelton and J. Summerson, *A Description of Maps and Architectural Drawings in the Collection made by William Cecil First Baron Burghley now at Hatfield House* (Roxburghe Club, 1971), pp.28–30.

22 Teeling, *Orgaan der Vereniging van Technische Ambtenaren van het Kadaster,* VII, p.161.

23 Shelby, *John Rogers,* pp.146–9; Skelton and Summerson, *Description of Maps,* p.29; Harvey, *History of Topographical Maps,* p.162.

24 L.F. Salzman, *Building in England down to 1540* (Oxford, 1952), pp. 14–20; cf. the surprisingly limited use made of plans and drawings by Italian architects in the sixteenth century (J.S. Ackerman, 'Architectural Practice in the Italian Renaissance', *Journal of the Society of Architectural Historians,* XIII pt 3 (1954), p.8).

25 Winchester College muniments, 22820; Skelton and Harvey, *Local Maps and Plans,* no.11.

26 Corporation of London Records Office, Bridge House deeds, Small Register, ff.8r–11r; Skelton and Harvey, *Local Maps and Plans,* no.23.

27 London, Drapers' Company archives A.V.180, 183 (I am grateful to Mrs E. Cottrill, former Archivist to the Company, for telling me of these plans and supplying me with photocopies); Public Record Office M.P.I. 68(2); Oxford, Bodleian Library MS. Top. Berks. c.28 (R) (reproduced in Colvin, *History of the King's Works,* 3, pl.20).

28 B.L. Cotton MS. Aug. I Suppl.7; its origin and date are discussed by Colvin, *History of the King's Works,* 3, pp.349–51 (where a redrawn version is reproduced on p.350).

29 B.L. Cotton MSS. Aug. I Suppl.1, I.ii.11, I.i.84; their date and authorship are discussed by Shelby, *John Rogers,* pp.34–46, 158–63, and they are reproduced there as pl.10–12. The likely scale of these plans, sixteen feet to one inch, occurs also on Hatfield House C.P.M. II.34 (fortifications at St Mary's, Scilly Is.) and possibly on B.L. Cotton MS. Aug. I.i.67 (Sandown Castle, Deal), while Hatfield House C.P.M. II.28 (Carlisle) is drawn at half this scale, thirty-two feet to one inch.

30 H. de la Croix, 'Military Architecture and the Radial City Plan in Sixteenth Century Italy', *Art Bulletin,* XLII (1960), pp.273–4.

31 De la Croix, *Art Bulletin,* XLII, pp. 264–8; J.R. Hale, 'The Early Development of the Bastion: an Italian Chronology, c.1450–c.1534', in *Europe in the Late Middle Ages,* ed. J.R. Hale, J.R.L. Highfield and B. Smalley (London, 1965), pp.466–94.

32 De la Croix, *Art Bulletin,* XLII, pp.268-73; H. de la Croix, 'The Literature on Fortification in Renaissance Italy', *Technology and Culture*, IV (1963), pp.38-9.

33 De la Croix, *Art Bulletin,* XLII, p.276.

34 Da Vinci's maps are nearly all in the Royal Collection at Windsor Castle; they are reproduced and discussed by M. Baratta, *I Disegni Geografici di Leonardo da Vinci conservati nell Castello di Windsor* (Reale Commissione Vinciana, 1941).

35 E.g. see those reproduced by Ackerman, *Journ. of the Soc. of Arch. Hist.,* XIII pt 3, figg.1–3, 6,8, and by De la Croix, Art Bulletin, XLII, figg.3, 4.

36 B.H. St J.O'Neil, *Castles and Cannon* (Oxford, 1960), p.61.

37 B.L. Cotton MSS. Aug. I.i.69, I.ii.16, I Suppl.8 and 9 (two fragments of a single plan); they are reproduced and their author's English career discussed by B.H. St J.O'Neil, 'Stefan von Haschenperg, an Engineer to King Henry VIII, and his Work', *Archaeologia,* XCI (1945), pp.137–55, pl.XI–XIII.

38 B.L. Cotton MS. Aug. I.ii.23; Shelby, *John Rogers,* pp.9–10; Colvin, *History of the King's Works,* 3, p.367.

39 B.L. Cotton MS. Aug. I.ii.67.

40 Hatfield House C.P.M. I.42, I.65; B.L. Cotton MS. Aug. I.ii.7 All three are discussed in Skelton and Summerson, *Description of Maps,* p.57.

41 B.L. Cotton MS. Aug. I.ii.74; Hatfield House C.P.M. Suppl.12 (described in Skelton and Summerson, *Description of Maps,* p.60).

42 'non più per pittor, ma per ingegniere s'accomodò a' servigi suoi'. Vasari's biography is in *Le Opere di Giorgio Vasari,* ed. G. Milanesi (Florence, 9 vols, 1878–85), Vol.5, pp.135–9; the quotation is taken from p.138. There are brief accounts of Da Treviso's work in England by R.W. Carden, 'The Italian Artists in England during the Sixteenth Century', *Proceedings of the Society of Antiquaries of London,* 2nd ser. XXIV (1911–12), pp.199–202, and A.E. Popham, 'Hans Holbein's Italian Contemporaries', *Burlington Magazine,* LXXXIV (1944), pp.13–14.

43 P. Pouncey, 'Girolamo da Treviso in the Service of Henry VIII', *Burlington Magazine,* XCV (1953), pp.208–11.

44 B.L. Cotton MS. Aug. I Suppl.11, a finely drawn map of the circuit of the town's walls with, apparently, an inner line of defence that is incompletely drawn, being shown partly by pecked lines; the only other features shown are some streams and paths outside the walls. Montreuil was never in English hands in the fifteen-forties but operations against it were contemplated, and in 1544 a siege was begun, then abandoned. Da Treviso (known as Jeronimo in many English records) had, about 1542, been employed in reconnoitring and, apparently, investigating the defences of towns on French territory; when, once the siege was begun in 1544, Montreuil proved to be better fortified and less easily approached than expected from his map, he said that he had been concerned only with the interior of the town and that when he was there one particularly effective bulwark by the Abbeville gate had not been built (*Letters and Papers of the Reign of Henry VIII* (London, 21 vols, 1864–1920), Vol.17, p.229; Vol.19 pt 1, pp.543, 549, 560).

45 'M. Ricchardo Ventuorth gentil'huomo della Maesta del Re de Inghıltera' (Tartaglia, *Quesiti* (1546), f.'58'v: the leaves are in fact misnumbered). The

edition of 1538 is mentioned by De la Croix, *Technology and Culture,* IV, pp.39, 49, but that of 1546 is the earliest in the British Library, the Bibliothèque Nationale and the *National Union Catalog* of books in the United States.

46 *L. and P. Hen. VIII,* Vols 1–4, contain quite a few references to him. Among the more significant are those on 1, pp.37, 124, 662, 931; 2, p.791; 3, pp.243, 1371; 4, p.3108.

47 B.L. Cotton MS. Aug. I Suppl.4; reproduced by Shelby, *John Rogers,* pl.8.

48 J.A. Pinto, 'Origins and Development of the Ichnographic City Plan', *Journ. of the Soc. of Arch. Hist.,* XXXV (1976), p.35.

49 The concepts and techniques underlying Da Vinci's map of Imola are discussed by Pinto, *Journ. of the Soc. of Arch. Hist.,* XXXV, pp.38–42; his survey of the city walls of Cesena is reproduced there, p.41, his measured sketches of streets at Imola by C. Pedretti, 'Rilievi sconosciuti di Imola in un foglio di Windsor', in his *Studi Vinciani* (Geneva, 1957), pp.217–21, pl.X.

50 *L. and P. Hen. VIII,* 20 pt 2, pp.9–10.

51 B.L. Cotton MS. Aug. I.ii.8; reproduced by Shelby, *John Rogers,* pl.18.

52 B.L. Cotton MS. Aug. I.i.56 (on its date see n.18 above). But while the longitudinal planks are lacking, its bridges do have lines representing side beams and dots representing nails, as on the Portsmouth map, features that also occur on the bridges drawn on B.L. Cotton MSS. Aug. I.i.61 (Harwich), I Suppl.11 (Montreuil: see n.44 above).

53 I am grateful to Mrs Sarah Tyacke for advice and very interesting information on this.

XV

Estate Surveyors and the Spread of the Scale-map in England, 1550–80[1]

[37] Not the least of the services performed by Maurice Beresford's *History on the Ground* in 1957 was to show what a valuable source of information we have in the estate maps of the late sixteenth century. The maps reproduced there – of Holdenby (Northamptonshire) by Ralph Treswell in 1580 and 1587,[2] of Toddington (Bedfordshire) by Ralph Agas in 1581 (Pl. I), of Whatborough (Leicestershire) by Thomas Clerke in 1586 and of Maids Moreton (Buckinghamshire) and Salford (Bedfordshire) by Thomas Langdon in 1596 – are all carefully and accurately drawn, from measured survey, to a consistent scale set out on a scale-bar, mostly with a pair of dividers drawn above it. They show us the villages and their individual houses and closes, the fields with their often intricate divisions and boundaries, the streams and bridges, the hedgerows and trees. They offer us the closest, fullest, most exact view we can get of the detailed landscape of Tudor England – and of medieval England as well, for we have nothing comparable for earlier periods. From the Middle Ages we have practically no local [38] maps at all – only some thirty-five older than 1500, most of them from the mid-fourteenth century onwards (Skelton and Harvey 1986). In the first half of the sixteenth century many more local maps were drawn – probably some 200 survive from between 1500 and 1550 – but, as Beresford himself points out, these early-Tudor maps were a far cry from the highly professional estate maps that he shows us from the 1580s and 1590s (Beresford 1957, pp. 46–7). Mostly they were picture-maps,

---

1 The article has been reset with minor changes, but the original page numbers (which have not been used for the index) have been entered in square brackets. The general development of mapping in 16th-century England is described, with particular reference to its use by the government, by Barber 1992 and, in outline but with many more illustrations of maps than can be shown here, by Harvey 1993. The growth of cartographic understanding, which underlies the argument of the present article, is discussed in detail by Morgan 1979.

2 Treswell is not named on either map, but of the other nine 16th-century maps in the same volume (Northamptonshire Record Office, Finch Hatton MS. 272) seven are specifically his, and as they are identical in style it seems likely that he drew the others too.

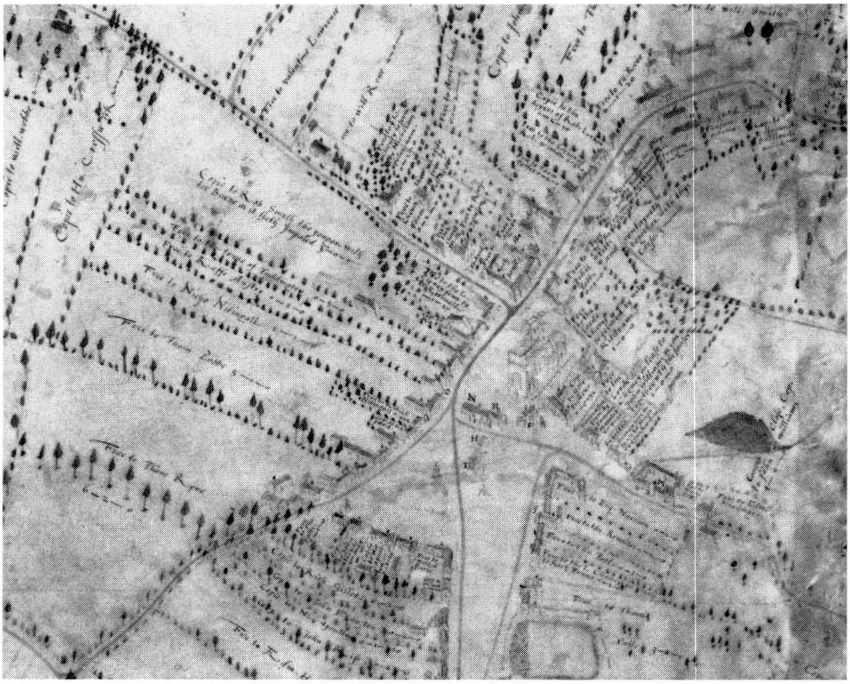

Plate I. Part of the map of Toddington (Bedfordshire) by Ralph Agas, 1581 (BL Add. MS. 38065 N)

bird's-eye views giving an impression of landscape as though viewed from a height, varying greatly in their faithfulness of representation and in their artistic skill but alike in showing features in elevation or perspective rather than in simple ground-plan and in making no attempt at uniform scale. The earliest English maps based on measured survey and drawn to scale date from about 1540 and from then until 1550 we have some forty scale-maps in all, most of them with the scale specifically stated in a written note or a scale-bar or both. About half are of places in England, half of sites in the lands around Calais that were then still in English possession. Nearly all are plans of actual or projected fortifications – town walls or castles – and probably all were drawn by military engineers. In 1550 drawing maps to scale was known in England only as a technique of military engineering, a technique probably imported from Italy like the styles of fortification shown on the plans and like some of the men who drew them (Harvey 1981; Merriman 1983).

Yet by the 1580s surveyors were drawing to exact scale estate maps of which some – like Ralph Agas's of Toddington – were of a size and complexity far beyond the early scale-maps of the military engineers. What was the sequence of development in those thirty years? How did scale-maps pass from the engineers to the surveyors, from plans of fortifications to plans of estates? More is involved than simply the spread of a technique: it was the spread of a new concept, the introduction of the cartography which places features within an accurately surveyed framework drawn in exact proportion throughout.

One of the biggest difficulties in answering these questions is that few local maps from the sixteenth century bear dates. Certainly many of the undated maps can be dated, often to within narrow limits, by internal evidence: the landowners they name, the buildings they show, and so on. But this involves, for any one map, a great deal of detailed investigation, and any overall survey of their cartographical development is bound to rely on whatever work has already been done in dating individual maps. Thus it was possible to establish a chronology for the introduction of the scale-map by the military engineers, only because of the extensive research on particular sites and particular maps that had been done by H.M. Colvin and his collaborators for volumes 3 and 4 of *The History of the King's Works*. However, for estate maps dating is less of a problem than for most other manuscript maps of the sixteenth century; most of the more elaborate estate maps either include the date of composition in the title or are attached to dated surveys. This means that we can be fairly sure of the chronology of the estate map drawn to scale – when it was introduced, how fast it spread. But we are much less sure of the cartographic background to this development because it consists of many maps of different sorts, carefully drawn or mere sketch maps, which from their appearance belong to the second half of the sixteenth century but which have not been more precisely dated. The suggestions made here about the lines of development behind the estate surveyors' adoption of scale-maps are based on the relatively few of these maps that are dated or easily datable. They may well need revision if the dates of a substantial number of the other undated maps are eventually worked out.

That so many of the estate surveyors' maps are dated reflects the fact that they were working in a well-established tradition not of maps but of written surveys. By the mid-sixteenth century English landowners had been having written surveys made of their estates for at least 500 years. These surveys were simply descriptions of the property: they might list the lands and other resources of the manorial demesnes, or name the local tenants with the rents and services due from each, or describe in detail the lands of a vill whether occupied by tenants or by the manorial lord. From the early thirteenth century precise measurement of lands often played a part in these surveys and we have occasional references to some particular estate official, perhaps with special skills or experience, going to a manor to measure lands for this purpose, work that could hardly be done by the local manorial juries which normally supplied the information for a survey (Harvey 1984, pp. 15–24; Skelton and Harvey 1986, pp. 12–14, 16). But by the end of the fifteenth century there is still no trace in England of the professional surveyors – land-measurers – who seem to have been already well established in Italy and the Netherlands (Skelton and Harvey 1986, p. 16). Although some notes and formularies were produced in the Middle Ages for drawing up surveys and for calculating areas of land, the first English treatise on surveying was John Fitzherbert's *Boke of Surveyeng*, published in 1523. It was followed in 1537 by Richard Benese's *Maner of Measurynge* and it seems to have been now, in the mid-sixteenth century, that surveying emerges in England as a craft or profession which may even have had some full-time practitioners.

This development had nothing to do with maps. The earliest English treatise on land measurement that refers to map-making was published in 1571: *A Geometrical Practise, named Pantometria*, by Leonard and Thomas Digges.[3] As Ralph Agas in the 1580s and 1590s was arguing vehemently that drawing estate maps was an essential part of the surveyor's craft, it is clear that this view was not shared by all his professional colleagues or those [39] who called on their services (Agas 1596, pp. 14–16; Skelton and Harvey 1986, p. 19). Of some forty-nine surveyors known to have practised between 1550 and 1580 – and, significantly, this is the earliest period from which their names survive – fifteen are known only from written surveys (Eden 1975–9).[4] John Norden was as certain as Ralph Agas that the surveyor ought to produce a map of the estate he surveyed, but in his *Surveyors Dialogue*, first published in 1607, he makes it clear that this is only a part, and not the most important part, of his business:

> *Farmer*. Then you say, that plotting (*i.e. drawing maps*) is the chiefe part of a Surveyors skill.
>
> *Surveyor*. I say not so, but I say, it is necessary for him that is a Surveyor, to be able to do it, and that he be painefull & industrious, & having this quality with the rest more necessary, he may be then called a Surveyour. (Norden 1607, p. 17)

---

[3] Benese 1537, sigg. A4r–F4v, may imply the use of sketch maps in his instructions for calculating the area of irregular plots of land, but certainly not as part of the finished survey.

[4] Written surveys were more often anonymous than estate maps; the proportion of non-map-making surveyors was probably far higher than these figures might suggest. An important article by Andrew McRae (McRae 1993) discusses the rise of professional surveyors and the reactions they provoked; it expands and elucidates some of the points mentioned here, and I am most grateful to Dr McRae for letting me see it before publication.

These other more necessary skills included not only the ability to measure lands and to value landed property, but a knowledge of common law, particularly relating to customs and tenures, and of the Latin and palaeography that would enable him to read ancient records (Norden 1607, p. 17).[5] The ideal surveyor was not just a technician; he was also a learned man, a point underlined by Agas's use of Latin quotation in his pamphlet of 1596 (Agas 1596, p. 8).[6] He was the outside expert called in from time to time by the landlord to inspect his estate and draw up a survey which was in effect a report on the state of the property. At the surveyor's behest the court of each of the estate's manors would be summoned in special session – the court of survey – to provide evidence, both by sworn testimony and from written records, about the local tenants' holdings and obligations. This, together with what he learned from his own inspection and measurement of the property, would give the surveyor the material for his survey. If we are to believe Norden (who in no way underestimated the surveyor's responsibilities or status) he would then advise the local farmer or bailiff on his management of the property, recommend improvements to increase profitability and so on (Norden 1607, pp. 185–244).

That drawing a map should come to be seen as a proper part of this work of survey is not surprising. We have seen that from the start of the sixteenth century there was a growing awareness of the use and value of maps, barely appreciated by anyone in the preceding centuries. This spread of map-consciousness continued apace in the second half of the century. No one has calculated how many English local maps survive from 1550–1600 to compare with the suggested 200 or so from 1500–50, but certainly the figure is much larger and 800 is a conservative guess. The printed catalogue of maps and plans in the Public Record Office has some 40 entries for maps of places in England which it dates between 1500 and 1553, over 180 for those of 1553–1603 (Public Record Office 1967), pointing clearly to continuing growth in the use of maps in the reigns of Mary I and Elizabeth I. Thus, courts of law seem to have become increasingly apt to order a map ('a true and perfecte platte' or similar phrase) to be drawn to show the terrain in cases of dispute over boundaries. In another sphere of activity, the interest taken in maps by William Cecil, Lord Burghley, is well known and many of those from his collection have notes in his own hand (Skelton and Summerson 1971; Barber 1992, pp. 68–77).

---

[5] Earlier, Worsop had divided the surveyor's skills into 'three principal parts: that is to say the Mathematicall, the Legall, and the Judiciall', respectively measuring and mapping, the business of the court of survey and knowledge of agriculture (Worsop 1582, sig. I3v).

[6] Agas was rector of Grassenhall (Norf.) from 1578 to 1583.

All the same, we should not overestimate the early Elizabethans' awareness of maps: familiarity with maps was rapidly growing, but they were not yet universally understood or accepted. Law courts may have had maps drawn to illustrate disputed bounds, but parties to a case seem seldom to have taken the obvious step of including in the interrogatories put to local witnesses a question about their accuracy. In a case where this was done, at Dedham (Essex) in 1574, only two of ten deponents answered the thirteenth of the sixteen questions put on behalf of the plaintiff: 'Item whether a certeyne chart or platt discribinge aswell the Landes in variaunce as dyvers other Landes medowes wais and lanes therunto adjoyninge ... be a trewe discription and settinge Forth of the grounde in variaunce or no'.[7] Even in 1607 we find the farmer in Norden's *Surveyors Dialogue* saying that 'we poore Country-men doe not thinke it good to have our Lands plotted out, and me thinks in deede it is to very small purpose: for is not the Field it selfe a goodly Map for the Lord to looke upon, better than a painted paper? And what is he the better to see it laid out in colours?' (Norden 1607, p. 15).

Nor need the use of maps imply the use or understanding of scale. The scale-map was spreading too but in a much more restricted way. As in the 1540s, of course, military engineers continued to draw scale-maps of projected and actual fortifications and other works. The maps of works at Berwick-upon-Tweed by Sir Richard Lee and Rowland Johnson between 1558 and 1564,[8] and at Dover for Thomas Digges in 1581[9] are examples. But the map might be extended to show not only the works themselves but the countryside around, setting them in their local context. The map of Dover drawn for Digges shows not just the harbour, the site of the intended works, but the castle, the town and fields behind. One of Rowland Johnson's plans for Berwick extends some 10 miles inland, showing this area at a smaller scale than part of the town itself (1400 instead of 480 feet to the inch).[10] A map of Thanet that has been dated 1548 may be seen as an early example of the same sort.[11] But we also find scale-maps of areas that were not drawn primarily to show fortifications. Some certainly had military implications and may represent an extension by engineers of the techniques and concepts of their large-scale plans. Thus a scale-map by Henry Bullock in 1552 shows an area some 10 miles square to the north of

---

[7] Public Record Office (hereafter PRO) DL 4/16/18 and MPC 1/276. It may be significant that the plaintiff was Christopher Saxton's patron, Thomas Seckford.

[8] PRO MPF 1/137; Hatfield House (Herts.), Cecil Papers Maps (hereafter CPM) I.22–25, 27, and II.30 (Skelton and Summerson 1971, nos 73–6, 78, 80). *Cf.* MacIvor 1965.

[9] BL Add. MS. 11815a. *Cf.* Macdonald 1937, pp. 115–21, plate II.

[10] CPM I.24 (Skelton and Summerson 1971, no. 80).

[11] CPM I.61 (Skelton and Summerson 1971, no. 40, plate 2).

Carlisle, marking different alignments of the border across 'the Debatable [40] Land between England & Scotland'.[12] A map of Sheppey, finely drawn about 1572 by 'IM' (he has not been further identified), shows the whole island in some detail, marking the lands of Lord Cheney and, by differing shades of wash, of particular farms; but it distinguishes the fort at Swaleness by showing it in ground plan, and in one corner of the map is a note about access for ships (Pl. II).[13] Some other scale-maps are of very much larger areas and clearly owe no particular debt to the plans that the military engineers were drawing to scale in the 1540s. Of the maps in the Public Records that can be associated with particular lawsuits between 1550 and 1580 the only one drawn to scale is, significantly, of an exceptionally large area: Ashdown Forest in east Sussex and its surrounds, extending north and south some 25 miles from Godstone to Lewes. It was drawn for Sir Richard Sackville about 1563 to show lands that he was disputing with Sir Edward Gage.[14] Our earliest map of County Durham, drawn specifically to scale, dates from about 1569–70.[15] It is one of several manuscript scale-maps of counties from about this date that were in Lord Burghley's collections.[16] It is in this same context of the use of consistent, specific scale in maps of this sort that we should see Christopher Saxton's maps of the English and Welsh counties, engraved between 1574 and 1578 – a point we shall return to.

This, then, is the cartographic background to the introduction of mapping to estate surveys. On the one hand the thirty years 1550–80 saw increasing awareness of the value of maps, an increasing tendency to draw maps – whether the merest sketch or a finely drawn artistic production – for all sorts of purposes. On the other hand only certain sorts of map were drawn to consistent scale: plans of fortifications and other works drawn by military engineers, maps of somewhat larger areas likewise drawn for military or other official purposes, and maps of whole counties or similar areas. Naturally enough, estate owners were involved in the general spread of map-making in this period, and we find maps of the whole or parts of various landed estates drawn, not to scale, for purposes that are mostly obscure. An example, probably of the 1550s, is a pen-and-ink sketch of the surrounds of Cotehele House (Cornwall) showing

---

12 PRO MPF 1/257.

13 PRO MPF 1/240 (Public Record Office 1967, p. 178).

14 PRO MPF 1/144 (Public Record Office 1967, p. 265).

15 BL Royal MS. 18 D.III, fos 69–70, discussed by Manley 1936.

16 R.A. Skelton suggests 'c.1570?' as the date for three manuscript county maps (Northamptonshire, Shropshire, Worcestershire) in the atlas, formerly Burghley's, that is now BL Royal MS. 18 D.III (Skelton and Summerson 1971, nos 4, 5, 56); all are specifically drawn to scale.

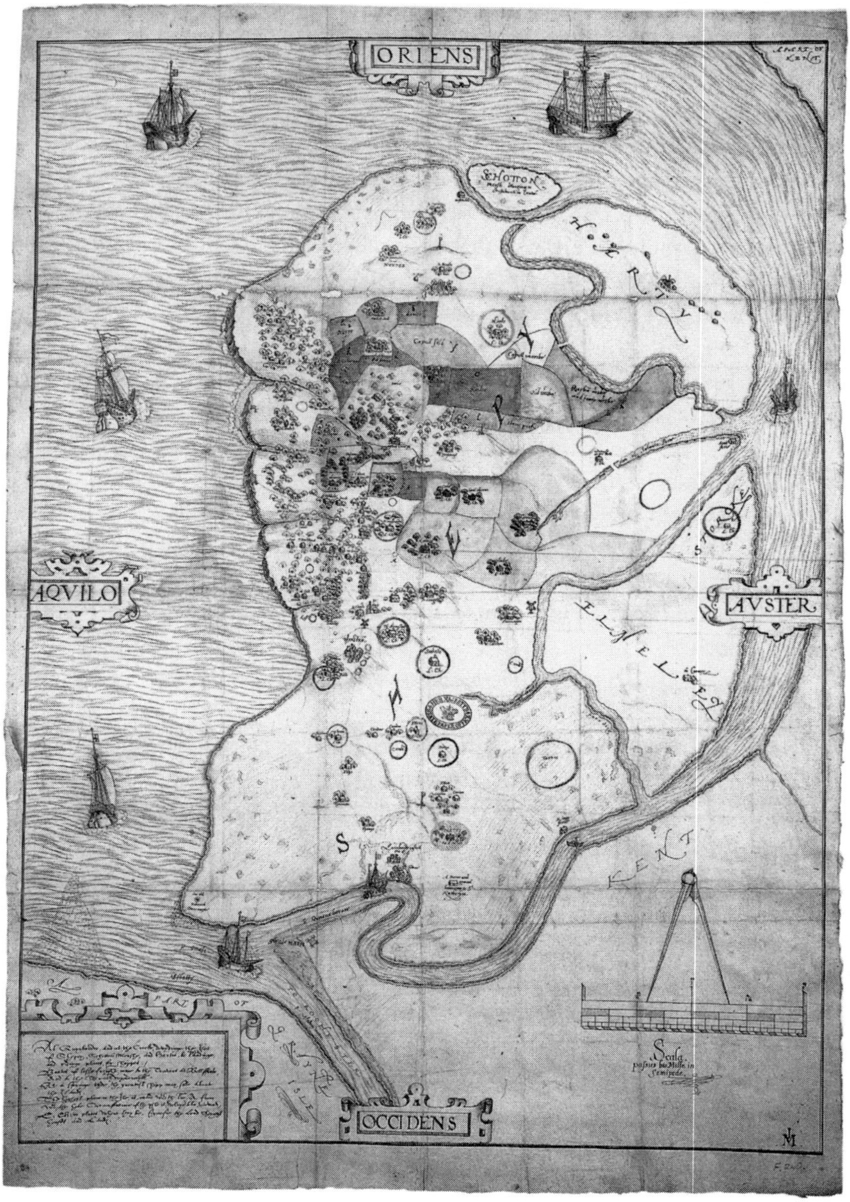

Plate II. The Isle of Sheppey (Kent) by 'IM', *c.*1572 (PRO MPF 1/240)

its mills and other properties.[17] Another, of 1569–78, is an elaborate picture-map of the manor of Sherborne (Dorset) and surrounding estates, showing the landscape in detail and with each manor painted in a different colour.[18] Such maps, however, had no known connection with estate surveys or with surveying – the precise measurement of lands. Surveyors may, even in the fifteenth century, have often drawn rough sketch plans as a way of assembling information (Skelton and Harvey 1986, pp. 17–18),[19] but, if so, very few survive and they certainly formed no part of the finished written survey.

Interestingly, one of the two witnesses at Dedham in 1574 who answered the question about the map's accuracy was Robert Mawe of Rendlesham (Suffolk) who had been making a survey of the manor; his answer implies that he was not the map's author, but we have sketch maps from Dedham dated August 1573 which are very likely working notes for Mawe's own written survey for they give not only the areas and occupiers' names for the plots of land but also page references to unidentified records.[20] We find an early link with estate surveys in a map of lands in Elmswell, Wetherden and Woolpit (Suffolk), probably drawn between 1569 and 1575: written on most of the pieces of land it shows are the area, a reference to the relevant page and entry number in what was evidently a terrier or other survey, and the name of the owner or occupier, but as these names include the abbot of Bury St Edmunds this survey must have been an old one, compiled before the monastery's dissolution in 1539.[21] Evidence of surveying appears in a map drawn for a lawsuit over lands at Billington (Lancashire) in 1571; it is not drawn to scale, but on each plot it enters the area, either taken from an existing survey or measured specially on the ground.[22]

No doubt these hints of connections between surveying and mapping in the course of the 1570s can be multiplied, especially if more undated maps can be assigned to this period. But I know of no estate survey illustrated with maps earlier than those by Israel Ames of Belchamp St Paul (Essex) in 1576 (Pl. III)

---

17 Cornwall County Record Office, DD ME 2369, discussed by Ravenhill 1984.

18 BL Add. MS. 52522, discussed by Harvey 1964–5; by Cobbe 1973, pp. 646–7, plate 1; and by Barker 1991.

19 *Cf.* note 3 above.

20 PRO DL 4/16/18 and MPC 1/77, fos 4r–7v; Mawe's evidence showed that compiling his survey had involved a full search of court rolls and other records. Mawe too was later associated with Saxton (*cf.* note 7 above), acting as his surety in a bond in 1589 (Evans and Lawrence 1979, pp. 71, 167).

21 Archives of Lord Stafford, Swynnerton Park (Staffs.), discussed by N. Scarfe and P.D.A. Harvey in Skelton and Harvey 1986, pp. 349–50.

22 PRO MPC 1/48 (Public Record Office 1967, p. 478).

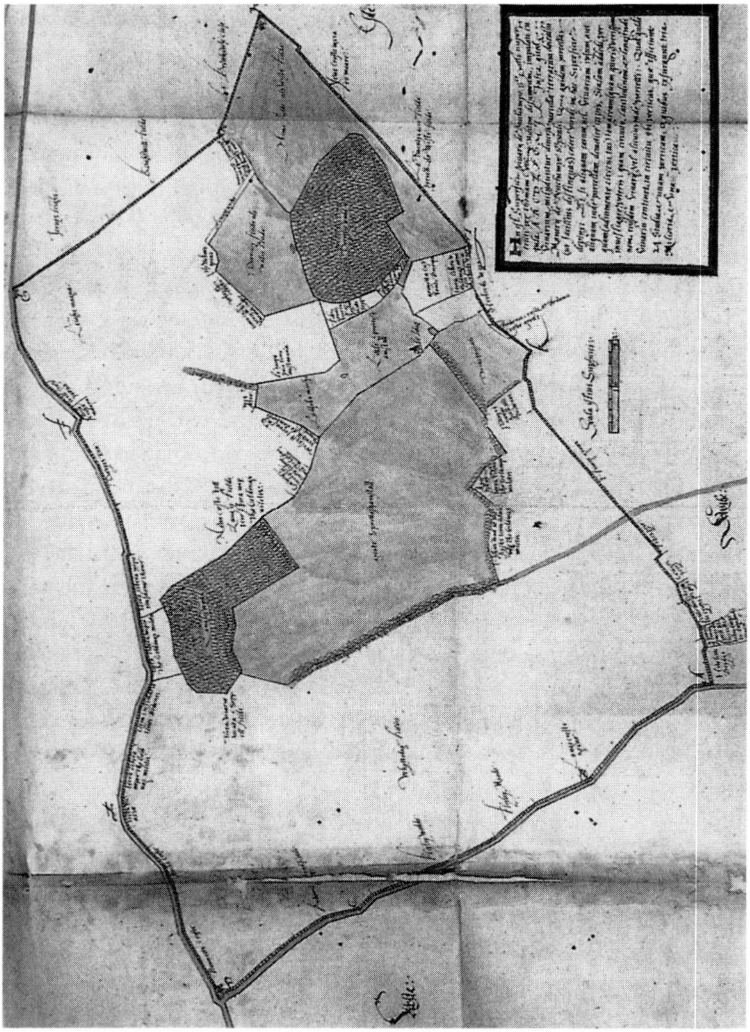

Plate III. One of the maps illustrating the survey of Belchamp St Paul (Essex) by Israel Ames, 1576 (Guildhall Library, London, MS. 25517/1, f.100)

and of Edmund Tirrell's Essex properties in 1579,[23] no map which effectively presents in graphic form the full details of an estate survey earlier than Ralph Agas's map of West Lexham (Norfolk) in 1575 and Ralph Treswell's map of Holdenby in 1580.[24] Almost certainly earlier precursors can be found. But it seems to have been only in the 1580s that it began to be understood that an estate survey might involve making maps. These maps were drawn to scale, specifically so, like Ames's maps of Belchamp St Paul and of Tirrell's lands at Plumberow[25] and the maps of West Lexham and Holdenby: there was no intermediate stage of non-scale-maps associated with estate surveys. This is a point we shall return to, but we may note here that the careful measurements in the field needed to work out the areas of land recorded in the survey would, with little extra work, be a sufficient basis for a scale-map of the estate.

Behind this lay the techniques and instruments that had come to make such accurate surveying possible. Constructing a map by triangulation had been described in 1533 by Gemma Frisius (Jemme de Fries) of Louvain in his *Libellus de locorum describendorum ratione*, published in both Latin and Dutch; the first account in English was given in 1559 by William Cuningham in his book *The Cosmographical Glasse*, which includes by way of illustration a diagram showing a triangle measured in Norfolk between Norwich, Swardeston and Wymondham (Cuningham 1559, p. 140). Various instruments for surveying and drawing maps were described by Leonard and Thomas Digges in *A* [41] *Geometrical Practise, named Pantometria* (1571) and by William Bourne in *A Booke called the Treasure for Traveilers* (1578). As Norden later pointed out, 'there are but two principall instruments, fit indeede for the plotting of grounds', others being simply variants of them (Norden 1607, p. 127). One [42] was the plane-table, which by its sightings from point to point enabled the map to be actually drafted in the field. The other was the theodolite which gave precise measurements of angles from which the map would be constructed afterwards in the study. Norden's account of making an estate map takes for granted that the plane-table will be used, presumably because it was the quicker and easier method (Norden 1607, pp. 128–42). Agas on the other hand argued eloquently in favour of the theodolite (Agas 1596, pp. 3–13).

---

23 Guildhall Library, London, MS. 25517/1; BL Harley MS. 6697.

24 Holkham Estate records, Map 87a (a photocopy is BL Maps 188 n.1(10)), and Northamptonshire Record Office, Finch Hatton MS. 272, fo. 59 (Beresford 1957, plate 19).

25 BL Harley MS. 6697, fo. 19; this is the most elaborate of the eight plans in the survey and the only one on which the scale is stated. It is interesting that in a contemporary copy (Guildhall Library, London, MS. 25517/2) of Ames's Belchamp St Paul survey the maps, less carefully drawn than in the original, lack the notes of scale.

Agas was a perfectionist, Norden was not (a glance at their maps will show the difference), but satisfactory scale-maps could be drawn by either method.[26]

Successive treatises reflect the broad chronology outlined here. In 1556 Leonard Digges, in *A Boke named Tectonicon*, tells how to measure land (as well as wood, stone and glass) and describes an instrument for doing it (Digges 1556, sigg. B1r–C4v, F1r–F4v); the book's title includes 'surveyers, landmeters' among those who will find it useful, but nowhere is it suggested that maps will be drawn from these measurements. In *Pantometria* in 1571 Leonard and Thomas Digges give directions not only for measuring distances but also for drawing scale-maps; here, however, both the text and the accompanying woodcuts constantly refer to military operations and never to estate surveying. Valentine Leigh's treatise of 1577, *The Moste Profitable and Commendable Science, of Surveying of Landes, Tenementes and Hereditamentes*, while dealing mainly with the form of the written survey and of other manorial records in the surveyor's purview, says that 'the Surveiour should have some skill in measuryng of Lande' and offers instruction in how to do it, but says nothing of drawing maps (Leigh 1577, sigg. G4r, O1r–Q4r). William Bourne in 1578 does not seem to envisage the use of his methods by surveyors: his instructions for measuring distances and heights emphasise the value of this in military operations and the instrument he describes for making scale-maps (a form of plane-table) is said to be particularly useful for generals and captains. When, later in the book, he discusses measurement of land he does not mention mapping (Bourne 1578, books 1, especially f. 26r, and 3). The first work to assume that maps will be drawn by an estate surveyor was by Edward Worsop in 1582: its title was *A Discoverie of Sundrie Errours and Faults daily committed by Landemeaters, ignorant of Arithmetike and Geometrie*, but even the surveyors he criticised would 'knowe howe to take a plat by any one instrument' – their fault lay in their knowing nothing of the principles underlying their techniques of mensuration and mapping (Worsop 1582, sig. E3r,v). In 1596 Ralph Agas recounted [43] how, about twenty years earlier, he had been employed to trace the bounds of a foldcourse and locate its boundary marks by compass direction; he did this successfully and accurately with his theodolite, 'whereby I considered of what force a bounder by plat might be in time to come, which carrieth the hedges at a haire breadth by a circular division of infinite parts'. The point is that drawing a scale-map had been no part of that particular job, but it had struck Agas what a good idea it would have been to have had one (Agas

---

[26] As Professor Ravenhill points out to me, given British weather Agas might be said to have been a realist too. Woolgar 1985 reproduces and discusses working draft maps, now in the archives of Corpus Christi College, Oxford, that were produced by both techniques in the early seventeenth century.

1596, p. 16).[27] The treatises confirm the evidence of the maps themselves that mapping became a part of estate surveying quite suddenly in the late 1570s.

To account for this it is not enough to look simply to a spread of knowledge of mapping techniques and instruments among surveyors. The principal pioneers of mapping in estate surveys were (despite Worsop's strictures) learned men who will long have been aware of what maps had to offer and how to draw them. We should look rather to those for whom the maps were made, the landowners who were prepared to pay extra fees and costs to the surveyor for the sake of having a map. To the landowner only a scale-map would be worth paying for. The map that was not drawn to scale would show him the relative positions of his fields and other lands – but all this at least the resident landowner would know already. In the words of Norden's farmer, 'is not the Field it selfe a goodly Map for the Lord to looke upon, better than a painted paper?' Such a map would add little or nothing to the written survey. But the scale-map would. With it the landowner could see at a glance, more clearly than from the written survey, the relative sizes of his various pieces of land; it would show, as the written survey could not, where they could be subdivided for sale, for letting or for his own cultivation and precisely how much land would be involved in each subdivision (Agas 1596, pp. 14–15). But why did landowners suddenly come to appreciate these and other merits of the scale-map?

We are familiar with maps drawn to a consistent scale; the idea calls for no particular explanation and we are apt to take it for granted. This was not the case in sixteenth-century England. When the earliest scale-maps – those of the military engineers in the 1540s – specify the scale used they do not just say what it is; they give what amounts to an explanation of scale for the user unfamiliar with the concept. Thus the 1548 map of Thanet tells us that 'In thys plat 3 inches contayneth one mile'.[28] Soon this was no longer necessary. The map of Sheppey in about 1572 has simply a scale-bar with the legend 'Scala passus bis Mille in Semipede'.[29] But the scale-maps of small areas that were drawn in the twenty-five years after 1550 were for military or other official use. They were drawn for well-educated men who would understand the use of [44] scale as a technique peculiar to this form of mapping. It does

---

[27] I cannot agree with Taylor 1947, p. 132, that the passage implies that a map was actually drawn (to me the most natural interpretation is that it was not), still less that this was 'almost certainly' Agas's map of the manorial foldcourse at Tangham (Norfolk) that survives as BL Egerton MS. 2789B; except that this was a foldcourse similarly defined by a written survey in a book there is nothing to support this identification.

[28] CPM I.61.

[29] PRO MPF 1/240.

not follow that an understanding of scale was widespread. Indeed, the treatises show just the opposite. William Bourne in 1578 devoted an entire chapter to explaining how to make 'a Tronke or skalle of measure in a mappe or Carde', showing with diagrams how dividers can be applied to the scale-bar to measure the distance between any two points on the map (Bourne 1578, book 1, f. 29(bis)r,v). In fact, the dividers drawn or engraved on so many sixteenth-century maps are not there for decoration. They are graphic instructions for use: this is where you put your dividers if you want to measure distances on the map. And in Worsop's work of 1582 we find a very revealing conversation (the whole book is cast in the form of a discussion between the surveyor and four other characters). The surveyor is asked 'What meane you by this worde (Scale) which I see in so many places of your booke?'; he replies 'The Scale is a measure used in platting, taken at the will of the plat maker, eyther greater, or lesse, to set foorth the plat measured in true proportion, and Symetrie, upon paper, or any other superfice'. Another listener says, reasonably enough, 'I understand not this definition', and Worsop then spends a couple of pages, with pictures of scale-bars and dividers, in a more lucid explanation. 'I nowe perceive', says the first questioner, 'what is meant by this worde scale. I remember I have seene the like lines, and compasses set in mappes, but I never understood what they meant till nowe' (Worsop 1582, sigg. B2v–4r).

Clearly the employers of Agas, Ames and Treswell, probably many other landowners as well, had achieved this understanding some years earlier than Worsop's questioner. How? Even after its publication in 1578 we can hardly suppose they were all reading Bourne's *Treasure for Traveilers.* It is far more likely that they had been introduced to the idea of scale by Saxton's county maps. The earliest of these are dated 1574, the latest 1578, and it is assumed that the individual maps could be bought separately both before and after the issue of the whole series as an atlas in 1579. Certainly they were produced in quantity and found a good market; the relatively large number of copies that survive today is sufficient evidence. Those who bought them are likely to have been just that class of landowners who were the potential employers of estate surveyors (Morgan 1979, pp. 133–4, 145–9). Each of Saxton's maps bears a scale-bar: 'Scala [45] miliarium' or, in one case, just 'Scala' (Skelton 1970, pp. 7–8, 14–16). These maps, on which their purchasers would soon learn to work out the distances between places in their counties, may well have done more than anything else to introduce the English gentry at large to the idea of the scale-map. With this preparation it became easy, or at least possible, for the enterprising surveyor to sell the landowner the idea of having a scale-map of his own estate. It may not be going too far to say that but for Saxton's county maps in the 1570s there would have been no estate maps in the 1580s.[30]

Nor were estate maps the only new application of scale-mapping to small areas at this time. A long tradition of building plans had drawn outlines roughly to shape but not to scale; instead the dimensions of each room were written on the plan (Harvey 1981, pp. 38–9). In the course of the sixteenth century we find these plans approximating ever more closely to consistent scale; often it is difficult to tell whether a plan is drawn to scale, but with some inaccuracies, or merely drawn to shape with particular care and skill. From the 1540s onwards military engineers drew scale-plans not only of fortifications but of other buildings as well, the earliest known being plans of the royal manor house at Hull by John Rogers in 1542.[31] However, even the finely drawn plans by Henry Hawthorn for the works at Windsor Castle in 1576 were probably not intended to be scale-plans,[32] and outside the context of royal works the earliest surviving building plans drawn specifically to scale seem to be those by William Browne (with elevation) for the customs house at Hull in 1573[33] and by John Symonds for works at Cursitors Hall (London) about 1575 (Pl. IV).[34] Interestingly, Worsop in 1582 refers to the carpenter's use of plans with the scale shown in feet (Worsop 1582, sigg. B2v–4r). Turning to maps drawn for law suits, we find that three maps drawn for a court of the duchy of Lancaster in 1580–81 were drawn to scale, maps not of wide tracts of country like the Ashdown Forest map of about 1563, but of relatively small areas in Barnoldswick (West Riding of Yorkshire), Haslingden and Penwortham

---

30 It is interesting and may conceivably be significant that none of the early estate surveyors' maps that have been mentioned is demonstrably older than Saxton's map of the relevant county; his map of Norfolk is dated 1574 and of Essex 1576.

31 BL Cotton MSS. Augustus I.i.84, I.ii.11 and Supp.1, discussed by Shelby 1967, pp. 34–46, 172, plates 10–12.

32 PRO MPF 1/150, 254, 266 and SP 12/109, fo. 69, discussed by H.M. Colvin and J. Summerson in Colvin 1963–82, 3, p. 326, where they are described as 'among the most interesting architectural drawings of the period'.

33 CPM II.3 (Skelton and Summerson 1971, no. 155): I am grateful to Mr R.H. Harcourt Williams, who has kindly confirmed that the scale is actually noted on the drawing. It may be, however, that priority should be given to CPM I.9, a plan of the Royal Exchange, London, with a scale of feet along its right-hand edge, which cannot be dated more precisely than '1566 or later' (Skelton and Summerson 1971, no. 157).

34 PRO MPA 1/71 (Public Record Office 1967, p. 22); CPM II.21 (Skelton and Summerson 1971, no. 154; *cf.* Colvin 1963–82, 4, p. 152). It is interesting that CPM Supp. 36k, an elevation of a gateway at Theobalds (Herts.) drawn by Symonds in 1577, also has its scale specified: 'This is 4 fotes To The ynch' (Skelton and Summerson 1971, no. 170).

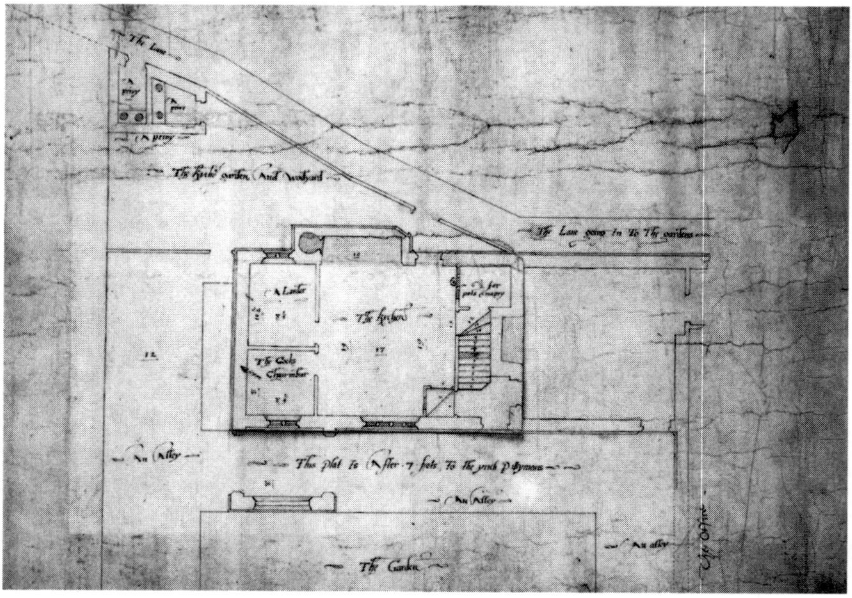

Plate IV. Cursitors Hall (London) by John Symonds, *c.*1575 (PRO MPA 1/71)

(both Lancashire) (Pl. V).[35] Uniform, specified scale was starting to be seen as a normal feature of any map or plan.

The course of development outlined here throws some light on the career of Saxton himself. The attribution to him of a copy, made in 1569, of a map of Belfast Lough, seems to be apocryphal (Evans and Lawrence 1979, p. 120). His earliest known works are thus his maps of the counties, 1574–8, and the two versions of his map of the whole of England and Wales of which the later dates from 1583. From this year onwards (see Appendix) he appears as the surveyor of many estates, particularly in Yorkshire; this work produced some twenty-five surviving maps of which the two earliest, of properties in Kent, date from 1590 (article XVI, plate) (Evans and Lawrence 1979, pp. 79–80). It has been thought surprising that we have no earlier estate maps by Saxton; one might suppose that his work on county mapping had been preceded by an apprenticeship in estate [48] mapping (Evans and Lawrence 1979, p. 9). In fact the pattern of his known work neatly fits the

---

[35] PRO MPC 1/91, 244, 245 (the two last were formerly numbered 244 (1) and (2); Public Record Office 1967, pp. 477, 489, 507). Similarities between these maps (as the scale-bars with 5-rod divisions and the accompanying notes) suggest that they were the work of a single map-maker.

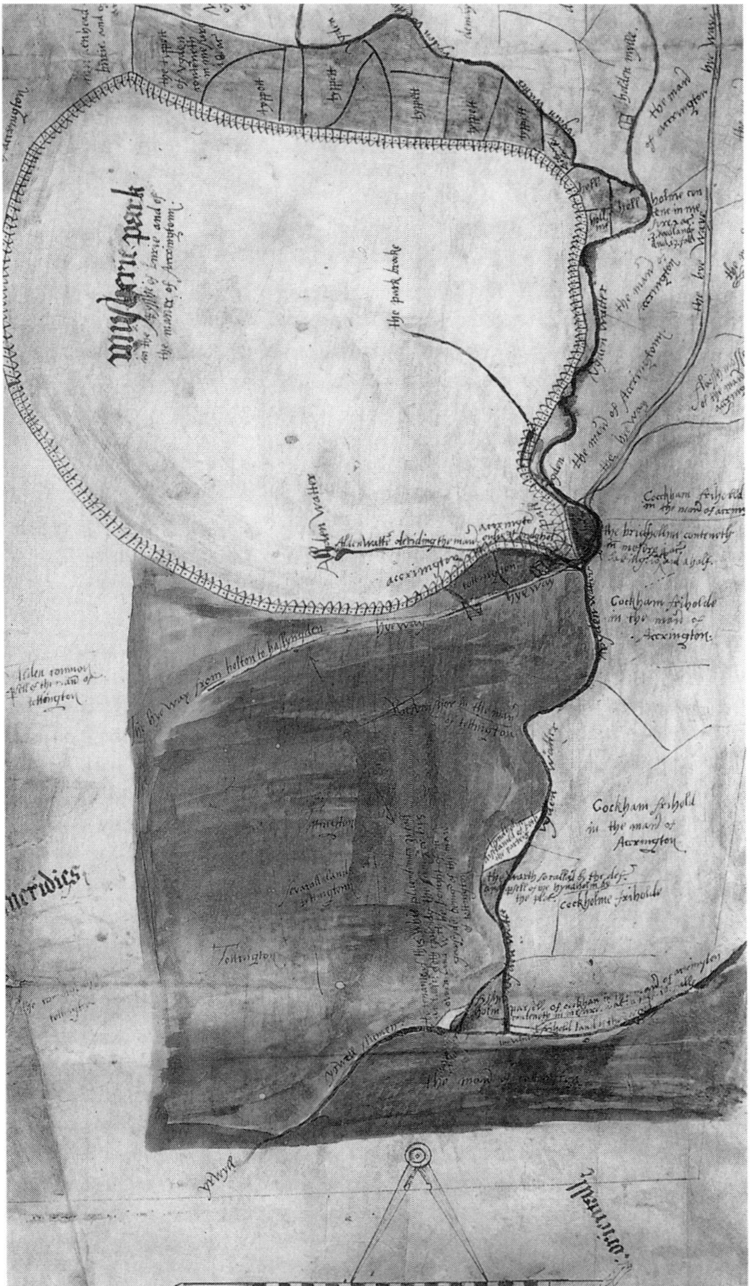

Plate V. Haslingden (Lancashire), 1581 (PRO MPC 1/245)

general chronology of surveyed maps in England; he may well have drawn estate maps earlier in the 1580s, but we probably have none earlier than his county maps simply because none was drawn.

Within this broad chronological framework there is still much to discover. A glaring need is for work on the written surveys of the sixteenth century: the development of the court of survey and the work of the surveyor as land-measurer before he became a cartographer, how far the growth of surveying as a distinct profession arose from the development of new techniques and instruments for measurement, how far it was the result of proprietorial and agrarian demands for more frequent, more detailed surveys.[36] We need more work too on the spreading use of maps – when, how and why – bearing in mind that the technical innovations are only part of the story. It was not just for these that the period 1550–80 was crucial in the history of cartography in England. It was as much because it was now that maps in general, scale-maps in particular, stopped being restricted to particular military, legal and other technical purposes and started to become familiar objects of everyday life.

Acknowledgements

I am grateful to Professor M.W. Beresford for permission to publish, in the Appendix, his note on an estate survey by Christopher Saxton, four years earlier than any hitherto noticed. I am grateful too to the late Dr J.B. Harley and to Professor W. Ravenhill, Mrs S. Tyacke and Dr C.M. Woolgar, who all kindly read a draft of this article and who made very helpful comments and suggestions, and to British Library Public Services which, through the kindness of Mr David Way, made available the photographs used for the illustrations.

Bibliography

Agas, R., 1596. *A Preparative to Platting of Landes and Tenements for Surveigh* (London).

Barber, P., 1992. 'England I: Pageantry, defense and government: Maps at court to 1550' and 'England II: Monarchs, ministers and maps 1550–1625', in *Monarchs, Ministers and Maps*, ed. D. Buisseret (Chicago), pp. 26–98.

[36] In our present state of knowledge we can only regard as interesting possibilities the suggestions that increasingly precise mensuration in surveying resulted from increasing enclosure of lands or from the distribution of monastic estates following the dissolution (Darby 1933; Skelton and Summerson 1971, p. 30). For the later history of the estate map, Bendall 1992 is crucial pioneer work.

Barker, K., 1991. 'An Elizabethan map of north-west Dorset: Sherborne, Yetminster and surrounding manors', in *Maps and History in South-West England*, ed. K. Barker and R.J.P. Kain (Exeter).

Bendall, S., 1992. *Maps, Land and Society: A History, with a Carto-Bibliography of Cambridgeshire Estate Maps, c.1600–1836* (Cambridge).

Benese, R., 1537. *The Maner of Measurynge of all Maner of Lande* (Southwark, [1537]).

Beresford, M., 1957. *History on the Ground* (London).

Bourne, W., 1578. *A Booke called the Treasure for Traveilers* (London).

Cobbe, H., 1973. 'Four manuscript maps recently acquired by the British Museum', *Journal of the Society of Archivists*, 4, no. 8, pp. 646–52, plates I, II.

Colvin, H.M. (ed.), 1963–82. *The History of the King's Works*, 6 vols (London).

Cuningham, W., 1559. *The Cosmographical Glasse* (London).

Darby, H.C., 1933. 'The agrarian contribution to surveying in England', *Geographical Journal*, 82, pp. 529–35.

Digges, L., 1556. *A Boke named Tectonicon* (London).

Eden, P. (ed.), 1975–9. *Dictionary of Land Surveyors and Local Cartographers of Great Britain and Ireland*, with *Supplement* (Folkestone).

Evans, I.M., and Lawrence, H., 1979. *Christopher Saxton: Elizabethan Map-Maker* (Wakefield and London).

Harvey, P.D.A., 1964–5. 'An Elizabethan map of manors in north Dorset', *British Museum Quarterly*, 29, pp. 82–4, plate XX.

Harvey, P.D.A., 1981. 'The Portsmouth map of 1545 and the introduction of scale-maps into England', in *Hampshire Studies*, ed. J. Webb, N. Yates and S. Peacock (Portsmouth), pp. 32–49.

Harvey, P.D.A., 1984. *Manorial Records* (British Records Association).

Harvey, P.D.A., 1993. *Maps in Tudor England* (London).

Leigh, V., 1577. *The Most Profitable and Commendable Science, of Surveying of Landes, Tenementes and Hereditamentes* (London).

Macdonald, A., 1937. 'Plans of Dover Harbour in the sixteenth century', *Archaeologia Cantiana*, 49, pp. 108–26, plates I–IV.

MacIvor, I., 1965. 'The Elizabethan fortifications of Berwick-upon-Tweed', *Antiquaries' Journal*, 45, pp. 64–96, plates XXIII–XXXII.

Manley, G., 1936. 'The earliest extant map of the county of Durham', *Transactions of the Architectural and Archaeological Society of Durham and Northumberland*, 7, pp. 278–87.

McRae, A., 1993. '"To know one's own": estate surveying and the representation of land in early modern England', *Huntingdon Library Quarterly*, 56 (1993), pp. 332–57.

Merriman, M., 1983. 'Italian military engineers in Britain in the 1540s', in *English Map-Making 1500–1650*, ed. S. Tyacke (London), pp. 57–67, plates 17–30.

Morgan, V., 1979. 'The cartographic image of "the country" in early modern England', *Transactions of the Royal Historical Society*, 5th ser., 29, pp. 129–54.

Norden, J., 1607. *The Surveyors Dialogue* (London).

Public Record Office 1967. *Maps and Plans in the Public Record Office*, vol.1.

Ravenhill, W., 1984. 'The Plottes of Morden Mylles, Cuttell (Cotehele)', *Devon and Cornwall Notes and Queries*, 35, part 5, pp. 165–74, 182–3.

Shelby, L.R., 1967. *John Rogers: Tudor Military Engineer* (Oxford).

Skelton, R.A., 1970. *County Atlases of the British Isles 1579–1703* (London).

Skelton, R.A., and Harvey, P.D.A. (ed.), 1986. *Local Maps and Plans from Medieval England* (Oxford).

Skelton, R.A., and Summerson, J., 1971. *A Description of Maps and Architectural Drawings in the Collection made by William Cecil First Baron Burghley now at Hatfield House* (Roxburghe Club).

Taylor, E.G.R., 1947. 'The surveyor', *Economic History Review*, 1st ser., 17, pp. 121–33.

Woolgar, C.M., 1985. 'Some draft estate maps of the early seventeenth century', *Cartographic Journal*, 22, pp. 136–43.

Worsop, E., 1582. *A Discoverie of Sundrie Errours and Faults daily committed by Landemeaters, ignorant of Arithmetike and Geometrie* (London).

Appendix

Professor M.W. Beresford writes:

> W.R. Shepherd's *History of Kirby Underdale* (Batley, 1928), pp. 59–64, has an account of a pre-enclosure survey of the open fields at this East Riding village in 1583. According to a terrier 'of 1590' which he cites, the survey was followed by enclosure at the hands of eight commissioners. He then quotes from a document of 20 March 1591 'the nayme of the Survayer was Christoffer Saxton a westoran (*West Riding?*) man that mesuryd the land'. If the survey was made in 1583 it is earlier than any cited in Evans and Lawrence 1979, pp. 79–80.

Note

An earlier estate map than those noted here has since come to light, as noted at the beginning of article XVII: Queen Camel (Somerset) by Grindall Painter in 1573 (Somerset Record Office, DD/MI, c/186).

XVI

A MANUSCRIPT ESTATE MAP BY CHRISTOPHER SAXTON

CHRISTOPHER SAXTON is best known for his maps of the counties of England and Wales, the first of their kind, which he published between 1574 and 1578. It is not certain how far these were based on original surveys, how far on existing manuscript maps, but they were sufficiently accurate to serve as the basis of nearly all county maps for the next hundred years, while as late as 1770 maps were still being printed from some of the original plates. Saxton was a young man when he carried out this work—he cannot have been more than thirty-six when it was completed[1]—but apart from combining his county maps into a large single map of the whole country, published in 1583, he undertook no more major projects of this sort, and the only other maps that he is known to have produced are manuscript plans of single towns and estates. In all, fifteen of these are recorded,[2] of which only thirteen survive, seven in the Public Records and the others in various public or private collections. Hitherto, there has been no example of Saxton's unprinted work in the British Museum, and the recent gift by Mr. C. E. Kenney of his plan of an estate in Sittingbourne and Murston (co. Kent) is a notable accession to the Department of Manuscripts.

The map, which has been numbered Additional MS. 50189, is on a single sheet of parchment measuring 29 in. × 25½ in. (a portion is reproduced as Pl. xxix). From a comparison with Saxton's maps in the Public Record Office, of which six appear on several grounds to be autograph, it seems unlikely that any part of the present map is in Saxton's own hand. It is presumably a fair copy made from his rough drawings and its various ornaments are reminiscent rather of contemporary printed maps than of the more austere style of Saxton's other manuscript maps. Like Saxton's other estate plans, however, it has a plain border containing the names of the points of the compass in capital letters. In the top right corner is the title in an ornamental cartouche: 'A Plate of the Mañers of Bayford and Goodmanston in the parishe of Sittingborne in the Countie of Kent. And made by Christofer Saxton in the month of September, Ano Dm̄i. 1590.'; below this is a summarized list of the areas of the estate, with their total. The scale, in the bottom left corner, is decorated with a pair of dividers in the same style as Saxton's county maps—that on the map of Hertfordshire is almost identical. These embellishments have been painted, and the map itself is coloured to mark roads (pale yellow-brown), water (blue), and the boundaries of the estate (pink), besides trees, buildings, and ships. The map gives the names and areas of the fields of the estate, and also shows some of the adjoining properties including part of Sittingbourne village. The owner of the manors of Bayford and Goodneston in 1590 was Sir William Garrard,[3] who was the son of one Lord

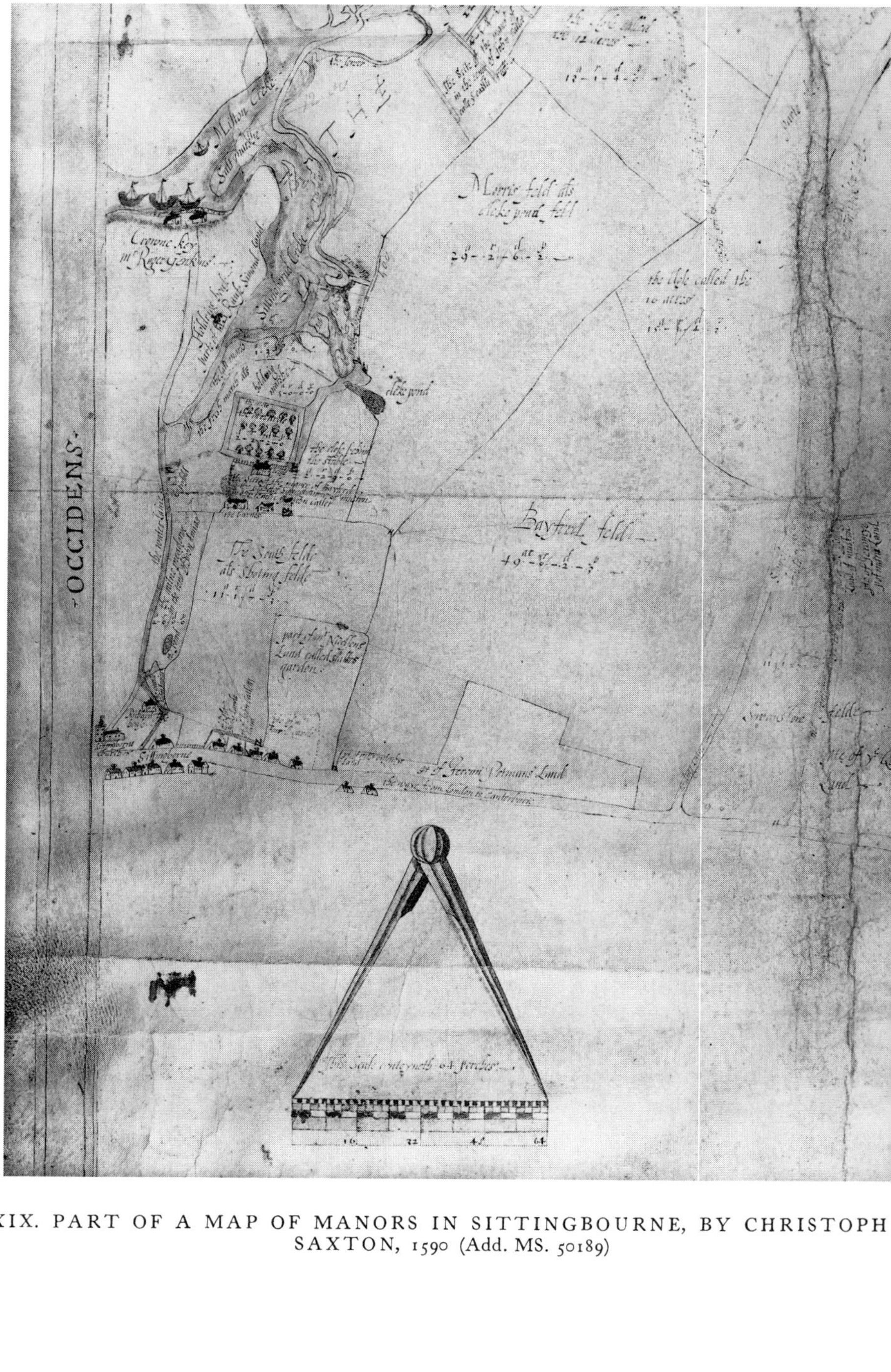

XXIX. PART OF A MAP OF MANORS IN SITTINGBOURNE, BY CHRISTOPHER SAXTON, 1590 (Add. MS. 50189)

Mayor of London and the son-in-law, the brother, and the great-great-uncle of three others, and it was presumably he who commissioned Saxton to make this map. It seems to have remained among the records of the estate; at some time, probably in the eighteenth century, a few lines and measurements were added to it in pencil, and in 1879 it was still in local ownership.[4] The condition of the map is fairly good, though at some time it seems to have been kept folded, with the result that some of the writing along the folds has been obscured.

The representation of Sittingbourne church is conventional, not pictorial; the actual building is more elaborate than that shown on the map and seems never to have possessed a spire. In view of this it is likely that the other buildings are also marked conventionally; even so it is interesting that half of the houses are shown as of the very simplest type of timber-framed construction, with their longest walls rising sloping from the ground and meeting to form the roof.[5] The cowls which surmount all the chimneys are also of interest. The form of the estate, made up of enclosed fields and extending into more than one parish, is typical of Kent, and so too is the use of the *daywork* as a measure of land equal to four perches or one-tenth of a rood[6]—all the areas are given in acres, roods, dayworks, and perches. Although the daywork in this sense is particularly associated with Kent and south Essex, at an earlier date it may have been more widespread, for in 1320 it occurs at Wellingborough (co. Northt.);[7] by the mid-sixteenth century it had died out in Essex, but in Kent it survived longer, and Anthony Denton's survey of the Kentish estates of Sir Roger Twysden in 1630 (Additional MS. 34155) similarly gives areas in acres, roods, and dayworks. In other parts of the country the daywork was also used as a land measure, but with different meanings: in seventeenth-century Warwickshire it appears as the equivalent of an acre,[8] while in Yorkshire it was equal either to three roods or to a local measure of nearly two-thirds of an acre.[9] It is curious, therefore, to find the four-perch Kentish daywork appearing on maps which Saxton subsequently made of lands in Yorkshire, at Bingley in 1592 and at Horbury and Netherton in 1598,[10] as well as in his survey of Burley-in-Wharfedale in 1602;[11] we may conjecture that he was himself introducing to his native Yorkshire the measure which he had found in use in south-east England.

Apart from the areas of two small pieces of land, added by a different but contemporary hand, the map seems to have been the work of a single draughtsman; there is no evidence of his identity. His work lacks the fine precision of, for instance, Ralph Agas's map of Toddington (co. Bedf.) in 1581 (Additional MS. 38065), but it was not until the late eighteenth century that this was considered essential to maps of this sort. More serious are some evident mistakes in the figures on the map. Thus, the area of *Morris feld* appears in the list at the side of the map as 29 a. 2 r. 0 d. 2 p., but on the map itself as 29 a. 2 r. 6 d. 2 p., while 0 and 6 have probably again been confused in the area of a piece of freshwater marsh, given as the impossible 22 a. 3 r. 9 d. 6 p. Even allowing for this, the total area of freshwater marsh shown on the map does not correspond to that given in the list at the side, while as the list stands the total acreage of the estate has been added incorrectly. Faulty copying seems the most likely cause of these discrepancies.

[1] An account of his life is given by Sir G. Fordham, 'Christopher Saxton, of Dunningley. His life and work', *Miscellanea*, Thoresby Society, 1928, pp. 356–84.

[2] They are listed by G. R. Batho in *The Geographical Journal*, cxxv (1959), 73–74.

[3] E. Hasted, *The History and Topographical Survey of the County of Kent*, 1778–99, ii. 613.

[4] W. A. Scott Robertson, *Sittingbourne, and the Names of Lands and Houses in it: their Origin and History*, 1879, pp. 10, 22.

[5] This type of house is described by H. M. Colvin in *Medieval England*, ed. A. L. Poole, 1958, i. 80–81.

[6] H. L. Gray, *English Field Systems*, 1915, pp. 300–1; J. L. Fisher in *Transactions of the Essex Archaeological Society*, N.S. xxiii (1942–5), 92–96.

[7] *Wellingborough Manorial Accounts*, ed. F. M. Page, Northamptonshire Record Society, 1936, p. xxiv.

[8] *Ecclesiastical Terriers of Warwickshire Parishes*, ed. D. M. Barratt, Dugdale Society, i (1955), 39–40. It is probably in this sense that the word is used in the Staffordshire examples of 1534 and 1792 given in the *Oxford English Dictionary* and the *English Dialect Dictionary*.

[9] *English Dialect Dictionary*.

[10] Public Record Office, M.P. B 30, M.P. C 9.

[11] It is printed by Fordham, op. cit., pp. 381–4.

XVII

English Estate Maps: Their Early History and Their Use as Historical Evidence*

[27] Origins: The First Estate Maps

Estate map is here taken to mean a plan of landed property, drawn not for a particular occasion or for some closely defined purpose but for general reference. It was a work the estate owner could consult for detailed information about the lands it showed; or he might point to it with pride, seeing it as a graphic epitome of his property, wealth, and social position. Often it was clearly designed for display, beautifully coloured and elaborately ornamented. Often signs of wear, and many added corrections and annotations, testify to long service as a functional tool of estate management.

Such estate maps appeared quite suddenly on the English scene.[1] There was no long course of development in which the genre gradually evolved, moving from the rudimentary to the elaborately detailed and sophisticated; those who made them did not feel their way towards plans of this kind. One of the most ambitious of all English estate maps before the nineteenth century, Ralph Agas's map of Toddington (Bedfordshire), dates from 1581, only a few years later than the earliest we know.[2] These date from the mid-1570s: the map of Queen Camel (Somerset) by Grindall Painter in 1573, the map of West Lexham (Norfolk) by Ralph Agas in 1575, and the maps by Israel Amyce of the St. Paul's Cathedral estate at Belchamp St. Paul (Essex) in 1576[3] (article XV,

---

* The article has been reset with minor changes, but the original page numbers (which have not been used for the index) have been entered in square brackets. In these notes the following abbreviations are used: BL (British Library), PRO (Public Record Office), WCRO (Warwickshire County Record Office).

1 The first few pages, an account of the first appearance of English estate maps in the 1570s, are abridged from P.D.A. Harvey, 'Estate surveyors and the spread of the scale map in England, 1550–80', *Landscape History* 15 (1993): 37–49. P.D.A. Harvey, *Maps in Tudor England* (London, 1993), 79–93, sets estate maps in the context of the general development of mapping in the sixteenth century.

2 BL Add. MS 38065.

3 Somerset Record Office, DD/MI, c/186; Holkham Estate records, map 87a (photocopy BL Maps 188 n.1[10]); Guildhall Library, London, MS 25517/1 (one of these maps by Amyce

plate III). Other early estate maps include the series produced by Ralph Treswell between 1580 and 1587 of lands of Sir Christopher Hatton in Northamptonshire (figs 7, 8).[4] Others, some perhaps earlier than 1575, will probably come to light. But it is clear that the estate map in England emerged fully fledged, almost certainly in the 1570s, and that from the start it was produced by various mapmakers; it was not the invention of a single innovator.

It did however have antecedents in the plans of the whole or parts of landed properties that were drawn for some particular, immediate purpose. These had a long history behind them; indeed there are even a few medieval English examples, such as the thirteenth-century diagram of Wildmore Fen (Lincolnshire) that sets out the pasture rights agreed between Robert Marmion and Kirkstead Abbey.[5] After 1500, like all forms of local [28] map, maps of landed properties became much more common and we have many examples, particularly from the mid-sixteenth century onward. Some are simply sketches. A plan of Sir Richard Edgcumbe's house and its surrounds at Cothele (Cornwall) in the 1550s was drawn to show how access would be given to the local mills when a new deer park was created. It was a pen-and-ink drawing showing the river and the relevant roads and bridges, with rough outlines of church, houses, gates, and trees. Others include detailed information from written surveys.[6] Plans of parts of the fields of Dedham (Essex), drawn in 1573 perhaps in connection with a lawsuit, include a mass of notes giving the area and the tenant of each tiny plot of land and even page references to the survey that must have supplied the information.[7] Others again are carefully and elaborately drawn and coloured. One of an area of marshland and its surrounds at Formby (Lancashire) was made probably for display in a court of law in 1557–58; the clumps of marsh flowers and the house, churches, mills, and woods are carefully executed but are all painted large and bold, to be seen from a distance.[8]

---

is illustrated in article XV, plate III). An earlier map that fits our definition of an estate map was drawn between 1444 and 1446 to show Edmund Rede's manor of Boarstall (Buckinghamshire), but this was an isolated production without successors (R.A. Skelton and P.D.A. Harvey, eds., *Local Maps and Plans from Medieval England* (Oxford, 1986), 211–19, plate 18). See colour plate 58 in Harvey, *Maps in Tudor England*, for a map by Israel Amyce.

4 Northamptonshire Record Office, Finch Hatton MS 272. Israel Amyce's survey with maps, made in 1579, of the lands in Essex of the late Edmund Tirrell (BL Harley MS 6697) scarcely meets our definition of an estate map, for it was made so that the lands could be divided between Tirrell's heirs.

5 Skelton and Harvey, *Local Maps and Plans*, 71–81, plate 3.

6 W. Ravenhill, 'The Plottes of Morden Mylles, Cuttell (Cotehele)', *Devon and Cornwall Notes and Queries* 35, pt. 5 (spring 1984): 165–74, 182–3.

7 PRO MPC 1/77, fols.4–7.

8 PRO MR 1/2.

These earlier ad hoc plans of landed properties, made to meet some particular need, differ in another very significant way from the estate maps that first appeared in the 1570s. None is drawn to [29] scale: each serves its intended purpose without any consistent relationship between distance on the map and distance on the ground, whether stated or implied. The Queen Camel map of 1573 was not drawn to scale, but other early estate maps are all scale maps. Certainly there had been earlier scale maps in England. The first date from the early 1540s: plans, mostly of projected fortifications, drawn by military engineers who probably brought the technique from Italy.[9] Over the following forty years quite a number of surviving scale plans of military works were drawn; there are particularly notable series from Berwick-upon-Tweed[10] and Portsmouth.[11] There was also a modest extension – no more – of the use of scale in two other sorts of map. One was a straightforward development of the plan of fortifications: military works would be shown in the context of what might be quite a large area of the surrounding countryside, and the whole map would be drawn to scale. Thus a superb scale map of the Isle of Sheppey (Kent), drawn about 1572 and signed with the unidentified monogram IM, was probably produced for military purposes and includes a minute plan of the fort at 'Swale Nesse', but it is very much more than a simple plan of fortifications (article XV, plate II).[12] The other scale maps of this period were of very much larger tracts of country, 200 square miles or more, drawn for a variety of purposes: for military planning, like one of part of the Anglo-Scottish border in 1552; for use in a lawsuit, like one of Ashdown Forest (Sussex) about 1563; or for general reference, like one of County Durham in 1569–70.[13] The value of consistent scale was appreciated sooner in these maps of relatively large areas than in maps and plans of more limited scope, apart from those drawn by military engineers.

---

[9] L.R. Shelby, *John Rogers: Tudor Military Engineer* (Oxford, 1967); P.D.A. Harvey, 'The Portsmouth Map of 1545 and the Introduction of Scale-Maps into England', in *Hampshire Studies*, ed. J. Webb, N. Yates, and S. Peacock (Portsmouth, England, 1981), 32–49; M. Merriman, 'Italian Military Engineers in Britain in the 1540s', in *English Map-Making*, ed. Sarah Tyacke (London, 1983), 57–67.

[10] PRO MPF 1/137; R.A. Skelton and J. Summerson, *A Description of Maps and Architectural Drawings in the Collection Made by William Cecil, First Baron Burghley, Now at Hatfield House* (London, 1971), nos. 73–6, 78, 80.

[11] D. Hodson, *Maps of Portsmouth before 1801* (Portsmouth, England, 1978).

[12] PRO MPF 1/240; see Harvey, *Maps in Tudor England*, colour plate 22.

[13] PRO MPF 1/257, MPF 1/144; G. Manley, 'The Earliest Extant Map of the County of Durham', *Transactions of the Architectural and Archaeological Society of Durham and Northumberland* 7 (1936): 278–87.

The printed county maps of Christopher Saxton fall neatly into this pattern. They are scale maps of substantial areas: every one has a scale bar. The individual maps bear dates from 1574 to 1578, and the complete collection was published as an atlas in 1579.[14] The number of surviving copies shows that the maps sold well, and they must have done much to familiarize Englishmen with the value of maps in general and with the concept of scale in particular. It is not impossible that they played a large part in the introduction of estate maps; that is to say, landowners commissioned maps of their properties because they had bought or seen Saxton's maps of their counties. The coincidence of dates is interesting and may be significant.

[30] Whatever its cause, we should clearly look to the landowners as well as to the surveyors for the impetus that produced the sudden emergence of estate maps in England. Ralph Agas wrote in 1596 how the value of a map in setting out a surveyed boundary had suddenly occurred to him some twenty years earlier;[15] but from what we know of the men who produced the first estate maps – their education, their interests, their writings – we would have expected them to have been long aware of maps, aware even of the possibility of mapping to scale. There was no development in techniques or instruments that made it more practicable to draw estate maps in the mid-1570s than it had been fifteen or twenty years before. All that was new was the idea, the application of mapping to estate management; and what happened in the 1570s was probably that the surveyors who were to produce the maps succeeded in convincing the landowners who were to pay for them that this idea was a good one.

We should remember that every estate map had two parents: the commissioning landowner who fathered it and the mapmaking surveyor who brought it to birth. We can only guess the part that each played in the origin of estate maps in England, but in following their development down to the early eighteenth century, we can distinguish fairly clearly the role of each parent.

Development: The Role of the Surveyor

By the 1570s estate surveyors in England had a long history behind them. Fully measured surveys of estates first appear in the early thirteenth century, and by at least the early fourteenth we find some men who, while far from being professional surveyors, were apparently considered to have special knowledge or skill in measuring and describing lands. The products of their work were written surveys, formal descriptions of estates that the owners would use in the work

---

14 R.A. Skelton, *County Atlases of the British Isles, 1579–1703* (London, 1970), 7–16.

15 Ralph Agas, *A Preparative to Platting of Landes and Tenements for Surveigh* (London, 1596), 16.

of management, precisely defining their property and revealing its potential, providing a yardstick with which actual production and profit might be compared. These written surveys might be long and elaborate, giving details of sometimes thousands of individual plots of arable, besides meadow, pasture, and woodland. Mapping played no part in medieval surveying; it is just possible, no more, that fifteenth-century surveyors occasionally sketched rough maps as an adjunct to their working notes, but a map never formed part of the final survey. The work of the estate surveyor was to measure and describe lands, not to draw maps.[16]

This was the case until the 1570s. From the beginning of the sixteenth century the understanding and use of maps spread fast, and we have seen that sketch maps of pieces of landed property might be made for some particular purpose. No doubt estate surveyors made some maps of this kind, just as they probably made use of sketch maps in drafting their surveys. But the work of a surveyor was still to measure and describe lands in written surveys: he was no more a mapmaker than the landowner who employed him. We see this in published treatises on surveying. The earliest of these, John Fitzherbert's *Boke of Surveyeng* (1523) and Richard Benese's *Maner of Measurynge* (1537), like their medieval predecessors that circulated in manuscript, make no mention of drawing maps: they are concerned with measurement, particularly how to measure areas of variously shaped plots. The same is true of mid-sixteenth-century treatises on measuring distances and areas, by Leonard and Thomas Digges in 1556 and 1571, by Valentine Leigh in 1577, and by William Bourne in 1578. Those that mention mapmaking – and those of [31] 1571 and 1578 refer to making scale maps – see this as part of military science, not as anything to do with the estate surveyor. The earliest book that envisages mapmaking as part of a surveyor's work was published in 1582: *A Discoverie of Sundrie Errours and Faults Daily Committed by Landemeaters, Ignorant of Arithmetike and Geometrie*, by Edward Worsop.

It was thus as measurers and describers of lands, not as mapmakers, that surveyors began in the mid-sixteenth century to be seen as specialists having particular skills and expertise. This reflects the development of new techniques and new instruments for mensuration, and from then on many surveyors were highly qualified and experienced professionals. But much work was still done by part-time, even amateur, practitioners, and in the eighteenth century some treatises were still recommending methods of measuring that were far from adequate.

By the end of the sixteenth century even the most professional still saw mapmaking as only a part, and by no means the most important part, of an

16 Skelton and Harvey, *Local Maps and Plans*, 12–18.

estate surveyor's work. The most map-minded might see it as an essential part. Agas wrote: 'No man may arrogate to himselfe the name and title of a perfect and absolute Surveior of Castles, Manners, Lands, and Tenements, unlesse he be able in true forme, measure, quantitie, and proportion, to plat [i.e. draw maps of] the same in their particulars *ad infinitum*.'[17] But the very fact that Agas had to assert this suggests that there were those who thought otherwise. Even so prolific and proficient a mapper of estates as John Norden made it clear in his *Surveyors Dialogue* (1607) that mapmaking was not 'the chiefe part of a Surveyors skill': other abilities were 'more necessary'. Besides actually measuring and describing land he had to interrogate local inhabitants about bounds and rights, he had to be able to examine records, including those in Latin and in antique handwriting, he had to value lands before they were leased or sold, he had to advise the landowner how to run his estate and how to get the most profit from his property.[18] In Norden's eyes the surveyor was a highly skilled professional, entitled to much respect from estate owners and others, but not simply because he ought to be a competent cartographer. It should be added that Norden wrote as a propagandist: he probably idealized the estate surveyor's skills and exaggerated his importance and status.

Since estate maps were made by surveyors, whose work was based on the measurement of land, it is easy to see why, right from the start, they were always drawn to scale. To the landowner, once he had mastered the idea, the scale map offered much more than a map drawn without consistent scale. And for the mapmaker there would be no difficulty: he would already have by him all the information needed for a scale map, which would simply be a graphic representation of what he had had to do anyway. Some estate maps include pictures of the surveyor at work in the field, underlining their origin in the careful measurement of land. An early and particularly interesting example is on a map of 1659 from Bryanston (Dorset) (Fig. 1). Here, as so often on early estate maps, the outline plans of the fields are enlivened by typical scenes: herds grazing, men at work ploughing, sowing, and harvesting, juxtaposed in cheerful disregard of the due order of seasons. At one edge of the map is another scene, the surveyor's team in action, half a dozen strong, with their names. 'Mr Stratton', the surveyor, stands with his stick, supervising the operation, 'Mr Fisher' is holding a pole, 'Mr Andrews' is carrying what is probably a plane table, 'John Heath' holds the end of a chain, and so on.[19] Measurement and estate mapping were soon seen as parts of a single process, so that certainly

---

17 BL Lansdowne MS 165, fol.91.

18 John Norden, *The Surveyors Dialogue* (London, 1607), 17.

19 Privately owned (copy, Dorset Record Office, photocopy 500).

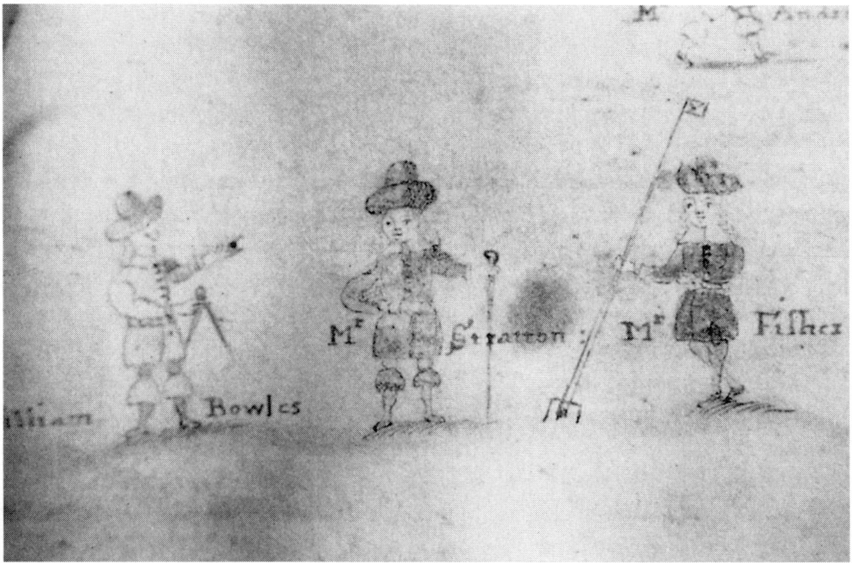

Fig. 1 N. Stratton, detail from a map of Bryanston (Dorset), 1659 (Dorset Record Office, photocopy 500)

by the [32] late seventeenth century the estate surveyor would commonly be thought of as a mapmaker. The connection that we take for granted between surveying and mapping was already established.

This appears clearly in two developments that occurred early in the history of estate mapping in England. One was that surveyors began to be brought in to make scale maps in circumstances that earlier had called for nothing more than a simple sketch map, drawn without measurement to no consistent scale, a map that did not need the special skills of the surveyor. We see this in maps made in connection with lawsuits. In the records of the duchy of Lancaster is a notable series of sixteenth-century maps that were made for production in the duchy's own courts – often on the orders of the court itself – in suits concerning its tenants. The earliest recognizably drawn to scale are a group of three dating from 1580–81, of lands in Accrington, Penwortham (both Lancashire), and Barnoldswick (West Riding of Yorkshire). They all have scale bars but are so crudely drawn and painted that we might suppose this to be no more than a gesture toward the idea of the surveyed map.[20] However, a document attached to the Accrington map implies that, although the whole

[20] PRO 1/MPC 245, MPC 1/244(1), MPC 1/91.

map covers about 1,000 acres, only the particular pieces of land involved in the case, some 50 acres in all, were measured and drawn to scale; the rest must have been sketched in to set these lands in their local context. Of the lands at issue 'the moste parte' took some four or five hours to survey on 29 May 1581, and the finished map was returned to the court the following day. Probably the surveying was fairly rudimentary.[21] All the same, it is very interesting that, just as estate maps were coming rapidly into fashion, measurement on the ground and consistent scale began to be demanded in this other sort of map as well.

An interesting group of maps of Woodmancott Farm near the Candovers (Hampshire), property of Winchester College, neatly illustrates the course of [33] this development. Only one is dated, but they all belong to the late sixteenth century, and their chronological order probably corresponds to their undoubted typological order. First are two similar but independent plans of the farm, not drawn to scale, which the catalogue of the college archives suggests were drawn in connection with lawsuits.[22] Then comes a general estate map of 'the manor of Candever', a much wider area including the farm; it was drawn to scale, with a scale bar, and is signed by Ralph Treswell in 1588.[23] Finally, there are two copies of a map of the farm alone, probably again for a lawsuit, but this time drawn to scale; with them are a set of six of the surveyor's drafts from which the map was constructed.[24] The estate map, drawn to scale from the surveyor's measurements, set a new standard in mapping of properties, to which other sorts of map now began to conform.

The second development was that, once estate surveyors had successfully introduced the map as a way of setting out the information from a general survey, giving it graphic realization, drawing maps came more and more to be seen as a normal part of their work. Thus they would often produce scale maps when they were employed on other lesser [34] jobs for the proprietors of landed estates: measuring a piece of land that was to be sold or divided, for instance, or calculating the effects of some projected improvement. Probably

---

[21] PRO MPC 1/245. The document tells how those commissioned to correct an earlier map, found to be faulty, had discovered it was so badly wrong that they had to draw a new one; one of their number withdrew from the operation because they were excluding from the survey a particular piece of land not specifically named in their commission. This piece of land in fact appears on the map – unavoidably, given its position in relation to the pieces specified. It follows, though this is not made clear on the map itself, that only the specified pieces were actually measured and plotted to scale. This is confirmed by the speed of the operation and by the map's sketchy appearance overall.

[22] Winchester College Muniments, 21446.

[23] Winchester College Muniments, 21443; 21445 (dorse) is a contemporary copy.

[24] Winchester College Muniments, 21444a,b; 21447a–f.

this came about quite quickly, but the chronology has yet to be investigated. Often it is not easy to tell why a particular map was made: the colouring or other conventions may not be explained, and where a map covers a considerable area of property, we can sometimes only guess that it was produced for some ad hoc purpose and not as an estate map for general reference.

This uncertainty can be illustrated by some maps from south Warwickshire at the end of the seventeenth century and the beginning of the eighteenth. One by John Garfield about 1700 shows the river at Stratford-upon-Avon with a substantial area on either side, including meadows, an orchard, the parish church, and houses (Fig. 2). Some lands are outlined in various colours, but their significance is not explained and there is no title, though one seems to have been erased from a panel in one corner. We can only say that it is likely, but not certain, that this map was drawn for some particular occasion of which we know nothing. We can be more certain about four plans of lands at Whitchurch and Crimscote drawn by the same John Garfield, one in 1687, one probably about the same time, and the other two in 1709 and 1712. They are functional outline plans, each showing only one or a small group of plots, covering from 8 to 20 acres in all, with notes of their areas and just a few other details. Why each was drawn we again do not know, but here there can be little doubt that it was for some immediate purpose.[25] On the other hand, a plan by James Fish of the churchyard at Snitterfield in 1697 [35] has a key that makes it quite clear why it was drawn: it was to show a proposed extension to the churchyard (Fig. 3). Probably all these maps served some equally clear and immediate purpose, now hidden from us, reflecting the way that surveyor and landowner might now use scale maps in the everyday work of running an estate.

The surveyor might have been a part-time occasional practitioner or a full-time professional. In either case he normally worked in a fairly limited area, and we find that in any generation the estate mapping within a single county is dominated by a very few names. Surveyors were already established in this pattern of work by the late sixteenth century, so we find it in the earliest estate maps. The work of Christopher Saxton provides a notable early example. Most of his fifty-odd known estate maps and surveys are of places in Yorkshire, but about a dozen are from other parts of eastern or south-eastern England, one is from Shropshire, and one from Lancashire. The seventeen maps or surveys that his son Robert is known to have produced are all from Yorkshire.[26] Again, Thomas Kington and George Withers are two names that recur in the

---

25 WCRO CR 936, box 15/7.

26 I.M. Evans and H. Lawrence, *Christopher Saxton: Elizabethan Map-Maker* (Wakefield, England, 1979), 79–80, 122.

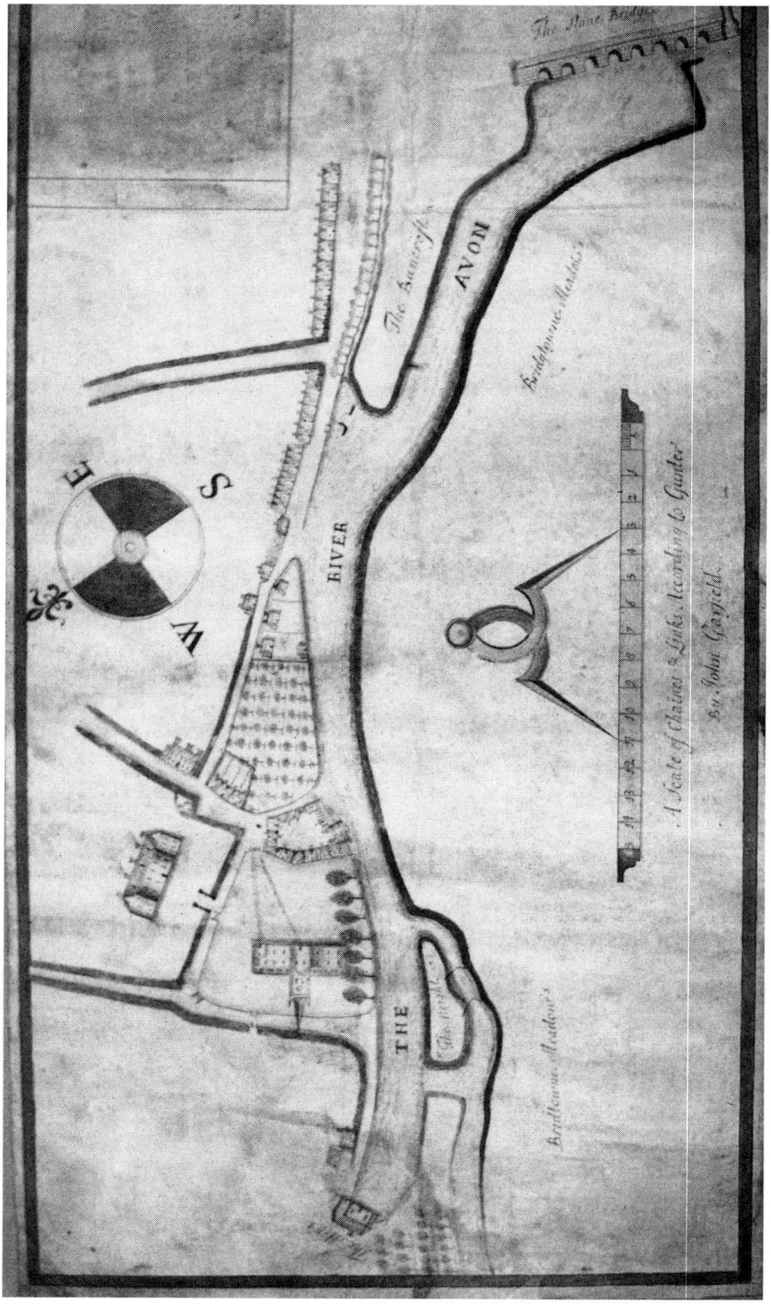

Fig.2 John Garfield, the river at Stratford-upon-Avon (Warwickshire), c. 1700 (Warwickshire County Record Office, CR 85, bundle 4)

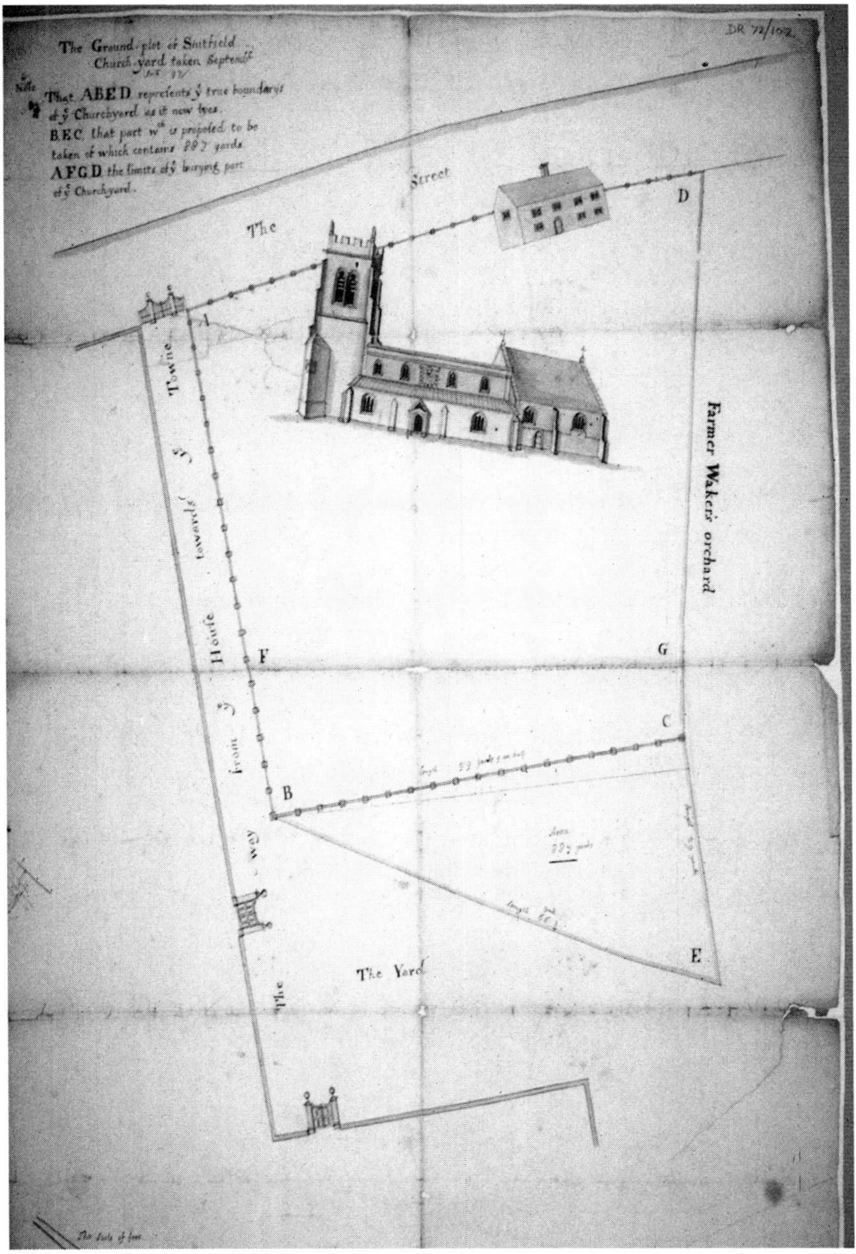

Fig.3 James Fish, plan of the proposed extension to the churchyard at Snitterfield (Warwickshire), 1697 (Warwickshire County Record Office, DR 72/102)

estate maps from early-seventeenth-century [36] Hampshire. But none of these achieved the local dominance that we find in the work of James Fish in Warwickshire over a long period, from the 1680s to the 1730s; some thirty of his maps survive from all parts of the county. In addition we have a few maps that he made in other counties, mostly adjoining Warwickshire but including Somerset (one map of Sutton Bingham, near Yeovil). This pattern was probably typical of the well-established professional surveyor at this time: most work would come from what might be quite a small number of landowners in his own area, but either because they owned some distant properties or because personal contact or reputation brought commissions elsewhere, some of his work would be done far from home. Whereas Fish worked all over Warwickshire – throughout his career he was based at Warwick – two of his contemporaries operated in smaller areas. John Garfield worked, as we have seen, in the south of the county, and Robert Hewitt in the north; each produced some eight or nine surviving maps.[27]

In the course of the seventeenth century, as making maps became more and more a normal, accepted part of a surveyor's work, there developed a general proficiency, an increasingly professional approach to producing them. This appears in various ways. One is in greater uniformity of style. The earliest estate maps differed a lot one from another. The overall impression of the maps of Ralph Agas, for example, with their careful finicky lines, little use of colour washes, and lettering in red or green, is quite different from the contemporary maps of John Norden, probably just as carefully measured and drawn but with bold overall colouring and with lettering in black ink. It would be wrong to suggest that later surveyors' maps all looked alike: there were many individual quirks of convention and style. But by the end of the seventeenth century the differences were less extreme than they had been a century earlier.

One interesting instance of the professionalism of surveyors who were producing maps regularly is the use of printing blocks for conventional signs and decorative features on the maps they drew. An early and rather crude example is a map of Thankful Frewen's estate at Northiam (Sussex), drawn by Giles Burton in 1635: the circle around the number given to each field, the scrolls containing their areas, the houses, even horses and cattle and a man carrying a staff, have all been stamped on the map from engraved wood-blocks.[28] We find the same technique on a map of Sir John Mill's manor of Langley in Eling (Hampshire), drawn by Nicholas Ayling in 1692, but here the printing is confined to the decorative elements (Fig. 4). The elaborate floral

---

27 WCRO, card index of surveyors.

28 BL Maps MT 6 b.1(30).

border is printed, using probably seven blocks, as are the ships on the water and the thirteen figures of deer in the New Forest; nine different blocks were used, showing deer walking, browsing and so on, which suggests that Ayling was accustomed to having landowners in the forest as his clients. The blocks are finely engraved, and (unlike Burton's map) it is not at once obvious that these features were printed. On the other hand, the houses and other non-decorative features on Ayling's map are all drawn by hand.

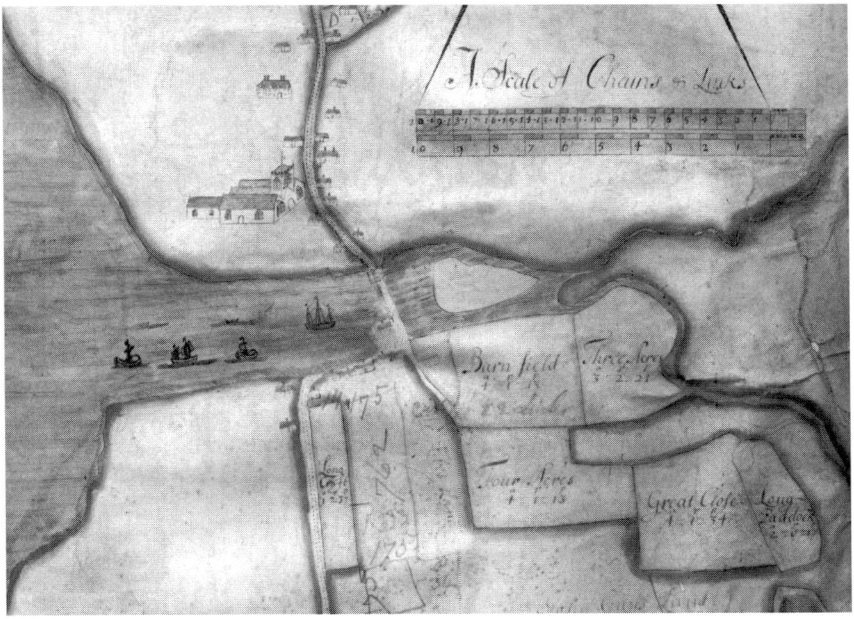

Fig. 4 Nicholas Ayling, detail from a map of the manor of Langley (Hampshire), 1692 (By permission of the Warden and Scholars of Winchester College, muniments, 21335)

Yet however skilled surveyors became as cartographers, however much drawing maps was seen as a normal part of their business, still in the early eighteenth century, just as 150 years earlier, the primary work of a surveyor was to measure and report on land. He might present the results of this work in a map or in a written report; how it was done, which method was chosen, was between him and his client. In fact, throughout this period there is great variety in the roles of the written survey and the estate map. All the information gathered [38] might be entered on the map, which would then entirely replace the written survey. This seems to have been the case in Ralph Agas's 1581 map of Toddington (Bedfordshire), drawn on twenty pieces of parchment stuck together to form a single sheet measuring 8.3 by 11.5 feet, which was then attached to a roller; at the edge of the map are lists and summaries that supplement the notes of area and tenure written on each plot of land, and there was probably no other product of the survey – all the information is contained on the map itself.[29] The same is true – to take another early example – of the anonymous map of Bolton Percy (West Riding of Yorkshire) about 1595. This is on thirty-eight sheets bound as a volume, but here detailed information about the areas and tenants of the lands is given in lists beside the maps, not on the map surface.[30]

Often, however, both a written survey and a map or maps were produced, and the relationship between them could differ greatly. The written survey might simply explain or elaborate on the information given on the map. We see this in two very different examples from Warwickshire. James Fish's magnificent map of the earl of Rochester's estate at Kenilworth in 1692, some 6 feet square, is accompanied by a forty-four-page survey book that summarizes and tabulates information already given on the map but also adds the rents due and copies of relevant medieval title deeds.[31] The small but handsome map of Dr. Thomas's lands at Atherstone-on-Stour that John Doharty drew in 1738 has on a separate piece of parchment what amounts to a key to the map: it lists the names, areas and use of the plots of land that on the map are simply given numbers.[32] On the other hand, the map might do no more than illustrate what was primarily a written survey. A written description of first the demesne and then the tenants' lands of the manor of Bisterne (Hampshire) in 1591 occupies twenty-seven leaves of a bound volume and is followed by a single opening on which are drawn nine plans, in varying states of completion, of particular parts of the estate.[33] From Ansty (Warwickshire) we have a survey written in 1599 and a map that may well have been drawn up at the same time by way of illustration but that gives none of the information about the areas or tenure of plots that is given in the survey.[34] Also from Ansty is a book of some ten leaves setting out the precise measurements and other details of the enclosure and redistribution

29 BL Add. MS 38065.

30 Borthwick Institute of Historical Research, York, PR.BP 15.

31 WCRO CR 143, CR 143A.

32 WCRO CR 1945, CR 936, box 3.

33 Hampshire Record Office, 23 M 54/1.

34 WCRO CR 285/3, 56. The survey is dated April 1598, forty-first year of Elizabeth I; but April 1598 was in the fortieth regnal year. The date of the court entered immediately above the survey is Thursday, 19 April in the forty-first year, certainly 1599, when 19 April fell on a Thursday.

of lands from common fields in 1602, the work of Richard Bankes, 'Practitioner [39] in the Matthamatticks'; no map accompanies it.[35] Of course, where a map and a written survey were drawn up together but – as so often – on separate sheets or books, there is always the risk that the two might become divorced over the years and one of them lost from sight or destroyed; maps often have a different archival history from the other records of an estate.[36] We might then wrongly suppose that the survivor of the pair, whether map or written survey, was the sole product of the surveyor's work.

However, written surveys without maps were being produced throughout the period. One undoubted example, made in 1721, is of the lands of Richard Mariett and his tenants in Preston-on-Stour (Warwickshire), a roll of six pieces of parchment, 11 feet long, describing in turn over three thousand strips of common-field arable or other land;[37] even in the early eighteenth century the estate surveyor would not be thinking of a map as the only way to present his work. On 4 May 1685 James Fish wrote from Warwick to Davenport Lucy, then staying in Covent Garden in London, about work he had been doing on Lucy's lands at Hampton Lucy and Charlecote (both Warwickshire):

> The last weeke I was preingag'd in business that I could not defer, without prejudice to the Gent': by whome I was imploy'd, and having no more time then this day to peruse any thing of Hampton Lucy or to cast up that part of Charlcote that I have allready survey'd I humbly beg yor: pardon for the imprefect Account I have here inserted viz:

	A	R	P
Charlcote Field is about	500:	00:	00
with 5 little Mead: as apart thereof containing about	015:	00:	00

> There's about 60 Acres of ye field next Newbold that is very course Ground, the rest by experienc'd and impartiall men is valu'd a^{t} as [40] good a rate as any common field land thereabouts and lies so accommodated with all Conveniences that makes it capable of good improvement by inclosing.

---

35 WCRO CR 285/57.

36 The map and survey book from Fish's survey of Kenilworth in 1692 (above, n. 31) are a good example. The map was deposited in the WCRO in 1950 by the earl of Clarendon, descendant of the earl of Rochester for whom the survey was made; the book came to the WCRO four years later from a different source.

37 WCRO CR 936, box 15.

Continuing, he mentions that he has made a 'collection' of the tenants and their tenures and will transcribe it if wished, also a rental of the estate and an inventory. He concludes: 'The Totall of y[r] whole Estate viz: Hatton Fullbrooke Hampton Lucy Charlcott & Hunscote is about 4500 Acres In my next I shold give a more particular account and insert those Remarques I have made from y[e] opinion of experiencd men tho they all owne themselves farr short of yo[r] great judgment in these affaires.' In all this there is not a word about drawing a map.[38]

On other occasions Fish did draw maps of parts of the Lucy family's property; they knew perfectly well the value and use of maps in estate management, but it was not only in order to draw maps that one employed a surveyor. There was other work for him to do as well. Some other landowners probably never saw any need for the trouble and expense of having maps made of their estates. After all, as John Norden made the tenant farmer say in his *Surveyors Dialogue* of 1607, 'is not the Field it selfe a goodly Map for the Lord to looke upon, better than a painted paper? And what is he the better to see it laid out in colours?' (15). Some landowners would have agreed with this point of view. Others did not. As we have seen, every estate map had two parents: the surveyor who gave it to the world and the landowner who was responsible for its conception. The landowner's role in the early development of the estate map in England was just as important as the role of the surveyor.

Development: The Role of the Landowner

The number of early estate maps differs greatly from one county to another. For Warwickshire in the Midlands a full census of manuscript maps has been made, and another is in progress for Hampshire in the south.[39] From Warwickshire there are forty estate maps earlier than 1701 in repositories accessible to the public, from Hampshire only sixteen; and from County Durham in the north of England there is only one.[40] These three counties are of different sizes, and

38 WCRO L 6/1032.

39 The information for Warwickshire, collected primarily by D.J. Pannett, is available in a card index in the WCRO; the project is described, with an interesting and informative account of the history of estate maps in the county, in D. Pannett, 'The Manuscript Maps of Warwickshire, 1597–1880', *Warwickshire History* 6, no.3 (1985): 69–85. G.A. Rushton is collecting the information for Hampshire for publication in the Hampshire Record Series, and I am most grateful to her for making it available to me.

40 The one County Durham map is of Layton by Robert Farrow in 1608. It is in the Baker Baker papers in the University of Durham, Department of Palaeography and Diplomatic, 72/249, enclosure attached to fol.12.

if we adjust the figures to give the number of maps per 1,000 square miles the result is Durham 1, Hampshire 11, Warwickshire 42. It would be wrong to attach importance to the precise figures. In some cases there is room for doubt whether a map is or is not an estate map by our definition. The total numbers are small, and the simple accident of a map's having a better chance of survival on one estate than on another may affect the figures. And, of course, reckoning only those in accessible repositories may significantly distort the picture. All the same, the figures for these three counties are so very different that they must point to real differences in the number of early estate maps that were produced.

Much more work will have to be done in individual counties before we can fully appreciate, let alone explain, these differences – careful, detailed work of the sort that Sarah Bendall has carried out on the maps from Cambridgeshire.[41] But some preliminary comments can be made on the evidence from Durham, Hampshire, and Warwickshire. First, it is unlikely that the lack of maps from Durham points to a contrast between north and south, that understanding and use of estate maps spread late to the north of England. On the contrary, in neighbouring Yorkshire, estate maps appeared early on the scene; I have already referred to the work of Christopher [41] Saxton and to the map of 1595 from Bolton Percy. What is significant, however, is that most of County Durham was owned by a very small number of large landowners, and the archives of only two of them are in publicly accessible repositories. Warwickshire had a larger number of substantial landed estates, but if we had no maps from three of them, the Warwick Castle estates and those of the Leigh and Newdegate families, our figure of surviving and accessible maps from the county would be very different. Even allowing for the accident of loss or survival of particular estate archives and maps, it seems clear that probably the majority of landowners did not have maps made of their properties. We see this in the early estate maps made for Oxford colleges. Some, notably All Souls and Corpus Christi, commissioned considerable series of maps; others seem to have had maps made only as occasion demanded or not at all.[42]

The two major County Durham landowners whose archives are publicly accessible contain not a single early estate map and it is interesting that both these landowners are ecclesiastical: the dean and chapter of Durham and the bishop. One of the earliest of all English estate maps, of Belchamp St. Paul

---

[41] A. Sarah Bendall, *Maps, Land and Society: A History, with a Carto-Bibliography of Cambridgeshire Estate Maps, c. 1600–1836* (Cambridge, 1992).

[42] P. Eden, 'Three Elizabethan Estate Surveyors: Peter Kempe, Thomas Clerke, and Thomas Langdon', in Tyacke, *English Map-Making*, 68–84; C.M. Woolgar, 'Some Draft Estate Maps of the Early Seventeenth Century', *Cartographic Journal* 22 (1985): 136–43.

(Essex), was made for the dean and chapter of St. Paul's Cathedral in 1576[43] (article XV, plate III). All the same, a first impression suggests that the church was less inclined than other landowners to commission estate maps. Glebe terriers – detailed accounts of the lands held for the incumbent's support – which the parish clergy were required from time to time to submit to the archdeacon or bishop, practically always took the form of written surveys. Of some 120 Warwickshire glebe terriers dating from 1585 to 1675 not one is even accompanied by a map;[44] from County Durham one, for Stanhope in 1663, was submitted as a map, without a written description, and from Hampshire too there is just one map among the glebe terriers before 1750, a sketch plan of 1718 showing glebe land of a church within the city of Winchester.[45] The force of tradition and the difficulty of mapping small properties made up of scattered plots and strips must have contributed to this; but behind it may lie a particular reluctance on the part of church administrators to use maps in estate management. The point needs investigating.

So too does the relationship between estate map, written survey, and manorial court. From the sixteenth century onward we have records of manorial courts meeting in what was called a court of survey. This was held to compile a survey that listed the manorial tenants, what they held and by what tenure, and what they owed in annual or other payments. If we are to believe the treatises on holding manorial courts, which were still being published in the nineteenth century, neither mapping nor mensuration played any part in this operation: the information was gathered from the evidence of the tenants themselves and from deeds and other documents.[46] However, John Norden in 1607 envisages the court of survey as being summoned and held by the professional surveyor whom the landowner has brought in to survey the estate; in it he interrogates the tenants and examines the documents they produce – but then he goes on to measure and map the manor as part of the same process.[47] Some early estate maps and measured surveys were certainly drawn up as business of the manorial court. The survey already mentioned of Bisterne in 1591, with maps

---

43 Guildhall Library, London, MS 25517/1 (one of the maps showing part of the estate is illustrated in article XV, plate III)

44 D.M. Barratt, ed., *Ecclesiastical Terriers of Warwickshire Parishes* (Dugdale Society, 1955–71), 1: ix–xii.

45 University of Durham, Department of Palaeography and Diplomatic, DR terriers, Stanhope 1663; Hampshire Record Office, 21 M 65/E15/132. I am grateful to G.A. Rushton for information about the maps among the Hampshire glebe terriers.

46 E.g. G. Jacob, *The Complete Court-Keeper; or, Land-Steward's Assistant*, 8th ed. (London, 1819), 143–51. Cf. P.D.A. Harvey, *Manorial Records* (London, 1984), 61.

47 Norden, *Surveyors Dialogue*, 83–144.

attached, was drawn up in a court of survey; so was the survey of Ansty in 1599, which may or may not have had an accompanying map; so too was Ralph Agas's map of Toddington in 1581, which may well have stood on its own without a [42] written survey. However, most English estate maps have no apparent connection with a manorial court. Throughout the seventeenth and eighteenth centuries manorial courts themselves were in decline: their effective business diminished, proceedings became formalized, and many courts simply stopped being held. But even if the landowner held a court for the lands that he wished to have surveyed or mapped, it may well have seemed an unnecessary complication to involve it in the business. How far estate mapping and manorial courts went hand in hand, whether it was the decline of manorial courts that separated them, and whether they had ever really been closely connected are questions that future research may well answer.

In any case the owner of a mapped estate need not have been the lord of a manor. We have maps of estates of every sort from those of great lords down to those of quite humble tenants. On the one hand we have John Norden's survey in 1607 of the King's Honour of Windsor (Berkshire), an area measuring some 16 miles by 20: a parchment book, handsomely bound and with the royal arms at the front, containing twenty coloured maps with notes that include a table of the number of deer on the estate[48] (article XVIII, plate II). At the other extreme is a little book of six paper leaves containing plans by John Tipper of the 20 acres owned by Robert Smith of Coundon (Warwickshire) in 1694; on the first page is a pleasing pen-and-wash picture of his modest timber-framed house with attached stable (Fig. 5). Whether the estate was large or small, we still do not know where the idea for mapping it came from, whether it was the landowner who sought out [43] the surveyor for the purpose or the surveyor who took the initiative in selling his services to a wide range of customers.

For the landowner there were clear practical advantages in having a map of his properties. Some were described by Ralph Agas in *A Preparative to Platting of Landes and Tenements for Surveigh*, a twenty-page pamphlet published in 1596. One was that 'the bounders and abbuttals of every particular is so wholy put downe, as no booke may bee comparable with the same'; another that '[i]f you will sever any fielde or closase into two or more parcels: the Scale will readily bewray how many perches, & feet shall performe the same, and where may be the rediest cut'; and another that it eliminated the risk of losing lands because change of name or of use meant that they could not be identified in old written records (14–15). Certainly many landowners saw these and other advantages and made use of maps in the practical work of estate management.

[48] BL Harley MS 3749; one map is illustrated in article XVIII, plate II.

Fig. 5 John Tipper, page from a book of plans showing land owned by Robert Smith of Coundon (Warwickshire), 1694 (Warwickshire County Record Office, CR 299/574)

Annotations on an anonymous map of lands belonging to Lord Spencer at Radbourne (Warwickshire) in 1634 show clearly how estate maps might be used over a long period.[49] In 1736 it was revised by Henry Beighton, who added a new scale and some new field names and areas. One or more other hands have pencilled subdivisions on some of the closes, adding numbers or 'Not my Lords' or other notes, including some on land use, 'Plow', 'Probably plow', 'Mow'; some more field names have been altered, and the line of a canal entered across the property. Finally, two hands of the late eighteenth century have added more notes on the closes and, at the bottom, a table listing the closes, their areas, and their tenants. All these annotations show that the map was in administrative use for at least 150 years. Estate maps were indeed made to last. Most were drawn on parchment, and on the written survey accompanying John Doharty's map of Dr. Thomas's estate at Atherstone-on-Stour in 1738 the name of the farmer is on a detachable slip that could be replaced without disfiguring the document; map and survey alike were intended to outlive the term of a single lease.[50]

But to the landowner the estate map offered more than simple practical utility. Consider three late-seventeenth-century maps of the whole or [44] parts of Wedgnock Park (Warwickshire), which belonged to Lord Brooke. One, by James Fish in 1696, is a simple outline of three plots of land with their names and areas, uncoloured and drawn on paper.[51] It is wholly functional and may indeed have been drawn for some immediate ad hoc purpose, not for general reference. Another, again by James Fish in 1696, is a more elaborate map of a mostly wooded area, a good deal more detailed, drawn on parchment and coloured with grey wash – but still a functional and unpretentious map; there is no decoration, and the title and key are enclosed simply in borders of ruled lines.[52] The third, by John Wilson in 1682, is an altogether more splendid production. Drawn on two sheets of parchment glued together, it uses colour for many features. In one corner is an elaborate cartouche for the title: a marble tablet with a nymph on either side and Lord Brooke's coat of arms with coronet and with two swans as supporters (Fig. 6). Cartographically, none of these three maps is particularly complicated: mostly they show simple outlines of fields and [46] woods. The difference among them is in the presentation: the greater or lesser formality and ornamentation of the surrounds. There is no difficulty in finding other estate maps of every level of elaboration.

49 WCRO CR 732.

50 WCRO CR 936, box 3.

51 WCRO CR 1886/M568.

52 WCRO CR 1886/M2.

Fig. 6 John Wilson, detail from a map of Wedgnock Park (Warwickshire), 1682 (Warwickshire County Record Office, CR 1886/M3)

Clearly John Wilson's map of 1682 was meant for display – like many other estate maps. In the novel *Il gattopardo* (chap. 1), Giuseppe Tomasi di Lampedusa's mid-nineteenth-century Sicilian Principe di Salina has a room decorated with mural pictures of the family estates, a graphic symbol and reminder of his power and wealth. In their more modest way many English estate maps served exactly this purpose. Nor need this be seen as a less practical, less functional purpose than the use of the maps in the detailed work of running the estate. They were meant to be looked at and to impress, to excite the beholder's admiration and to satisfy the owner's pride of possession without even the trouble of travelling to the spot. They were a statement of ownership, and of the status that came with ownership of land – and this was as true of Robert Smith with his 20 acres at Coundon as of Lord Brooke or the ruler of the kingdom.

This then was the value of English estate maps to those landowners who commissioned and paid for them, those who used and enjoyed them. They are of no less value to historians, to those of us who preserve them and who still use and enjoy them today.

The Historians and the Fields

Topographical history, the history of individual places, has been one of the great strengths of English historical scholarship from the time of William Lambarde, whose *Perambulation of Kent* was published in 1576, down to our own day, which has just seen the appearance of the 200th volume of the monumental *Victoria History of the Counties of England.* Central to this strength is the history of landed properties: how estates were built up and divided, how they passed from owner to owner across the generations, their size, their composition, their value. Legal requirements, family piety, antiquarian curiosity, and historical inquiry have over the centuries all contributed to the amassing of a [47] vast amount of information about estates, incorporated in county histories and many other published works. However, it is only in the present century that this great tradition of research, with all its emphasis on the exact definition of rights, tenures, and amounts of land, has turned to precisely where properties lay, and thus to the sort of information that estate maps are peculiarly suited to provide. This is strange, for some of the tradition's early exponents were well aware of the value of maps and were even interested in the techniques of drawing them; a notable example is William Dugdale, whose maps of the hundreds were published in his *Antiquities of Warwickshire* in 1656.[53] With few exceptions it is only recently that topographical historians have begun to use estate maps systematically for the details of local landownership and local history.

This means that it was historians working on more general themes who led the way in using estate maps as a source of evidence, and the first to do so were the historians of field systems. The pioneer was Frederic Seebohm. In his book *The English Village Community*, published in 1883, he described the three-field system of cultivation that had operated before enclosure: the individual held his arable in strips, scattered among those of his neighbours, in open fields that were cropped following an agreed uniform rotation and were used as common pasture when they were not sown. He saw these arrangements as reflecting the economic and social pattern of the medieval and earlier periods. In the book's frontispiece is a map that is the direct ancestor of innumerable textbook and other illustrations of the three-field system: a map of Hitchin (Hertfordshire)

[53] P.D.A. Harvey and H. Thorpe, *The Printed Maps of Warwickshire, 1576–1900* (Warwick, 1959), 11–15.

in about 1816, before enclosure, showing the large fields that were the cropping units and the innumerable arable strips of the individual landholders. He took his map from one drawn to accompany a tithe apportionment, but if he had chosen another locality for his example, it could as easily have been from an estate map. Beside this map is another, also of Hitchin, showing the distribution of the strips and plots of land of a single landholder about 1750. The 'Hand Map' (i.e. manuscript map) from which this was taken was probably an estate map.

Seebohm's interest in the way fields were organized before enclosure was followed up by many later historians. It was Seebohm who first discussed the historical implications of arable strip holdings in a three-field system. But the first historian to describe the variety of field systems in medieval England was H.L. Gray in his book *English Field Systems*, published in 1915. Maps accompanying tithe apportionments and estate maps were crucial to his work: he was in fact investigating the clear differences in field systems that early estate maps show. There was not only the three-field (or two-field) system that Seebohm described, the Midland system. There was the East Anglian system of holdings of strips and plots, concentrated in particular areas of open fields that were divided into so-called foldcourses for grazing sheep. There were the ancient enclosed fields of Kentish agriculture. There were the various combinations of closes with common fields held in strips, which were found in the other counties of southeast England. There was the system of heavily cropped infields and lightly cultivated rotating outfields, found in the areas of marginal cultivation beside upland moors. The outward differences between all these ways of organizing arable holdings leap to the eye on estate maps, and Gray made full use of them in his work. Presumably for clarity, he reproduced none in full, but used instead drawings taken from them; among them was Ralph Agas's 1575 map of West Lexham (Norfolk), already mentioned among the earliest English estate maps that have so far come to light.[54]

[48] Gray, like Seebohm, extended his investigations to the period after the Middle Ages, and many of the estate maps that Gray called in evidence were very much later than 1575. All the same, they provide invaluable information about what happened in the Middle Ages. It is not that nothing changed, that we can assume that in any one place the layout of fields was the same in the twelfth century as in the seventeenth. We cannot. The point is rather that the map is so clear and precise that it leaves open no question of the field layout at the time it was made; it thus acts as a key to a whole mass of contemporary and earlier written surveys and other documents from the same place and similar places. Without the map these records would be hard to explain and open to

---

54 H.L. Gray, *English Field Systems* (Cambridge, Mass., 1915), 317.

varying interpretations. The map explains them, and with their help we can then carry back the picture the map gives into much earlier periods when estate maps were unknown. The earlier written records may show us that we have to modify the picture given by the later map; but at least we have a starting point firmly based in the local terrain for our understanding of the medieval fields of the place in question – and, by cautious analogy, of other places in the same area for which early estate maps are lacking. The map primes the pump, enabling us to use other evidence that would otherwise be obscure.

This use of estate maps as a key to written surveys was pioneered by Gray but has since been used by many historians in work on particular places and on English field systems in general. A notable example is the seminal work of C.S. and C.S. Orwin. The first edition of their book, *The Open Fields*, was published in 1938; in it they discuss the actual methods of agriculture – especially the techniques of ploughing – that everywhere lay behind the strip cultivation of open fields. They then look at what happened in practice in one particular place, Laxton (Nottinghamshire), and use as a constant point of reference a magnificent estate map with accompanying written survey, revised by Mark Pierse in 1635 from a map and survey produced by Francis and William Mason ten years earlier. The map is a unusually fine one, very detailed and embellished with scenes of work in the fields, hawking, and hunting; the Orwins reproduced it (redrawn from the original) in full. With its help they explored other records, earlier and later, of the fields of Laxton, a place particularly suited for this study since a form of openfield farming has, uniquely, continued there down to the present day.

In all this the estate map is doing for the historian exactly what Ralph Agas in 1596 said it would do for the landowner: it records 'bounders and abuttals' better than any written document can do, and it enables us to trace lands through changes of use or changes of name that have made the older written records difficult to understand. Moreover, Agas points out, once the information gathered by the surveyor has been presented in the form of a map, the map can be used to draw up a written survey: 'upon the perfecting of any such surveigh, you may make a faire parchment booke with a large margent, you may enter, & ingrosse the same from the saide Plat'.[55] This is perfectly true. But what Agas does not point out, though it is implicit in what he says, is that the opposite is not the case – and it is this that makes the map so valuable to the historian. In practice it is not generally possible to construct an accurate map from the information given in even the most detailed written survey. At first sight one might suppose it to be a fairly simple task. A survey like that

---

55 Agas, *Preparative to Platting*, 14–15.

of 1575 from Neithrop (Oxfordshire), which describes some two thousand strips in the fields, arranges them in the furlongs (or [49] flatts, shots, or wongs: groups of parallel strips) in which they lay and, for every furlong, names the furlong, boundary, or other feature that adjoined it on each side.[56] A jigsaw puzzle to be sure, but one capable of patient solution. As F.W. Maitland put it in 1898, in writing of a mid-fourteenth-century survey that listed about a thousand pieces of land in Cambridge West Fields, 'In each field it describes the various furlongs or shots in such a manner that an ingenious man, who had time to spare and a taste for the Chinese puzzle, might depict them on a map.'[57] The passage conjures up a splendid picture of the Victorian gentleman, idly beguiling the time by playing with a medieval survey in this way. In fact he would have been lucky if he had succeeded, as anyone who has ever tried it will know. The survey gives bounds and areas but it does not give angles or shapes, and few furlongs were rectangular. Fitting them with confidence into even the known facts of the local landscape can be all but impossible. Professor D.G. Kendall, who developed a computer program to achieve this construction of a field map from a written survey, suggests that if, as may happen, the survey gives on average only three of the four abuttals for each furlong, then the 'margin of the possible' has been reached.[58] Where agrarian historians have a map, none of these problems arise: they start with the layout of the fields at one particular time laid out firmly before them and can interpret the written surveys and other records in its light.

The detailed layout of the fields is effectively information that only a map can give, and how they were worked is often clearer from a map than from any written source. There are many details of tenure and land use that surveys and other written records will provide but that are much more easily seen on an estate map. Varying measures are an example. A map by George Withers in 1623 of a property in Weston Patrick (Hampshire) gives the area of each parcel of the estate.[59] For agricultural land and woodland (marked with trees) only one amount is entered, but for copses (marked with alternate trees and shrubs) there are two, 'lande measure' and 'Wood measure'; on one plot the former is just over 50.9 acres, the latter just over 42.5. The woods, as so often, were measured with a longer perch than the other lands – 18 feet instead of the standard 16.5 feet. The copses, semi-woodland, were measured with both.

---

56 Bodleian Library, Oxford, MS Top. Oxon. c 454.

57 F.W. Maitland, *Township and Borough* (Cambridge, 1898), 56.

58 D.G. Kendall, 'The Recovery of Structure from Fragmentary Information', *Philosophical Transactions of the Royal Society of London*, ser. A, 279 (1975): 559–75.

59 Hampshire Record Office, 11M 49/509.

The map displays graphically, in a way the written record cannot, the contrast between the two measures.

The same map by George Withers outlines the various plots of land in colours that are explained in a key: red (which over the centuries has changed to purple) is freehold land, green leasehold, and yellow copyhold. This again is information that a written survey could provide just as fully and accurately. But on the map we at once see the relative sizes and positions of the areas in the three different sorts of tenure, and therewith any implications for the layout of the estate, for its history, and for the history of the locality. Relationships leap to the eye, whereas painfully plotting on a map the information as given in a survey might be thought unlikely to be worth the trouble. This is often the case when a map distinguishes different parts of an estate. A map of Lord Brooke's manor of Cestersover, near Monks Kirby (Warwickshire), by James Fish in 1691 outlines the fields in colour to show which lands belonged to each of four tenants; here again, the significance of the division of the four farms and their sizes and locations will at once be apparent to the historian working on the area.[60]

Sometimes, though, the colouring on an estate map raises more questions than it answers. William Folkingham, in 1610, described what colours to use for different sorts of land on a map and how to [50] make them. Thus 'Arable for Corne may be dashed with a pale Straw-colour compounded of Yellow Oker, and White leade, or of Pincke and Verdigreece'; meadows were to be light green, pasture a deeper green, heaths and fens 'deader Greene', trees 'a sadder Greene', and so on.[61] It would be interesting to discover how closely contemporary maps conformed to this. There seem to have been few generally accepted conventions, and we may be at a loss to understand the colours on a map. Another Warwickshire map of 1691, of a farm in Butlers Marston by an anonymous surveyor, uses green, brown, yellow, and purple wash to colour the fields; there is no key, and we can only guess what the colours show, perhaps differing land use.[62] Here the written record may explain the map, but it is far more often the other way around. To the historian of English fields and their exploitation, the estate map is a crucial source of evidence.

---

60 WCRO CR 1886/M12.

61 William Folkingham, *Feudigraphia: The Synopsis or Epitome of Surveying Methodized* (London, 1610), 57.

62 WCRO CR 2103/1.

The Historians and the Landscape

The estate map's importance has been underlined by a more recent approach to agrarian and rural history, that of historians who use the landscape of today as evidence of the past and who at the same time uncover the processes by which that landscape has come into being. They follow the same technique as the historians of the fields who use early estate maps to throw light on other documents – but in their case these documents include the modern landscape itself. An early investigation of this sort concerned the ridge-and-furrow grasslands that are still found in many parts of England: former arable, now grassed over but retaining the shape of the broad ridges created for surface drainage when the land was last under the plough. These can be used to identify the detailed layout of the medieval fields, but with certainty only where other evidence confirms that this really was the ancient [51] pattern – and the best evidence of all is an early map that shows the strips. Land normally under grass may have been ploughed up for a short time in the Napoleonic Wars or at some other late date, so that the pattern of the ridges we see may be no older than the early nineteenth century.[63]

A pioneer in this work was M.W. Beresford. Especially important, not least for their use of estate maps, are his *History on the Ground* (1957) and, in collaboration with J.K.S. St. Joseph, *Medieval England: An Aerial Survey* (1958). In the first of these works Beresford reproduced the whole or parts of eight sixteenth-century estate maps, not redrawn as in the books of Gray and the Orwins, but direct from photographs. This was not, of course, an innovation. Full reproductions of estate maps had been made long before, to illustrate works on particular places[64] or in collections for their antiquarian or wider interest; J.L.G. Mowat's edition of *Sixteen Old Maps of Properties in Oxfordshire in the Possession of the Colleges of Oxford, Illustrating the Open Field System* was printed in 1888, in the wake of Seebohm's work. But what Beresford did for the first time was to combine the evidence of these maps with the evidence of other documents and of the landscape to show how places had changed over the centuries and how relics of their history can be found in the landscapes of today. The history of boundaries, of fields and villages, and of deserted sites, the creation of market towns and of gentlemen's parks, were all viewed with the help of these early estate maps. In all this Beresford was using the maps as a way into past landscapes, a way of seeing what places really looked like when the maps were made.

---

63 M.W. Beresford, 'Ridge and Furrow and the Open Fields', *Economic History Review*, 2d ser., 1 (1948–49): 34–45; E. Kerridge, 'Ridge and Furrow and Agrarian History', *Economic History Review*, 2d ser., 4 (1951–52): 14–36.

64 E.g., the map of Dewsbury (West Riding of Yorkshire) by Christopher Saxton, 1600, in S.J. Chadwick, *Dewsbury Moot Hall: Account Rolls and Court Roll* (Leeds, 1911), facing p. 1.

This, of course, they do supremely well and they can thus be used, in a way we can use no written source, to trace change in landscape. We need look no further for examples than maps showing field systems before enclosure; but there are many others. Windmills are a feature of the rural landscape and economy that may easily be known only through their appearance on estate maps: they were apt to be short-lived, and the sites they occupied might be chosen from a great variety of suitable places on any one property. Thus one at Wykham, outside Banbury (Oxfordshire), appears as a landmark on one of the road strip maps in John Ogilby's *Britannia* in 1675 and again on an estate map of 1688, which is our only evidence for its precise location; as it does not appear on an estate map of 1746, we can be fairly sure that it had vanished by then. Without these maps all knowledge of the mill's existence would have been lost, apart from records of Windmill Quarter and Windmill Grounds as local field names.[65] Occasionally a map shows a settlement that has disappeared, even marking the site of a village or hamlet already abandoned when the map was made. A classic instance discovered by Beresford is on Thomas Clerke's map of Whatborough (Leicestershire) drawn for All Souls College in 1586, which names an area in the centre of the map 'The place where ye town of whateborough stoode'; the last houses of the village had been abandoned a hundred years earlier.[66] More often we are shown houses that were standing when the surveyor made the map but that have since been deserted. A map of Coughton (Warwickshire) in 1695 shows four houses, with their plots of land behind, in an area that was later taken into the grounds around the neighbouring Coughton Court; on the ground, only the raised areas where the houses stood are now visible. Another now vanished feature on the same map is the moat around Coughton Court itself.[67]

Sometimes maps provide evidence of the actual process of change in the landscape. Beresford reproduces and discusses two maps of Holdenby [52] (Northamptonshire), which show how, between 1580 and 1587, the grounds around the mansion were developed and some 600 acres of arable, meadows, and woods were converted into a park.[68] Both are in the volume of surveys and maps of Sir Christopher Hatton's estates drawn up by Ralph Treswell, and were presumably drawn first to plan and then to record the changes – even as a deliberate 'before and after' comparison. Two other maps in the same volume record change even more precisely (Figs 7 and 8). They show lands at Kirby, near Corby (Northamptonshire), in 1585 and 1586; between the two there had

65 *The Victoria History of the Counties of England: Oxfordshire* (London, 1972), 10: 71.

66 M.W. Beresford, *History on the Ground* (London, 1957), 95, 116–23, plate 11.

67 WCRO CR 1998/M7A,B. I am grateful to M.W. Farr for this information.

68 Beresford, *History on the Ground*, 211–14, plates 19, 20.

XVII

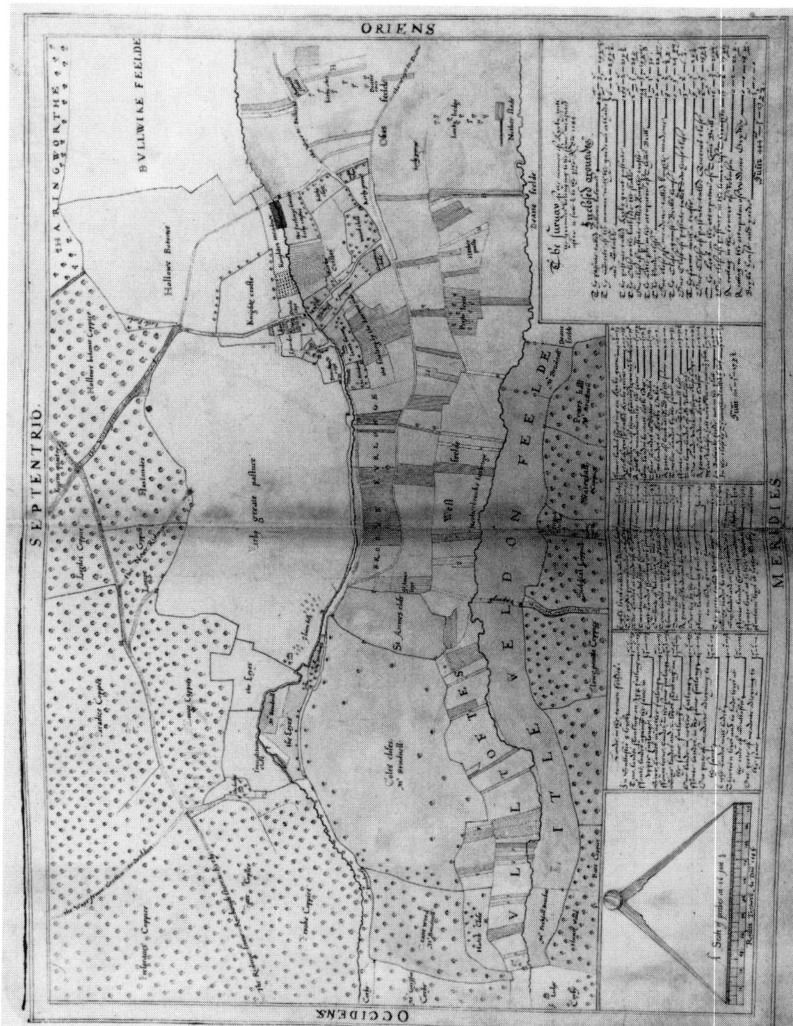

Fig. 7 Ralph Treswell, lands at Kirby (Northamptonshire), 1585 (Northamptonshire Record Office, Finch Hatton MS 272, fol. 5)

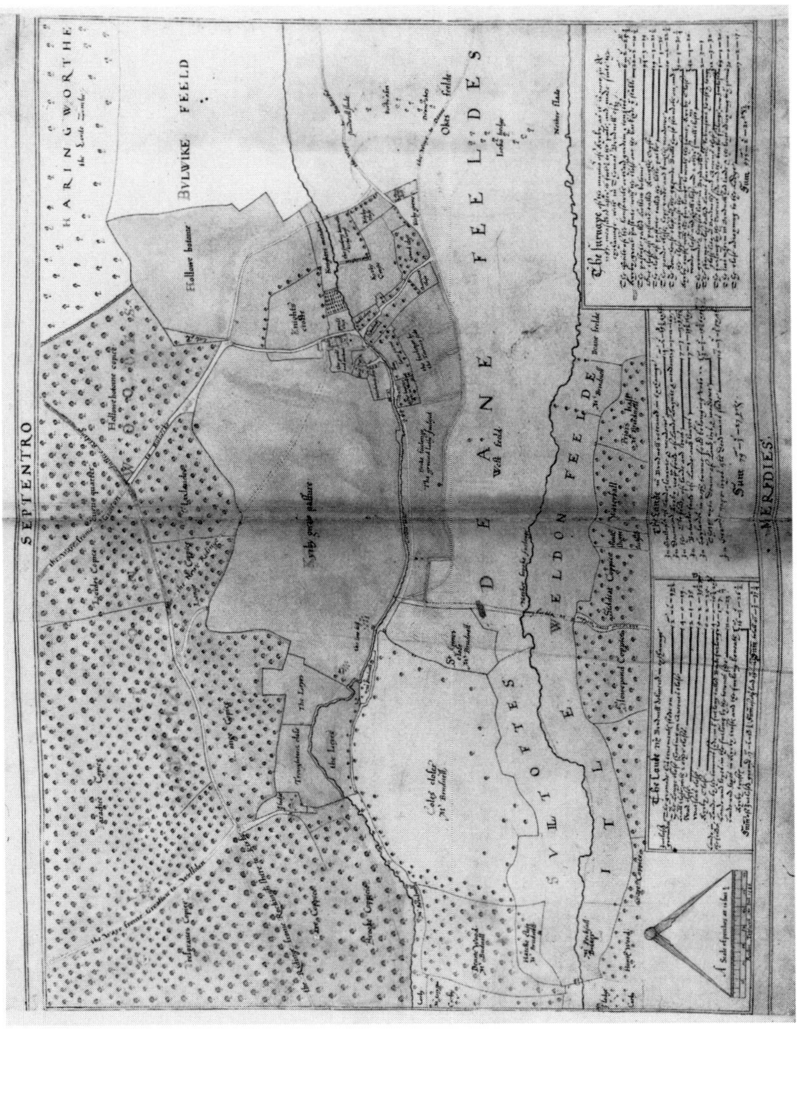

Fig. 8 Ralph Treswell, lands at Kirby (Northamptonshire), 1586 (Northamptonshire Record Office, Finch Hatton MS 272, fol. 6)

been enclosure and an exchange of open-field arable, recorded on the maps by the disappearance of many strip holdings and by details of what had happened entered in a panel on the second map. This apparently intentional record of change is most unusual – it was only the richest landowner who could afford to be so prodigal with his surveyor's time and skills. But successive maps from a single estate quite often demonstrate the sequence, even the process, of change, and even a single map need not be a static document, a record of a single point in time. We have already seen how the map of Radbourne (Warwickshire) in 1634 has annotations in a series of different hands, showing that it was being used in managing the property down to the late eighteenth century. These notes and alterations themselves record detailed change: the subdivisions on some of the fields, notes on the way the land was to be used, and some new field names, such as Barn Ground on Lower London Field – we remember Agas's comment that a map, unlike the written survey, avoids [53] problems caused by changes of name. Correctly interpreted, this one map can tell us much of the history of the property over a century and a half.

Besides providing a record of change the map places the past landscape before us with an immediacy and comprehensiveness equalled by no other source. No other mapmaker left quite so little to the imagination as Thomas Hornor, who in the early nineteenth century produced estate maps on a system that he called 'panoramic chorometry'. On the map was superimposed a near-vertical view of every feature, yet the map as a whole is not the bird's-eye view that it seems at first sight, for the basis was a true ground-plan with consistent scale, and each feature is a separate perspective picture, drawn to a separate horizon. The result is a kind of isometric aerial view, detailed and even beautiful; but it was clearly slow and costly to produce. The same technique was used by some artists in the sixteenth and seventeenth centuries for engraved views of cities, but Hornor's attempt to apply it – in full colour – to estate maps seems to have been unique.[69] All the same, some of the finer eighteenth-century surveyors approached fairly close to Hornor's method in the amount of detail on their maps and in showing it in near-vertical actual pictures instead of conventional symbols.[70]

Given the ability to read a map and a modicum of imagination in interpreting it, we can gain almost as vivid a picture of past landscape from many estate maps. For examples let us look at two of the most remarkable of the English estate maps of the sixteenth and seventeenth centuries.

---

[69] R. Hyde, 'Thomas Hornor: Pictural Land Surveyor', *Imago Mundi* 29 (1977): 23–34.

[70] E.g., the engraved map by John Rocque of the gardens of Richmond Old Park, 1734 (copy, BL Maps K.41.16f).

The map of Lord Cheney's manor of Toddington [54] (Bedfordshire) was drawn by Ralph Agas in 1581.[71] It is on twenty pieces of parchment which have now been separated but which together form a map that measures 11.3 by 8.5 feet. Starting at the village, we see that the church and all the houses are drawn not conventionally but from life, in some detail, as though in oblique bird's-eye view. We should not rely on the map's evidence for all minutiae, such as the position of every door and window; but we need not doubt the accuracy of the overall picture, for not only does Agas's work show every sign of meticulous care, but it was done for those who would have known intimately every nook and corner of the place. We see the market cross and the maypole and get an impression of the sizes, shapes, and general appearance of the houses and of the distribution of trees (article XV, plate I). From what other source could we get so detailed a picture of an Elizabethan village street? Moving beyond the village area, the map gives us a superb view of the agrarian landscape. Arable strips and closes, meadows, woods with deer in them, trees and hedges along the boundaries are all shown. There is of course a conventional element – Agas cannot be expected to show us the exact position of every tree, and he is only telling us that this is where trees were to be found. But his representation is wholly pictorial, so that one can easily imagine oneself walking along the paths, seeing what one would see on the ground; in some ways, indeed, one sees rather more on the map, for it identifies the area, the tenant, and the form of tenure of every plot of land. Moving to another part of the map we see Lord Cheney's mansion in bird's-eye view, clear enough to show its detailed architecture; it was a three-storied house, with gatehouse and turrets, built around a central courtyard with another large courtyard behind. A broad avenue of trees leads up to the yard in front of the house, and beside the house are formal gardens and an orchard. [55] Beyond are areas of water, the Black Pond and the Mill Pond, and a loaded two-wheeled cart with a team of horses crossing a field – to the cartographer a mere embellishment, but to the historian a further source of evidence. The whole map provides a mass of information on – quite apart from anything else – what Toddington looked like four hundred years ago.

We see this again in the map that James Fish produced in 1692 of the earl of Rochester's estate at Kenilworth (Warwickshire).[72] Drawn on six pieces of parchment glued together, it is much smaller than the Toddington map, but measuring some 6 feet square it is still an ambitious and impressive work. In the lower left corner is Kenilworth itself, a small town rather larger than Toddington and spread out spaciously over quite a wide area. Beside it is a

[71] BL Add. MS 38065.

[72] WCRO CR 143A; the accompanying survey book is CR 143.

panel giving a key to the names of its streets: New Way, Pepper Alley, Rosemary Hill, and so on. The scale is smaller than on the Toddington map, but we still see a great deal of detail: individual houses with their red roofs, what are probably inns with signs outside, the church with its steeple and a weathercock on top. The smaller houses are drawn conventionally, but the larger ones may well be from life. At one end of the town is the castle. On the map itself it appears within its outline site drawn in a weird semi-perspective; but beside the map are two more conventional pen-and-wash views of the castle, one from each side, with a horse and cattle in the foreground. This marks a new fashion, common in the eighteenth century, of embellishing estate maps – or at least those intended for display – with separate views of parts of the property. Those of Kenilworth Castle are especially interesting because some forty years earlier, after the Civil War, it had been slighted – made unusable for military purposes – and the great pool or mere beside it drained. The map marks the outline of where the pool had been, and the views of the castle show how little visible damage the slighting had produced – the castle was in fact still used as a residence (Fig. 9). They also show its vast size; during the slighting and draining Colonel Hawkesworth, in charge of operations, had made a home for himself in the gatehouse alone.[73]

Beyond the castle we move into the countryside, and the map again gives us an extraordinarily detailed picture of what was to be found. Roads [56] with trees and hedges on either side, streams crossed by bridges or fords, ponds, gates into the fields – all of them individually owned closes, for unlike Toddington a century earlier there were no common fields with strip holdings in late-seventeenth-century Kenilworth. One has to look closely at the smallest details of the map to realize how much it tells, how from every point one can envisage what one would actually have seen on the ground. In the farther parts of the estate were woodlands, marked with trees that are no more than conventionally placed but that reinforce the overall pictorial quality of the map. Here and there in the landscape are isolated farms, with their gardens and orchards beside them, and around them are the lands belonging to them – as we could not tell on the ground but can see easily enough on the map, for every tenant farmer's lands are outlined in a different colour. Nor could we see on the ground what is very clear on the map, the area and name of every field: Coneyberry Hill, Teutawing Close, Boggy Meadow. In fact the survey book accompanying the map tells us practically nothing the map does not except the rent paid to the earl by each farmer.

---

[73] P.K.B. Reynolds, *Kenilworth Castle, Warwickshire* (London, 1948), 3–4; *The Victoria History of the Counties of England: Warwickshire* (London, 1951) 6: 137–8.

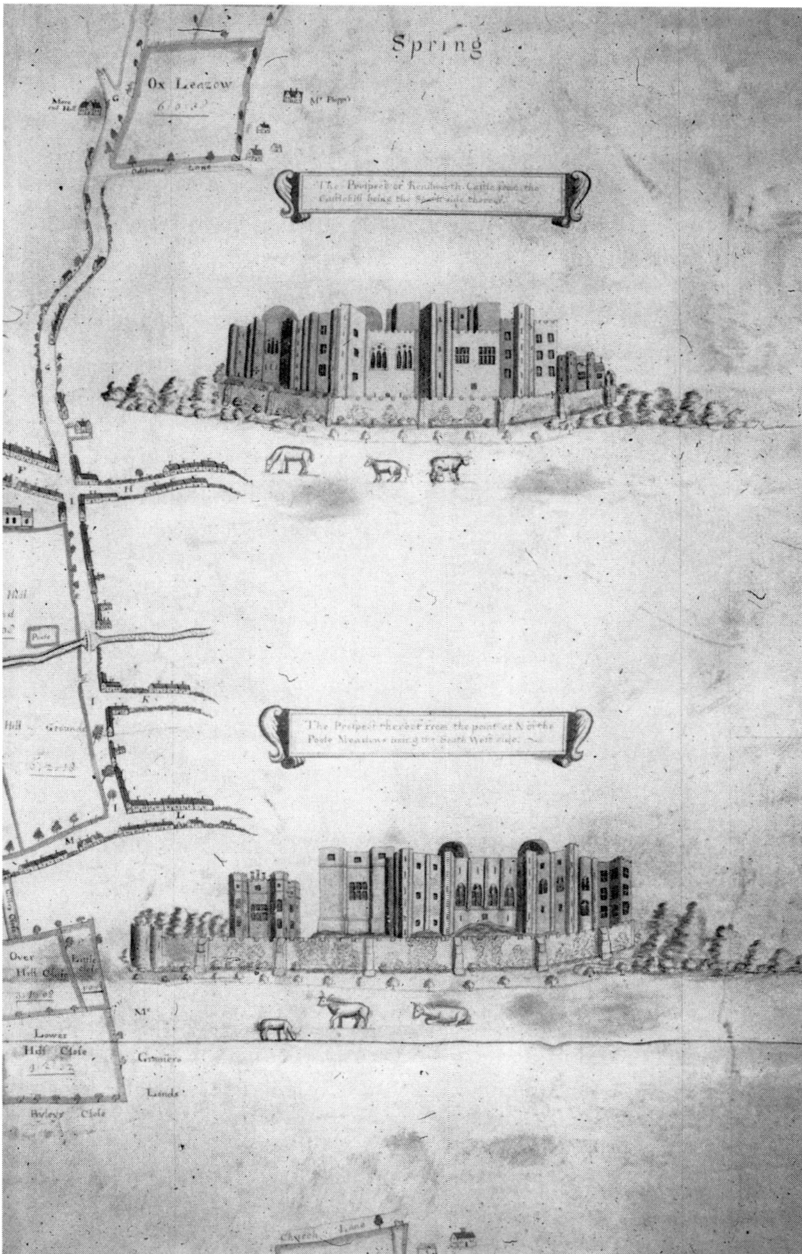

Fig. 9 James Fish, detail from a map of Kenilworth (Warwickshire), 1692 (Warwickshire County Record Office, CR 143A)

One could write a whole book about what is shown on each of these two maps, of Toddington and Kenilworth. Both are highly pictorial, and one needs little imagination to tell what one would see as one walked along every road and path: every cottage and outbuilding, every stream, gate, row of trees, and so on. Because in imagination one can move around the area in this way, the map sets us free in the past landscape in a way that a painting, which offers vivid realism but is tied to a single viewpoint, does not. The maps of Toddington and Kenilworth are of course exceptional works – but only in their size and comprehensiveness. Their accuracy and their detail can be paralleled on many smaller, less ambitious productions. Properly viewed, almost any estate map can give one insight into past landscape.

Moreover, once one becomes accustomed to the genre, once one gets one's eye in, so to speak, almost every estate map has some peculiarity, some unusual feature of particular interest. An example is the map of Brixworth (Northamptonshire), drawn by the vicar, Richard Richardson, in 1688. Unlike most estate maps it includes notes of the types of [57] soil in different parts of the area: 'Red Land' and 'Clay Land'. An even stranger feature is that it shows the standing places of ricks, marked with green-outlined rectangles. Again, Richardson was clearly interested in natural history, and the map includes notes on the wildflowers: 'Violae primulae', 'Serpyllum his locis nascitur' (i.e. thyme grows in these places), and so on.[74] Of course this is not a typical estate map – but very few estate maps are. There is something unexpected and interesting to be found on almost every one.

The Historians and the Landowners

When we use any historical document, we must always ask ourselves why it was written, for this was almost certainly not to give us information about the age that produced it but for some immediate, practical purpose that affected its form and content. If we treat it simply as an objective data source, we are likely to be led astray; we must make use of what it tells us in the light of what we believe it was trying to do. This is as true of estate maps as of any other sort of record. They are not general topographical maps like the official large-scale maps of today. They may incidentally show us how fields were cultivated before enclosure, what past landscapes looked like, and a host of other things, but this was not why they were drawn. Even when we use them for so seemingly neutral a purpose as a reconstruction of the landscape, we must be on our guard. The map was drawn for the landowner who, for one thing, is bound to have

---

[74] Northamptonshire Record Office, map 1555.

expected from his surveyor a reasonably flattering picture of his property. We may expect to find on the map the castle, the church, the neat row of houses on the village street, the hedges and fences carefully kept. What we cannot expect to find but must supply from other sources of evidence – even our own imagination – are the ruts and boggy patches in the roads, the sorry disrepair of the bridges and gates, the derelict outbuildings, the dung heaps.

Moreover, the map was usually drawn to show nothing except the property of the landowner. Sometimes this limitation is obvious. We see it at once in a map drawn by Thomas Hill in 1681 for the dean and chapter of Canterbury (article XVIII, plate I).[75] It shows their property at Vauxhall (Surrey), beside the River Thames: a series of closes in the hands of various tenants, together with houses lying along streets that are shown enlarged in a special inset. Beyond the area where their land lay, the map continues but shows only outstanding features, apparently included just to set the property in its local context: Spring Garden, Lambeth Palace and the row of houses joining it to Vauxhall, and on the other side of the river the houses of Westminster and the abbey behind. It is very clear that these features have all been drawn impressionistically: they form no part of the surveyed map. But much less clear is the status of closes belonging to other landlords that lay interspersed among the Canterbury lands; their owners are named, but their areas are not given. If the surveyor established the correct relative positions of the dean and chapter's pieces of land, he must have made some measurements of these intervening closes as well; but we cannot be sure that this was done, still less that the bounds between these other lands have been even sketched in from the ground – it could well be that they were put in simply to give the map an appearance of completeness and bore no relation to reality at all. This was the normal practice in entering neighbouring properties on a map of an estate that formed a single block of land: roads and houses might be sketched in approximately the right positions, but field boundaries would probably be merely drawn from imagination to produce a more [58] or less convincing pattern. This is almost certainly, for instance, the procedure followed on John Norden's maps of the Honour of Windsor in 1607 (article XVIII, plate II).[76]

Bearing in mind, then, why estate maps were made, it follows that they will tell the historian most reliably about the estates themselves: how they were run and how they were perceived by their owners. We have already seen that a map will often have quite a different archival history from the other records of an estate. Because they are picturesque or quaint, or even beautiful, estate

---

[75] BL Add. MS 34790.

[76] BL Harley MS 3749; one map is illustrated in article XVIII, plate II.

maps are often divorced from the context of the other records, even from the written surveys that they were meant to accompany. They may have been hung up as decorative pictures; they may have been preserved when other records were thrown away, or sold when other records were kept. Thus, almost all the estate maps in the British Library were acquired as single items, whether by gift or through the sale room. This means that we see them apart from the other records created in running the estate and cannot appreciate what part they were meant to play, how they were used in the work of management. The historian cannot, in other words, effectively draw on this important selection of estate maps as evidence for one aspect of the past on which they are peculiarly fitted to give information; and it is just that aspect on which we need information if we are to understand precisely why and how estate maps came to be made. Elsewhere, however, plenty of maps survive within their original archival context, often, indeed, still in the hands of the estates for which they were created. Sarah Bendall, in her work on the Cambridgeshire maps, is leading the way in investigating the role of estate maps in actual estate administration.[77] This may well produce results of great interest.

On the other hand, it need not be unhistorical to set the map apart from the other records of an [59] estate. It seems unlikely that when the earl of Rochester got his map and survey book of Kenilworth from James Fish in 1692 he handed them over to his steward to keep along with the accounts, the leases, the rent-rolls, and all the other estate papers. What such a map as this has to tell us is not so much how the landowner ran his estate as how he perceived it. We should see it as part of the baroque pomp with which the great surrounded themselves at this time, at one with the Palladian mansions, the liveried servants, the funeral monuments. Nor need it be only the great whose estate maps should be looked at in this way. A lesser man might view his estate map rather as someone today might view his bank statement when it shows a healthy balance. It was not for administrative efficiency that another Warwickshire landowner, Robert Smith of Coundon, had his 20-acre property mapped in 1694 (Fig. 5). This too has something to tell the historian.

---

[77] Above, n. 41.

XVIII

The Documents of Landscape History: Snares and Delusions[1]

[47] In 1960 I completed a doctoral thesis on the medieval history of Cuxham in Oxfordshire, a village with peculiarly good manorial records of the thirteenth and fourteenth centuries. When, unbelievably, the long-awaited moment arrived of writing the last sentence I could think of nothing to say. Everything sounded either pretentious or banal, or both. Finally, overcome by the sense of despair that I had felt throughout the whole work, I wrote (without second thoughts, to judge from the original draft):

> Only one general conclusion can safely be drawn. ... It is that no information from manorial records can be safely taken at face value, while least of all can we attempt to argue from their omissions. This is a negative and perhaps a commonplace conclusion, but its reiteration may not be without value.

This reflected years of grappling with records which said, for instance, that more acres might be harvested than had been sown, that the increase in volume on malting depended as much on the auditor's whim as on natural processes, that the tax-assessor had one idea of how much seed-corn was on the manor, the manor's own reeve quite another. Others working on agrarian records of the Middle Ages have shared my gloom: Professor Christopher Dyer has remarked that 'Reading fifteenth-century manorial records is an exercise in distinguishing between theory and reality' (Dyer, 1980, p. 162).

There are as many problems in using documentary sources for the history of landscape as in drawing on them for any other sort of information. There is the simple problem of understanding the technical or archaic words that they use. A quickset hedge is not a hedge that grows quickly, though there have been historians who thought it was. There are genuine problems of interpretation. The green corn fields through which Shakespeare's lover passed with his lass were unenclosed open fields, each extending over several hundreds of acres, for this was what the word *field* usually meant before the eighteenth century.

---

1 The article has been reset with minor changes, but the original page numbers (which have not been used for the index) have been entered in square brackets. Based on a lecture given at the Tenth Anniversary Conference of the Society for Landscape Studies, 23rd September, 1989.

Usually, but not always, for occasionally we find field (or its Latin equivalent, *campus*) used in its modern sense of a much smaller, enclosed area: the region, the date, the context of the particular document all have to be taken into account in deciding what it means. Some words are entirely ambiguous. In medieval records the Latin *fossata* can mean a ditch or a dyke, a moat or an embankment; the context seldom shows which, and usually it is anyone's guess, yet the difference is crucial if we are working on landscape reconstruction. Sometimes records express themselves with a precision and exactness well understood in their own time but now lost. When Domesday Book measured the woodlands of some counties in terms of the number of pigs they could support, this told contemporaries very much more than it tells us, for we can only guess at how much eleventh-century woodland was needed to sustain an eleventh-century pig.

In fact the compilers of Domesday Book were not interested in the actual acreage of the woodland; what they wanted to know was its economic potential: its value to its possessor, or to the king if it should ever come into his hands. This is made clear by the woodland entries for some other counties, in south-east England. Here the amounts of woodland are expressed not by the number of pigs they could support but by the number of pigs the manorial lord received as rent by those whose pigs were admitted to feed there. These pigs given in rent would be a fixed proportion of the total number and later records suggest that this proportion varied from place to place – it might be, say, one pig in seven or one in ten. Where Domesday Book tells us about woodland in a way that we can easily translate into terms of landscape – woods so many leagues long, so many leagues broad – that is our good fortune, but it is only incidental to the purpose of the record. This introduces a general moral: that whatever the purpose of any record or relic of the past, it was not to tell us today what the landscape of the past looked like. If we are to use its evidence intelligently – that is, accurately – we must first discover, if we possibly can, what that purpose was.

This, of course, applies to taking information on any subject, not just landscape, from any historical record; but it is extraordinary how often this elementary truth is overlooked, how often historians treat their sources as though they were encyclopaedias, an information retrieval service, designed to tell future generations the objective [49] truth about the age that produced them. Given this, it is not extraordinary at all that historians sometimes mislead both their readers and themselves. Sometimes the way the record's purpose affects the information it gives is obvious. One needs no particular expertise as an administrative historian to see why the amount of corn that the tax-assessor recorded on a manor may differ from the amount in the owner's own records;

but if we have only the tax-assessor's record and not the owner's there is a great temptation to treat it as a correct account.

The student of landscape history has at least this advantage, that the past ages that produced the written records, at least until the eighteenth century, had not the slightest interest in landscape. We can read the whole corpus of the Icelandic sagas without getting a hint of the extraordinary landscape, unlike any other in Europe, against which their dramas were played; all the chronicles of the Crusades without guessing that the landscape of the Near East differs appreciably from that of western Europe. If people in the past were uninterested in landscape they will not have been trying to mislead each other (and thus us) about it in their written documents. The historian of landscape is less likely than the economic historian to meet deliberate disinformation in the scraps of evidence that come his way.

But scraps of evidence are all he is likely to get, and one particular snare that lies in wait for him is the omissions in the record. The boundaries of estates that are set out in some tenth- and eleventh-century charters give us invaluable glimpses of the Anglo-Saxon landscape, but tell us only what lay on the edges of the property, leaving us to fill in, at best by intelligent guesswork, what lay in the middle. Here the omission is obvious, but often it is much less so and if we are to detect it we have to keep in mind the golden rule: always ask what the document's original purpose is likely to have been. Thus Domesday Book, entering properties under a particular place, often does not trouble to tell us that some of them actually lay somewhere else, like the houses of Staines that were really in London: the rents were paid at Staines and that was all that mattered to the compilers (Darby and Campbell, 1962, p. 131). Again, a late-medieval estate survey may be entitled and set out in such a way as to suggest that it describes all the lands in a group of fields, while actually describing only the manorial demesne, the home farm, simply leaving out the intermingled strips of the owner's tenants and neighbours. These are medieval examples, but the risk is the same in written records of all periods: the silent omission of information that is irrelevant to the writer's purpose but not to ours.

From the sixteenth century onwards we have a new source that is not only especially important for landscape history but might seem to meet this particular difficulty of omissions: the map. The writer can silently omit information from a written document, but to leave a feature off a map is as positive a statement as to put it in. This is true, provided again that we can discover, and always bear in mind, the reason why a map was drawn. The large-scale map drawn not for a particular purpose but to serve as a general work of reference seldom appears before the nineteenth century. Earlier maps were almost always drawn with a particular object in mind, and this affects what they show and what they leave out. Often this is obvious. In 1681 Thomas Hill made a map of Vauxhall manor in Surrey, an estate of the Dean and Chapter of Canterbury (Pl. I). It is clear that the map stops short at the boundaries of the Canterbury property, though some features beyond are put in by way of decoration or to put the estate in its local setting: Spring Garden and Lambeth Palace, as well as Westminster Abbey and some houses on the farther side of the Thames.[2] But a map's omissions may not be obvious at all. In 1607 John Norden produced a set of maps of the honour of Windsor: a magnificent bird's-eye view of Windsor Castle and sixteen plans making up a map of the whole of Windsor Forest, from Windsor to Guildford (Pl. II). The Forest, carefully surveyed and drawn, was the subject of the map, but where the map reaches the edge of the Forest it does not stop there but continues beyond, in much more detail than the surrounds of the Vauxhall map, showing rivers, roads, houses and fields. But we should not assume that the bounds of these fields are actual boundaries, mapped in the same way as the boundaries around and within the Forest itself. Norden's only concern in putting them in was, like Thomas Hill at Vauxhall, to place the estate in its local context. He wanted to show that it was adjoined at these points by areas of enclosed fields but he certainly did not survey them and the bounds shown on the map may well be entirely fictitious.[3]

Here we can distinguish the real from the conventional on the map largely because we know why the map was drawn: to provide a complete map of the Windsor Forest estate for its owner, King James I. Without this knowledge it is much harder, even impossible, to tell which features on the map are real, which may be conventional or imaginary, what are its actual, but concealed, omissions. The names of the local manorial lords on a map of the area around Sherborne, Dorset, show that it was drawn between 1569 and 1574. Different colour washes distinguish the lands of each manor and a great deal of other detail is drawn in, including many field boundaries.[4] But we do not know why the map was drawn: for whom or for what purpose. Consequently we cannot tell which features of the landscape are likely to have been carefully and accurately portrayed, which are likely to be sketched in to fill a gap. The field boundaries may be real or they may be conventional. Here, as always, knowing what our

---

2 British Library, Additional MS. 34790.

3 British Library, Harley MS. 3749.

4 British Library, Additional MS. 52522. *Cf.* Harvey, 1965; Cobbe, 1973, pp. 646–7, plate I. New light is thrown on this map by Katherine Barker, 'An Elizabethan map of north-west Dorset: Sherborne, Yetminster and surrounding manors', in *Maps and History in South-West England*, ed. K. Barker and R. Kain, Exeter Studies in History 31 (1991), pp. 28–53. I am grateful to Mrs Barker for kindly letting me read the typescript of this most interesting article.

Plate I. Map of Vauxhall manor, Surrey, by Thomas Hill, 1681. British Library, Additional MS. 34790, reproduced by permission of the British Library.

Plate II. One of the set of maps of the honour of Windsor by John Norden, 1607. British Library, Harley MS. 3749, ff. 16v–16*, reproduced by permission of the British Library.

document is trying to do is a crucial first step in interpreting it and in assessing correctly the information it conveys.

But its omissions, obvious or concealed, are only one of the snares that await us in the map that is used as historical evidence. Take, for instance, [50] the Ordnance Survey map that has at the top the title 'Ordnance Survey Tyneside Sheet 88 1:50 000 Second Series Crown Copyright 1976'. One might be excused for supposing that it shows the area as it was in 1976 or (allowing time for publication) a little earlier. But a note in small print at the bottom reveals that the map as a whole was revised in 1973, that low-water mark was plotted from air photographs of 1963–9, that it was revised for major roads 'and other significant changes' in 1975 and again, for major roads alone, in 1977, and that the contours (which are not entirely unchanging) were taken from the Ordnance Survey's last one-inch map of the area, going back probably to the 1950s, possibly earlier. It is not just that the map does not show us Tyneside as it was in 1976. What it shows us never existed at all at any one time: it is a patchwork of pieces of information drawn from various sources at various dates. And if this is the case in a modern Ordnance Survey map (which scrupulously relates what it has done) it is equally true of older maps (which do not). The first volume of the Portsmouth Record Series, published in 1971, reproduces on the endpapers a manuscript map from the British Library, and as its title reads 'A SURVEY of PORTSEA ISLAND ... done in the Year 1716' it was not unreasonably entered in the list of illustrations as 'SURVEY OF PORTSEA ISLAND 1716' (Willis and Hoad, 1971, p. ix). That the general editor had been altogether naive appeared when the fourth volume in the Series was published in 1978, the catalogue of *Maps of Portsmouth before 1801* by Mr Donald Hodson, who showed that this map had indeed been copied, with its title, from a map of 1716 but had been revised to include later features and dated in fact from the mid-1740s (Hodson, 1978, no. 14c).

That every map has omissions is in the nature of cartography: it gives us information about landscape that is stylised in form and highly selective in content. In the early nineteenth century Thomas Hornor, an estate surveyor and mapmaker, tried to get over this difficulty by what he called 'panoramic chorometry'. At first sight his maps look like fully detailed bird's-eye views; in fact they are true isometric maps, with the pictures of landscape features superimposed. The idea was not copied: Hornor's maps were beautiful, for he was a good artist, but they were slow to produce and too expensive to attract many customers (Hyde, 1977).

Yet it was not only Hornor whose maps spare our imagination by giving us actual pictures of landscape. Many estate plans and other maps of the seventeenth and eighteenth centuries include pictures of some or many features.

Earlier, in the sixteenth century, we have many maps that are not so much maps as pictures: the paths of the map and the bird's-eye view had not yet diverged and these picture-maps were the origin of both. Two surviving relics of Tudor England carry me back, give me an actual or supposed vision of the sixteenth-century countryside, better than any other records, better than a whole library of books by historians. One is the owl's song at the end of *Love's labour's lost*:

> When icicles hang by the wall,
> And Dick the shepherd blows his nail:
> And Tom bears logs into the hall,
> And milk comes frozen home in pail ...

– everyone knows it. Sparing in words, rich in images, it is an extraordinary evocation of a small rural household of Shakespeare's time. The other is less well known. It is a map, or rather a picture with some cartographic features, in the archives of Winchester College. It is quite small, about one foot by two feet, drawn in ink with coloured wash (Pl. III). In the foreground is a road, identified as the [51] highway from Basingstoke to London. Behind it are four fields and a wood – all quite small: their areas are given, and they total 44 acres. In one of the fields is a barn, timber-framed and thatched, with a path leading to it. On the road are a horse and cart and a man leading three packhorses. The trees of the wood, the hedges and gates are all carefully, minutely, drawn.[5] It is not a conscious work of art – that was not its purpose – but it tells me about that landscape, sets me down in it, in a way no carefully composed picture could do. This is, of course, a subjective, personal reaction that others may not share. For me, however, it is a superb evocation of past landscape, unparalleled even among the many pictures – and landscape painting is, after all, one of the glories of British art – that more than any other record show what England really looked like in the past.

This is partly because the picture, unlike the conventional map, is not confined to more or less permanent features. It shows not just the road, the fields, the wood; it shows the cart and the packhorses as well. It peoples the landscape. Often this is important: if we are to gain a true understanding of past landscape we must know how that landscape was used. Another Portsmouth Record Series volume is a calendar by Dr R.J.B. Knight of Portsmouth Dockyard records from 1774 to 1783. For the endpapers Dr Knight found a painting by William Elliott showing George III's visit to the fleet at Portsmouth in May 1778. In the background is a magnificent panorama of the Dockyard, a forest of masts, while in the foreground in the harbour are ships and boats of

---

[5] Winchester College muniments 3233.

Plate III. Plan of fields near Andwell, Hampshire, early 16th century. Winchester College muniments 3233, reproduced by permission of the Warden and Scholars of Winchester College.

every sort. It offers a striking contrast to the plans of the Dockyard, even the contemporary three-dimensional model, that are shown in the same volume. But more than this, because the records of the Dockyard are so full, and because we know the exact day, even the time of day, represented in the painting, it is possible to identify each of the ships in dock, ships that mostly appear simply as distant masts (Knight, 1987).

Unfortunately for the historian not all paintings of landscape are so detailed or accurate. Even topographical pictures have their snares for the unwary. As a boy I went, successively, to six schools. Most had a school song, sung by the whole school at the end of term and on other solemn occasions. None was a great poem; indeed, from the literary standpoint each seemed (or seems now) even more awful than the one before. One offered a brief conspectus of the school's curriculum:

> From fields of Classic lore we glean the culture of the past,
> In Mathematics reason finds delight,
> Keen Science with her magic wand reveals the latent truth,
> And varied Art redresses what in nature seems uncouth ... [6]

Also when young I was privileged to know the artist Muirhead Bone – privileged because he was a quite exceptionally nice man. Once, when I was a student, I met him in Beaumont Street in Oxford. He was sitting at his easel, which he had set up on the pavement, and was drawing the scene looking towards Balliol and St John's. 'But,' he said, 'the Martyrs' Memorial isn't in the right place for the picture, so I'm moving it towards the centre, so that it closes the vista.' I was fascinated: here was my school song in action. Here was Art, in the person of Sir Muirhead, hard at work redressing nature, which had uncouthly put the Martyrs' Memorial some twenty yards south of where it should have been. Once again the crucial question is, what was the purpose of the document, in this case the picture? Why was it drawn? If it was drawn with the deliberate aim of recording the scene we can rely on it as evidence. Many antiquarian and topographical artists have worked with precisely this end in view: Stukeley, Grim, Scharf, the Bucklers, many others. But if, as more often, the intention is to make an interesting picture, to convey atmosphere, an evocative impression of scenery, then it may or may not reproduce the landscape with accuracy. Other sources show that the artist who painted George III's visit to Portsmouth was exact and accurate in what he drew; but if there was no other evidence for the

[6] The school was Bishop Feild College, St John's, Newfoundland. The song is printed in *The Feildian*, 36, No. 3 (Oct. 1942), inside back cover.

appearance of Portsmouth Dockyard in 1778 it would be unwise to rely on the painting for any detailed information.

These then are the snares – or some of them. Most can be avoided if we know, and always bear in mind, what the particular record was trying to do – and whatever this may have been it was never to tell the late twentieth century about the landscape of its own time. But my title speaks of snares and delusions. In moving from written records first to maps, then to pictures, we have become more and more involved not just with what was there – out there in the landscape of the past – but with what it looked like. And at this point, having successfully avoided the snares, we are ready to be trapped by the delusion: that we can view the landscape of the past, re-create it on paper or in the mind's eye, see it as the past saw it.

We do not see landscape as the camera sees it. If we did, the way it is presented to us by Turner or Van Gogh, Cezanne or John Piper would be meaningless. The camera has a part to play in our understanding of landscape – it has its own insights, its own interpretation to offer. As a record of landscape it has many unique virtues as well as some snares all its own. But whereas the camera's view of landscape is static, unchanging from one generation to another, ours is not. Consider our aesthetic reactions to landscape – what we do or do not consider beautiful. Some years ago *The Times* published a pair of major leading articles on the environment, on the appearance of Britain. The writer remarked that by the end of the eighteenth century our ancestors had created a landscape of great loveliness: never before or since has Britain's outward appearance been so pleasing. Many of us [52] would agree. But if we had been alive two hundred years ago we might well have found this idea puzzling. Some parts of its landscape the eighteenth century appreciated as much as we do. People then liked the neatness of the increasing number of enclosed fields. They liked their Palladian mansions. But not all will yet have followed Horace Walpole in seeing beauty in the nation's great Gothic monuments, let alone in the innumerable medieval village churches, and William Gilpin was only starting to instruct the most sophisticated in the beauty of wild natural scenery. There were far more timber-framed thatched cottages then than there are now, in towns as well as in villages, but they were likely to be viewed rather as we see high-rise flats: disagreeable in appearance and the home of social problems. And the new townscapes that the Cubitts and others were creating – the Georgian terraces and squares that we so much admire – might be seen as plain, monotonous and ugly. It will not only have been Smollett's Squire Bramble who saw the new houses of Bath as 'contrived without judgment, executed without solidity, and stuck together with ... little regard to plan and propriety' (Smollett, 1771, vol. 1, letter of 23 April). It is salutary, and intriguing, to ponder what features of our own landscape future ages will think we have failed to appreciate as we should. Perhaps power-station cooling towers, whose interesting curves dominate the surrounding countryside. Or the lines of electricity pylons, intricate patterns of metal between the graceful rise and fall of the wires. Or the bridges on motorways with their elegant shapes in concrete. Or the neat rows of semi-detacheds on a housing estate, each with its beautifully kept front garden. These are meant as serious suggestions.

It is impossible to see past landscape as the past perceived it, just as we cannot tell how future generations will view the relics of our own landscape that we bequeath to them. But it is not only the landscape historian who has problems of this sort. The musicologist, through long and skilled research, can reconstruct the instruments and style of music played in the past. We can hear the same sounds as those who heard the first performance of a Bach cantata at St Thomas's, Leipzig, or of a Gabrieli motet at St Mark's, Venice. But we hear them through different ears – just what those sounds meant to those who heard them, just *how* they heard them, we can only guess. The same problem arises if we try to appreciate the contemporary impact of a painting or any other work of art (Gombrich, 1972, p. 54). I live in Durham in a house built in 1901 and from my window, which looks out over the city, concealed here and there by trees, there is no single building or other feature clearly visible to the naked eye that was not already there when the house was built. The first owner of my house would today see nothing intrusive in the landscape, but I cannot suppose that his view of it was the same as mine. Some features of course were already old in 1901: Durham Castle assumed its present form in the early nineteeth century, the enclosed fields beyond the town are older, Durham Cathedral is older still, yet not as old as the line of hills that closes the horizon. Landscapes of the past are with us forever, for no generation can sweep away all that went before. But the past landscape has forever gone: *où sont les neiges d'antan?*

Bibliography

Cobbe, H., 1973. 'Four manuscript maps recently acquired by the British Museum', *Journal of the Society of Archivists*, 4, No. 8, pp. 646–52, plates I, II.

Darby, H.C., and Campbell, E.M.J., 1962. *The Domesday Geography of South-East England* (Cambridge).

Dyer, C., 1980. *Lords and Peasants in a Changing Society: The Estates of the Bishopric of Worcester, 680–1540* (Cambridge).

Gombrich, E.H., 1972. *Art and Illusion*, 4th edn (London).

Harvey, P.D.A., 1965. 'An Elizabethan map of manors in north Dorset', *British Museum Quarterly*, 29, pp. 82–4, plate XX.

Hodson, D., 1978. *Maps of Portsmouth before 1801*, Portsmouth Record Series 4 (Portsmouth).

Hyde, R., 1977. 'Thomas Hornor: pictural land surveyor', *Imago Mundi*, 29, pp. 23–34.

Knight, R.J.B., 1987. *Portsmouth Dockyard Papers 1774–1783: The American War*, Portsmouth Record Series 6 (Portsmouth).

Smollett, T., 1771. *The Expedition of Humphry Clinker*, 3 vols (London).

Willis, A.J., and Hoad, M., 1971. *Borough Sessions Papers 1653–1688*, Portsmouth Record Series 1 (Portsmouth).

INDEX